Sources and Studies in the History of Mathematics and Physical Sciences

Sources and Studies in the History of Mathematics and Physical Sciences was inaugurated as two series in 1975 with the publication in Studies of Otto Neugebauer's seminal three-volume History of Ancient Mathematical Astronomy, which remains the central history of the subject. This publication was followed the next year in Sources by Gerald Toomer's transcription, translation (from the Arabic), and commentary of Diocles on Burning Mirrors. The two series were eventually amalgamated under a single editorial board led originally by Martin Klein (d. 2009) and Gerald Toomer, respectively two of the foremost historians of modern and ancient physical science. The goal of the joint series, as of its two predecessors, is to publish probing histories and thorough editions of technical developments in mathematics and physics, broadly construed. Its scope covers all relevant work from pre-classical antiquity through the last century, ranging from Babylonian mathematics to the scientific correspondence of H. A. Lorentz. Books in this series will interest scholars in the history of mathematics and physics, mathematicians, physicists, engineers, and anyone who seeks to understand the historical underpinnings of the modern physical sciences. All works are reviewed to meet the highest standards of scientific literature.

More information about this series at http://www.springer.com/series/4142

Ruth Glasner • Avinoam Baraness

Alfonso's *Rectifying the Curved*

A Fourteenth-Century Hebrew Geometrical-Philosophical Treatise

Springer

Ruth Glasner
The Faculty of Humanities
Hebrew University of Jerusalem
Jerusalem, Jerusalem, Israel

Avinoam Baraness
The Faculty of Humanities
Hebrew University of Jerusalem
Jerusalem, Jerusalem, Israel

ISSN 2196-8810 ISSN 2196-8829 (electronic)
Sources and Studies in the History of Mathematics and Physical Sciences
ISBN 978-3-319-77302-5 ISBN 978-3-319-77303-2 (eBook)
https://doi.org/10.1007/978-3-319-77303-2

Mathematics Subject Classification (2010): 01A05, 01A35, 01A30

This Springer imprint is published by the registered company Springer Nature Switzerland AG
The registered company address is: Gewerbestrasse 11, 6330 Cham, Switzerland

To Gad Freudenthal for initiating this project and for his help and advice all the way to its completion

To Gita Gluskina for her excellent pioneering work, carried out in the hostile environment of the Soviet Union

Preface

Sefer Meyasher `Aqov (The Rectifying of the Curved, hereafter *SMA)* is an enigmatic Hebrew medieval geometrical-philosophical text that focuses on the foundation of geometry. The book is a mixture of two literary genres: philosophical discussion and formal, Euclidean-type geometrical writing. Parts I and II of *SMA* form one unit, in which the two genres are intertwined. Part I introduces Alfonso's agenda: to extend geometry by legitimizing the use of motion and superposition. Part II expands on some issues that were raised in Part I, especially the possible errors that may result from the use of motion in geometry. Part III is a compilation of 33 geometrical propositions (8 of which are missing)—some better known, some less so, some minor lemmas, other highly interesting on their own. Part III is presented as preparing for Parts IV and V, the proposed culmination of Alfonso's project. Of Parts IV and V, however, only the first 32 lines of IV are extant.

Sefer Meyasher `Aqov was discovered by Solomon Luria, and in 1984 it was published with a Russian translation, introduction, and notes by Gita Gluskina. Gluskina identified the author as Alfonso of Valladolid, the Christian name of the converted Jew Abner of Burgos, who was active in the first half of the fourteenth century. He lived all his life in Castile, rather far from the leading cultural centers of his time, but nonetheless at the crossroad of three cultures. He was raised in the Jewish tradition and was erudite in religious studies. Like many Jewish intellectuals in Spain at that time, he was versed in Greek-Arabic philosophy and science. He also had connections with some Christian nobles and towards the end of his life he converted to Christianity. Driven by his ambition to solve the problem of the quadrature of the circle, Alfonso acquired surprisingly wide knowledge and became familiar with several episodes in Greek and Arabic geometry that historians usually consider not to have been known in the West in the fourteenth century. *SMA* reflects his wide and deep erudition in mathematics and philosophy, and provides new evidence on cultural transmission around the Mediterranean.

Sefer Meyasher `Aqov has not yet received the attention it deserves. So far it has drawn mixed responses from historians of mathematics. Tony Lévy remarks that the author "manifeste plutôt les qualités d'un philosophe, polémiste de surcroît, que celles d'un mathématician créatif." Tzvi Langermann, on the other hand, considers

SMA as "the most interesting original Hebrew geometrical text of the medieval period." Both positions can be argued. The present volume offers a new edition with English translation, introduction and detailed, both mathematical and historical commentary. We hope that in its new form this unique text will now reach a wider audience and get the attention it deserves.

Jerusalem, Israel

Ruth Glasner
Avinoam Baraness

Acknowledgement

We are most grateful to Bijan Vahabzadeh, Roshdi Rashed, Shlomo Sela, Ahmad Igbaria, and Roi Wagner for their advice on several mathematical questions. We wish to thank Nurit Pasternak, Malachi Beit Arié, and Edna Engel for their help with the identification of the manuscript and the glosses. We wish to thank Ofer Elior and Hagar Kahana-Smilansky for reading the manuscript and for their helpful comments.

This research was supported by ISF grant no. 981/18.

Contents

Chapter 1
Introduction

1 The Manuscript

The text of *SMA* is preserved in a single manuscript in the British library (henceforth BL manuscript).[1] It is the sixth work in a codex of eight scientific books.[2] The focal subject of this codex is the astrolabe. The first group of texts (1–4) are copied in a single hand, Sephardic script. Texts 5–7, by Abraham Ibn Ezra, Alfonso, and Emanuel ben Yaʿaqov were copied in another Sephardic-Italian hand in the last third of the fifteenth century.[3] Text 8 by Yaʿaqov ben Makhir is copied in a third hand. The corpus belonged to an erudite owner who wrote commentaries and added diagrams in the margins. Nurit Pasternak suggested and Malachi Beit-Arié confirmed that this owner was Mordekhai Finzi.[4] Finzi was at the center of the mathematical and astronomical activity in the Mantua community, in the third quarter of the fifteenth century, and had Jewish and Christian scientific colleagues.[5] It seems that texts 5–7 were copied in Finzi's milieu and perhaps were ordered by Finzi himself.

The manuscript of *SMA* is incomplete. In the outline five parts are listed. Of these Parts I–II are complete, in Part III there is a big lacuna, and of Parts IV and V only the first 32 lines remain. The lacuna in Parts III and the interruption of the text in

---

[1] BL Add. ms. 26984 fols. 93b–128a; Margalioth catalog, vol. 3, p. 320, 1002/6; institute of microfilmed Hebrew manuscripts in Jerusalem F 5656. We rely on the work of Nurit Pasternak who examined the manuscript at the British library.

[2] It is made of paper and parchment. The parchment is Italian.

[3] Edna Engel suggested that the copyist was a pupil of Abraham Fariṣol, perhaps the one who copied the Parma manuscript that is photocopied in Engel (1992, 158).

[4] On the basis of comparing the marginal notes (not in SMA, but in the astronomical parts of the codex) to the known autographs of Finzi. Interestingly the line in the letter ק in Finzi's autographs is usually straight and in the marginal comments in the BL manuscript it is sometimes straight and sometimes curved, even in the same comment.

[5] Langermann (1988, 8); Wagner (2013, 195–6).

R. Glasner, A. Baraness, *Alfonso's Rectifying the Curved*,
Sources and Studies in the History of Mathematics and Physical Sciences,
https://doi.org/10.1007/978-3-319-77303-2_1

Part IV occur in the middle of pages, even in the middle of sentences, so it looks as if the *Vorlage* of BL manuscript was defective. The lacuna in Part III may be due to a missing folio or quire in the *Vorlage*. It is not impossible that the *Vorlage* of the BL manuscript was Alfonso's autograph. In some places the reader can discern hesitations, corrections, and additions in the text with which the copyist did not know how to cope. There is no way for us to know if Part IV was continued, and if anything of Part V ever existed. The quality of the copying is rather poor and there are quite a few errors, most often in letters that refer to diagrams. It seems that the copyist was not a mathematician and was not interested in the contents of the text.

Only one simple diagram (of a straight line) is included in the text,[6] and only once a space is left for a diagram.[7] In Part II there are added diagrams in the margins, as it seems in Finzi's hand.[8] Surprisingly, in Part III, which is mathematical throughout and cannot be read without diagrams, nothing is added in the margins.[9] We know, however, that Finzi read Part III. Tzvi Langermann has shown that Property III.29, which deals with the construction of the conchoids, was copied verbatim in a notebook of Mordekhai Finzi in his hand in or before 1473.[10] Perhaps Finzi read Part III from the manuscripts that came into his hands, copied one proposition, and then ordered his own copy. If this description is correct, the BL manuscript was probably copied between 1473 and Finzi's death in 1475. The cultural milieu in which it was copied was the lively Jewish community of Mantua of the early 1470s.[11] Finzi was the only person known to us (before the twentieth century) who was interested in Alfonso's book and who read at least parts of it.

2 The Russian Translation and Previous Research

The renewed interest in Alfonso goes back to the 1930s when the BL manuscript attracted the attention of the Russian historian of science Solomon Luria. In 1960, he received a copy from the British library, and asked Gita Gluskina to edit and translate the text.[12] Luria died in 1964. Gluskina did an excellent pioneering work in the hostile environment of the Soviet Union at the time. She identified the author, wrote two articles on the text, and eventually succeeded to publish a critical edition

[6] On fol. 122b. See picture in III.24 below.

[7] On fol. 109a.

[8] We are again indebted to Prof. Beit Arié for confirming this.

[9] There are a few corrections above the lines, e.g., in fol. 115a, but these could have been made by the copyist.

[10] Langermann (1988, 37–9; 1996, 34–5). See Oxford Bodl. ms. Mich 350, Neubauer 2052, IMHM 19337, fol. 92b. There are minor differences between Alfonso's text and Finzi's copy such as replacing the suffix ין by ים (e.g., יוצאין מקיפין by יוצאים מקיפים) or omitting a few words. Another text in Finzi's notebook is dated 1473.

[11] In 1473 the scholar and copyist Abraham Farissol who lived in Mantua at the time left for Ferara in 1473; Mordekhai Finzi died in 1475.

[12] Freudenthal (2005, Chap. IX, 1).

with introduction, notes, diagrams, and a Russian translation in 1983.[13] Gluskina's edition deserves much praise. Boris Rosenfeld added several mathematical comments to her edition, but the mathematical content of the book deserves more attention. So far only two passages of *SMA* were studied in depth by historians of mathematics: the attempt to prove the parallels postulate by Tony Lévy and the presentation of the squaring of the lunes by Y. Tzvi Langermann.[14] Avinoam Baraness, who worked on Part III, has already published his presentation of Propositions III.29–32 in an anthology edited by Victor J. Katz, and his analysis of the mathematical parts of SMA in his dissertation.[15]

3 The Author

The author of *SMA* refers to himself three times as "Alfonso." Gluskina identified him as Abner of Burgos, the Jewish sage of Castile (ca. 1260–1347) who converted to Christianity late in his life (about 1320) and became known as Alfonso of Valladolid.[16] She based her identification on the similarity of ideas, notably Alfonso's Platonism, and on stylistic resemblances.[17] About twenty years later Gad Freudenthal had some doubts and re-examined the evidence, including texts by Abner that were not available to Gluskina, and eventually found himself corroborating Gluskina's identification.[18] He offered additional stylistic arguments, and pointed out the interesting use of what he calls "the principle of charitable interpretation" by both the author of *SMA* and Abner.[19] When we began our study of this text, we still had doubts, because we had some reasons to suspect that the text was written in Italy.[20] It was eventually the reference to Euclid's *Elements* X (by no means a popular text) with the same purport in *SMA* and in the *Mostrador* that finally convinced us that Gluskina's identification is correct.[21]

---

[13] Gluskina (1974, 1978).

[14] Lévy (1992); Langermann (1996a).

[15] Baraness (2016, 2018).

[16] Gluskina (1978).

[17] E.g., the use of the Talmudic, but uncommon phrase התכבד בקלונו by both Abner and Alfonso. Gluskina 1978, 70–1.

[18] Freudenthal (2005, Chap. IX, 2–4).

[19] Freudenthal (2005, Chap. IX, 2–3). See Section I.3.1 below.

[20] An Italian origin could have brought us somewhat closer to understanding how Alfonso learned about the construction of the conchoid by Nicomedes (*SMA* III.29). This construction is known to us from Pappus' *Collection*, vol. 4. Sabetai Unguru (1974, 319–321) very cautiously, but reasonably offers that an archetype of Pappus' collection circulated in Italy before 1533. He argues that Witelo was familiar with parts of the *collection* and suggests that William of Moerbeke translated for him parts of it. The single manuscript of *SMA* and the only quotation from it known to us are from north Italy in the 1470s. The one explicit reference to a Christian scholar is to the Italian mathematician Campanus. Thus, *prima facie*, an Italian origin seemed to be likely.

[21] See Section 6, vii, example [i] below. See also commentary to Section II.5.5.

Not much is known about Abner's life before his conversion. Presumably he was a Rabbi and a head of a Yeshiva. His erudition in the traditional Jewish sources was praised by his contemporaries IsḤāq Pulgar and Moshe Narboni,[22] and is well attested in his writings.[23] After his conversion he participated in public debates and wrote polemical books. Notable among his extant books are *Mostrador de justiciar* (in Castilian) and *Teshuvot la-Meharef* (in Hebrew). Since Yitzhaq Baer's famous study on the history of the Jews in Christian Spain, Abner was considered mainly as a political highly polemical figure. Recent research throws some light on other aspects of his life and work.[24] His philosophical book *Minhat Qenaot* (in Hebrew) presents a strict deterministic philosophy, and another philosophical book that was lost probably offered criticism of Aristotelianism.[25] In the light of the recent research, the similarity between the intellectual profiles of Abner and the author of SMA comes to the fore.

- Ryan Szcpiech, who studied in depth Abner's major book *Mostrador de justiciar*, contends that several passages in this book "give a good indication of his knowledge of scientific and geometrical concepts, a knowledge which is in the background of Abner's works, but which is obscured by discussions of messianic and polemical topics."[26] In *SMA* scientific and geometrical knowledge is in the foreground and is very impressive, but here, too, the reader can also sense a theological or even "messianic" passion in the background.[27]
- The combination of wide erudition with occasional flaws in precision is also typical of both Abner and the author of *SMA*. Gershenzon comments that Abner "does not hesitate to quote material out of its context and to join sentences widely separate in the text." She even accuses him of "deliberately changing his source in order to make the desired point."[28] Szpiech confirms that Abner's "citations themselves are often distorted through translation and are often nearly incomprehensible on their own… He frequently mixes parts of one citation with parts of another, blending different texts together in a single passage."[29] The author of *SMA* apparently made a great effort to reach as many "sources" as he could find,

[22] See for instance Gershenzon (1984, 14).

[23] Ryan Szpiech (2006, Appendix 4, 645–657) compiled a complete list of the sources of Alfonso's polemical book *Mostrador de justicia* (extant only in Castilian) and has shown that the vast majority of the citations are from Jewish traditional texts.

[24] Several important dissertations have been writes: Gershenzon (1984); Sainz de la Maza Vicioso (1990); Hecht (1993); Szpiech (2006); Sadik (2011).

[25] See Sadik (2008).

[26] Szpiech (2006, 524–8). Szcpiech further points out that Abner's "attitude to his sources was ambivalent, split between a real objective interest in the ideas that he proffers, and an overriding concern with his messianic polemical arguments" (2006, 528).

[27] Most notably in I.4.2, I.3.1 and the poetical preface.

[28] Gershenzon (1984, 12–4).

[29] Szpiech (2010, 194–195); see also 14–5.

> yet his citations, especially of the scholars whom he criticized, are often careless, usually abridged. One sometimes gets the impression that he was eager to lead forward his own agenda, not always paying the full attention to the ideas of others.[30]

In what follows we will refer to the author of *SMA* as Alfonso and, only when the context requires, as Abner. *SMA* throws additional light on the complex personality of its author. The book summarizes a life-long quest on the subject of the quadrature of the circle. "To glorify his name and greatness by measuring the circle, which *from my youth to my old age* was the one thing I desired from the Lord" this is how Alfonso describes his goal in the poetical preface (I.1.0). Later he confirms again that he was thinking of the problem for a long time (I.3.1). The author's identifying himself as "Alfonso" indicates that the version of *SMA* that came down to us was written (or revised) in his Christian period, namely relatively late in his life.[31] We do not know whether Parts IV–V (or at least a significant part of them) were written and then were lost, or whether the writing of the book was interrupted by difficulties, old age, or death. There is no evidence that in the 125 years that separate Alfonso's death from the copying of the BL manuscript in Mantua the book drew any attention. There is no indication that it was translated either into Latin or into Castilian (as were other books by Abner) or that it was read by anyone in Hebrew before Finzi. A likely guess is that the book was not distributed because it was never completed. Still it is possible that the original included more than is contained in the BL manuscript. The outline of Part IV mentions that this part includes 16 propositions, and this indicates that at least the plan of Part IV was finalized.

4 The Geometrical-Theological Agenda of the Book

The mathematical agenda of *SMA* was very ambitious. In the outline and the introductory parts Alfonso declares his intention to extend geometry by providing tools for precise measuring of magnitudes by one another, whether rectilinear or curved. Measurement of any magnitude can be achieved by assuming indivisible parts, but Alfonso does not take this road. He is by no means willing to give up the Aristotelian-Euclidean conception of geometry as the science that studies continuous magnitudes. He observes the formal language and structure of classical geometry, but "upgrades" it by modifying two of its basic tenets:

[30] He probably would not have made a good chess player.

[31] It is possible that "Alfonso" was Abner's Spanish name already before his conversion, but Shalom Sadik, for several reasons, considers it as unlikely. We are grateful to Sadik for his advice.

[i] Euclid's requirement of homogeneity
According to Euclid's requirement of homogeneity a ratio (and hence equality) can be defined only for "magnitudes of the same kind," "which are capable, when multiplied, of exceeding one another."[32] This requirement was already overruled by Archimedes, who used in his proofs ratios between magnitudes that cannot be even locally superposed on one another (and thus cannot be said to exceed one another), such as the surface of a sphere and a flat circle. Obviously, the requirement of homogeneity cannot be satisfied by magnitudes that are not of the same dimensionality, and hence accords well with Aristotle's teaching that lines are divisible only into lines, areas into areas, volumes into volumes.[33] Lower-dimensional entities are regarded by Euclid and Aristotle as extremities, rather than as parts of magnitudes. Reconsidering this radical separation of magnitudes according to their dimensionality, Alfonso turns to Plato and to his idea of dimensional reduction, described in the *Timaeus*.

[ii] Euclid's notion of superposition
Euclid introduced the notion of superposition in *Elements* I, Common Notion 4 "Things which are superposed on one another are equal to one another."[34] Alfonso advances a new notion that he calls "mental superposition," an isometric superposition that can be used as a measuring device, but is still flexible enough to map curved on straight magnitudes. To this end he offers three additions to the axiomatic basis of geometry: he adds to Common Notion 4 the requirement that the superposition should be instantaneous (I.2.2), and introduces two new construction postulates, the postulate of imagining motion (I.5.1) and that of measuring by motion (I.5.3).[35]

With these modifications in the axiomatic basis of geometry, motion can bridge over the dimensional barriers, so that magnitudes can be measured by the motion of lower-dimensional magnitudes. Viewing motion as an intermediate existence between potential and actual, Alfonso turns it into an instrument of measurement. In Euclidean geometry a line is infinitely divisible potentially, but when actually divided it becomes two separate lines. Alfonso suggests a new sense of division in geometry, division by motion, that is neither potential nor actual: a magnitude can be divided by the motion of a lower-dimensional magnitude, which cuts equal sections (*shevarim*) one after the other, without actually separating them. The continuity and unity of the magnitude are maintained, and the task of dividing magnitudes

---

[32] *Elements* V definitions 3, 4.

[33] Aristotle, *Meta.* V.13, 1020a11–13. In the definitions of Part III, Alfonso Replaces the strict Greek language of homogeneous ratios by a more flexible language of geometrical operators. "I call an expansion [רבוי] of a straight line by a straight line the rectilinear right-angled triangle enclosed by the two lines". He by no means shifts to a symbolic algebraic language, but allows more flexibility within geometry.

[34] On the translation see Section I.1.1, note on the term רכב.

[35] The notions of Euclidean and mental superposition are discussed in more detail in the introductory note to Section I.2.

into *shevarim* this way can, according to Alfonso, be carried on indefinitely. The arguments in I.3.4 (Argument ii), I.5.3, and the extant beginning IV.1 suggest that Alfonso could have thought of measuring areas by the motion of lines, and volumes by the motion of planes. In the incomplete proposition IV.1 the extremity (i.e., one side) of a rectangle is moved maintaining a right angle with the basis so that *every height* on the basis becomes ontologically equivalent to the side, namely to an extremity. This is close to considering areas as infinite aggregates of lines, in a way that calls to mind that of Bonaventura Cavalieri, three centuries later.[36] Unfortunately the rest of Part IV is not extant, so we cannot know where Alfonso led his ideas about "evaluating and measuring" magnitudes by one another.

In Alfonso's new geometry superposition plays a major role and magnitude becomes a more flexible entity. Presumably, his revision of geometry is not independent of his theology-philosophy. Whereas in Aristotle's philosophy body is the manifestation of stability (whereas lower-dimensional magnitudes are conceived as "sections and divisions" of body), in Alfonso's theology every corporeal body has a subtle body (*cuerpo glorificado*) that has no quantity or dimensions.[37] It is God, not matter, that endows bodies with dimensionality.[38] Alfonso's notion of superposition is permeated with theological and philosophical meanings. The main theme of *SMA* is "mental superposition" (הרכבה שכלית), namely the isometric transforming of geometrical magnitudes on one another. The main theme of Alfonso/Abner's theology is *hitlabbshut*, a concept that stands for the Christian notion of incarnation,[39] but also for a more general notion of divine immanence. It has several manifestations in Alfonso's theology,[40] and is the backbone of his interpretation of the sacrament of the *Eucharist*.[41] It is not unlikely that the association between the geometrical notion of mental superposition and the theological notion of *hitlabbshut* contributed to the significance of both concepts in Alfonso's eyes. Perhaps he developed his own geometrical-theological theory of physical existence in his lost book *New Philosophy*.

---

[36] See citation in the commentary to IV.1.

[37] Sadik (2007, 136–7).

[38] Sadik(2011, 71–82) (Section 1.6.2.2). Allusions to this peculiar notion of dimensionality may be found in *SMA* as well: "He [God] determines the actual specific limits of distances that are potentially unlimited" (II.4.2); "first quantity is not an attribute of the first body" (II.1.7).

[39] Gershenzon 137. Baer (1958, 287) finds in the concept of *hitlabbshut* a mixture of Kabbalistic and Christian ideas.

[40] Gershenzon, Chap. V; Sadik (2011, Chap. 1.6.2).

[41] The metaphor of the mirror as an entity in which an object can be in many places without being anywhere (Szpiech 2006, 157) may help us understand Abner's imaginative and rather unique way of associating geometrical and theological conceptualization.

5 The Intellectual Environment of *SMA*

The Jewish Background

SMA is a product of the Jewish Iberian tradition.[42] Like other books that were written in the same intellectual milieu,[43] *SMA* is based on the mathematical vocabulary that was established by Abraham Bar-Ḥiyya two centuries earlier. Of the rich corpus that was created by Provençal Jews in the twelfth to fourteenth centuries only a few biblical commentaries are referred to in Alfonso/Abner's extant writings.[44] Of the many translations and commentaries on philosophy and the sciences that were produced beyond the Pyrenees Alfonso seems to be unaware. We found no indication that he used *any* Hebrew translation of either a philosophical or a mathematical text that was made in Provence (see Appendix below), nor any indication that he was familiar with original compositions (philosophical or mathematical) by Provençal Jews. In particular, he did not adopt the mathematical language that was developed in Provence in the thirteenth century by the Tibbonids in their translations of Euclid's *Elements* and other geometrical texts,[45] and adopted in the fourteenth century by Qalonimos ben Qalonimos (the translator of Archimedes' *The Sphere and the Cylinder*) and other Provençal writers. Jewish intellectuals in Christian Spain were familiar enough with the Arabic language not to be dependent on translations.

The Arabic Background

Alfonso's mathematical and philosophical education was based mainly on Arabic texts. Szpiech has shown that he was well versed in Arabic thought and was familiar with some works of the leading philosophers al-Fārābi, Ibn Sīnā, al-Ghazali, and Ibn Rushd, as well as of other Arabic thinkers.[46] Szpiech is not sure, however, whether Alfonso studied from Arabic or Hebrew texts: "While it seems unlikely that Abner/Alfonso knew no Latin or Arabic, he shows no indication of advanced ability sufficient to read extensively in the available philosophical and theological works in these languages."[47] Already more than half a century ago Georges Vajda questioned the assumption that Alfonso read al-Ghazali in

[42] See Szpiech ibid.

[43] This is notable in the mathematical texts that were written by authors who were brought up in Christian Spain: Yehuda ben Shlomo's *Midrash ha-Ḥokhma*, IsḤāq Isra'eli's *Yesod 'Olam* and IsḤāq al-AḤhadab' *Epistle of Number* (on the latter see Wartenberg 2007, 33).

[44] Abraham ben David of Posquières (Rabad), David QimḤi and Yosef QimḤi. Rashi (eleventh century France) is also quoted. See Szpiech (2006, 657, 659).

[45] Lévy (1997a, 185–6; 2005, 107–8).

[46] Szpiech (2006, 593–642).

[47] Szpiech (2006, 445).

Hebrew.[48] Steven Harvey showed that Alfonso "brings rather accurately" Alfarabi's summary of Plato's *Laws*, a text that was never translated into Hebrew.[49] The study of *SMA* similarly shows that Alfonso was acquainted with several texts that were available only in Arabic.[50] Moreover, Alfonso did not use texts that were translated into Hebrew (see Appendix). It is highly likely that he could read Arabic well, and that his mathematical and philosophical sources were mainly in Arabic. Recent research established that Arabic manuscripts (in Arabic script and in Hebrew script) circulated among Jews in Christian Spain (sometimes commissioned by Jews),[51] and that the knowledge of Arabic among Jewish intellectuals in Christian Spain—most notably, but not only among physicians[52]—was maintained in the fourteenth, even in the fifteenth century.[53]

The Christian Background

Although Alfonso converted to Christianity and had a sacristan clerical position in Valladolid, scholars have doubts that he acquired a good command of Latin.[54] Shalom Sadik argues that "the sources of the Christian Neoplatonic philosophy of Abner were not actually Christian, but derive mainly from Jewish or Arabic sources."[55] Sainz de la Maza suggests, nonetheless, that already years before his conversion Alfonso was welcome in the *Monasterio de las huelgas* in Burgos, and was supported by Dona Blanca (Blanche), the daughter of King Alfonso III of Portugal and Beatrice, the illegal daughter of King Alfonso X of Castile. Dona Blanca lived from time to time in the *Monasterio* until her death in 1321.[56] It is reasonable to assume that Alfonso, who was engaged throughout his life in the search for wisdom,[57] tried to gain access to the library of the monastery in his town.[58] Sainz is confident that Alfonso had connections with the monastery long before his official conversion,[59] either through his medical profession or, perhaps, through a job that he had as a copyist, illustrator, and binder of books.[60] Dona Blanca, known

[48] Vajda (1959).

[49] Harvey (2003, 66–7).

[50] E.g., by al-Nayarizi and Ibn al-Haytham.

[51] Langermann (1996b); Koningsveld (1992); Assis (2012); Hacker (2015).

[52] Sainz (1990, 153) has convincingly argued that Alfonso was a physician.

[53] Sirat and Geoffroy (2005).

[54] Szpiech (2006, 444–5); Sadik (2016a, 99).

[55] Sadik (2016a, 98, 102).

[56] Sainz (1990, 167).

[57] See Poetical Preface and Section I.3.1.

[58] Our attempt to find some of Alfonso's "sources" in the catalog of the Monasterio did not yield any results.

[59] Sainz (1990, 171).

[60] Sainz (1990, 190, note 86).

for her "*aficiones culturales,*" had in her possession some books that belonged to her grandfather King Alfonso X (*el sabio*), famous for his erudition.[61] Sainz suggests that she sponsored Alfonso during the process of his conversion to Christianity.[62]

Furthermore, it has been shown by Carlos Steel, Mark Geoffroy, and Guy Guldentops that Alfonso collaborated with Alfonso Dinis of Lisbon in the translation of Ibn Rushd's *Beatitude of the soul.*[63] Alfonso Dinis was a relative of Dona Blanca, and was also associated with the royal family of Portugal. He was a physician, later a bishop, who was interested in the sciences and theology. He had academic and noble connections in Paris, in Castile, and in Portugal.[64] Steel and Guldentops admit that they do not know what was the role of Alfonso of Valladolid in the joint translation of Ibn Rushd, but they suspect that he did most of the job, because of the poor quality of the Latin. Alfonso Dinis, they assume, should have written better Latin.[65] If they are right, i.e., if Alfonso of Valladolid was able to write Latin, even if poor, a fortiori he must have had reading skills in this language.

SMA provides no new evidence on Alfonso's command of Latin, but it suggests that he was familiar with some debates that occupied contemporary scholastic scholars, notably on divisibility.[66] This knowledge could have reached him in one way or another through the University of Valladolid, or perhaps through oral communication with Alfonso Dinis. Dinis stayed in Paris three times,[67] and when back home he shared his time between Portugal and Castile.[68] The joint translation of the two Alfonsos may have taken place between 1332 and 1342.[69] Alfonso Dinis was interested in astronomy,[70] and it is not unlikely that he updated his employee (or friend) Alfonso of Valladolid on the state of the mathematical and philosophical studies at the university of Paris.

6 Mathematical and Philosophical Knowledge in *SMA*

Alfonso had a wide knowledge in mathematics and philosophy, in some cases surprisingly so. He knew about some episodes in Greek and Arabic geometry that historians have not considered to had been known in the West in the middle ages.

---

[61] To our disappointment the catalog includes no books in Arabic. This does not necessarily imply, however, that Arabic books were not accessible in the monastery.

[62] Sainz (1990, 166–7).

[63] Geoffroy and Steel (2001, 87–91); Steel and Guldentops (1997, 87–9); Meirinhos (2007, 58–63).

[64] Meirinhos (2007).

[65] Steel and Guldentops (1997, 89).

[66] Sections I.3.5, I.4.3–4, II.5.5.

[67] He studied at the university of Paris medicine (1329 and 1332) and theology (1342–5). See Meirinhos (2007, 52).

[68] Meirinhos (2007, 50–51).

[69] Meirinhos (2007, 51).

[70] E.g., Merinhos (2007, 54).

In some cases his knowledge was based on texts that he had under hand (e.g., al Nairizi or Ptolemy), in some cases on second-hand or third-hand sources (e.g., in the case of Plato and the pre-Socratic thinkers), and in some other cases he seems to have learned from fragments or through oral communication (e.g., in the case of al-Tūsī). We do not know whether he ever left his homeland Castile, or whether he learned from encounters with people who traveled. In some of the propositions of Part III we have not been able to identify the sources. We can, however, confidently say that he tended to abridge, and sometimes to change the proofs that he used. He was quick to adapt what he learned to his own needs and style.[71]

In what follows we offer a list of the authors mentioned in *SMA* and a few additional authors not mentioned by name, with whose thought, we suspect, Alfonso was to some extent acquainted.

Greek Authors

Alfonso was highly interested in the Greek tradition, mathematical and philosophical alike. He castigated Aristotle for criticizing his predecessors too readily,[72] and made a serious effort to gain a more direct access to ideas of early Greek thinkers. This attempt was only partially successful. Alfonso learned about Platonic and pre-Socratic natural philosophy mainly through the mediation of the Aristotelian tradition. He gained access to some episodes in Greek mathematics, notably the squaring of the lunes by Hippocrates and the construction of the conchoid by Nicomedes; the route of transmission of this knowledge is as yet unknown.

Empedocles (אבן דקליס), Leucipus and Democritus (לוקיש, דימריטיש)
Sections I.3.1, I.3.4

These thinkers are mentioned by name in *SMA*. The reference to Empedocles is hardly informative. The reference to Leucipus and Democritus could have followed Ibn Rushd's middle commentary on Aristotle's *De Generatione and Corruptione*.

Hippocrates of Chios (אבקרט)
Sections I.2.3, I 2.5 (mentioned by name), III.23 (not mentioned by name)

In Part I Alfonso discusses Aristotle's criticism of Hippocrates' alleged suggestion to derive from the quadrature of the lunes the quadrature of the whole circle. Alfonso probably relied on Ibn Rushd.[73] In Part III Alfonso renders in detail two quadratures of lunes, *without mentioning Hippocrates' name*. These quadratures are

[71] As already noted, in Part III he fairly consistently modifies the arguments that he uses by shifting from the classical Greek language of proportion to a language of expansion (רבוי). On this term and its importance for Alfonso see the discussion in the introductory note to Part III.

[72] Section I.3.1.

[73] See commentary on Section I.2.3.

known to us through Simplicius' commentary on the *Physics* but we do not know what Alfonso's direct source was.[74] The Hebrew spelling is a transcription of the Arabic أبقراط.

Antiphon (אנטיפרס, אנטיפרש) **and Bryson** (ברוסן)
Antiphon Sections I.3.5, I.4.3, I.5.3, Bryson Sections I.1.3 and I.1.4

The attempts of these two sophists to square the circle were mentioned and criticized by Aristotle. Alfonso probably learned about them from Ibn Ruhsd's commentaries.[75] He had much respect to both of them and referred to Bryson as "The Sage."

Plato (אפלטון, once אפלאטון)
Nineteen references, in the outline, Sections I.3.1–5, I.4.2, and outline of Part IV

Alfonso has often been depicted as a Platonist, but his acquaintance with Plato's teaching was partial and deficient.[76] According to his own testimony (Section I.3), Alfonso was familiar with Plato's geometrical cosmology through the critical presentation of Aristotle, probably *via* Ibn Rushd's middle commentaries on *De caelo* and *De generatione et corruption*.[77] Part IV of *SMA* was meant to provide a sound mathematical basis for Plato's cosmology.

Aristotle (ארסטו, ארסטוי, in III.33 ארסטוטי)
Fourteen references, thirteen in Parts I–II, one in Section III.33

Alfonso's attitude to Aristotle *qua doxographer* is expressly critical. In 10 of the 14 references to Aristotle the context is Aristotle's presentation of the views of earlier philosophers (once in I.1.1, nine times in I.3). Alfonso explicitly criticizes Aristotle for underappreciating the views of his predecessors (I.3.1). His attitude to *Aristotle qua philosopher* is equivocal.

On the one hand, Alfonso is often perceived as a Platonist and a critic of Aristotle.[78] Shalom Sadik argued that in his theological writings Alfonso criticizes Aristotle's understanding of matter, dimensionality, and place.[79] In *SMA* too Alfonso adopts a philosophical conception of matter and dimensionality different from Aristotle's (II.4.2, II.1.7, I.4.4) and rejects Aristotle's conception of motion as being "from something to something" (III.33).[80] In *SMA* he challenges Aristotle on issue of

---

[74] Langermann (1996a). See commentary on Section III.23.

[75] See commentary on Sections I.1.3 and I.3.5.

[76] It has been suggested that in his other writings, when he mentions Plato by name Alfonso sometimes refers to Plotinus' *Enneads*, most frequently to the fifth Ennead. See Szpiech (2006, 662); Sadik (2011, notes 208, 257, 287, 331). This text was available in Arabic but not in Hebrew. See Zimmermann (1986, 112–13, 128–31). It does not seem that Alfonso relies on Plotinus in SMA.

[77] See commentary on Sections I.3.2 and I.3.4.

[78] Gluskina (1978, 69).

[79] Sadik (2007).

[80] *Physics* V.1. Alfonso refutes Aristotle's tenet that oscillatory motion must come to a rest at the endpoints (III.33).

homogeneity in geometry: [i] he objects to Aristotle's view that straight and curved magnitudes are incomparable entities (I.1.1), [ii] he objects to Aristotle's conception of one-, two-, and three-dimensional magnitudes as different types of entities.[81]

On the other hand, like the majority of medieval thinkers, Alfonso was trapped in the Aristotelian conceptual network. It was not unusual for medieval thinkers who criticized Aristotle's doctrines on certain matters, to accept uncritically his teaching on other matters. Alfonso, too, often draws on Aristotle, taking the Philosopher's tenets on several major issues for granted: [i] He accepts Aristotle's theory of continuity of *Physics* VI (I.3.2–4), in particular the thesis (*Physics* VI.1) that time, motion, and distance are isomorphic (I.4.2). [ii] He relies on Aristotle's (rather problematic) contention (*Physics* VI.5) that motion has no first instance (II.1.2). [iii] He strictly follows Aristotle (*Physics* III.8) in totally denying the possibility of actual infinity (II.1.7). [iv] He adopts the notion of minimal part that originated in Aristotle's *Physics*, and developed by later commentators, mainly by Ibn Rushd (I.4.2).[82] [v] He justifies the use of moving lines taking example from Aristotle (II.1.6).

Alfonso's presentation of Aristotle's views is often so brief that it is impossible to identify his direct source. In those cases that he provides some more detail, the impression is that he learned of Aristotle's teaching through the commentaries of Ibn Rushd.[83] There is no indication, however, that Alfonso read any of these texts in Hebrew translation.[84]

Euclid (אקלידס, a few times אקלידיס, אקלדס)
Twenty-four reference in Parts I–III

In the preface to Part III Alfonso announces his intention to summarize the mathematical knowledge needed in Parts IV and V, so that the reader will have to consult no text besides Euclid's *Elements*. Although he takes for granted the reader's acquaintance with Euclid's canonical book, it is doubtful that he himself had first-hand access to all the 13 books of the *Elements*.[85]

Our knowledge on the Arabic tradition of the *Elements* is still deficient. The first known Arabic translation of Euclid's *Elements* was made before 805 by al-Ḥajjāj Ibn Yūsuf ibn Matar and, according to some sources whose credibility is uncertain, was reworked years later by al-Ḥajjāj himself.[86] Of this translation we have today only fragments. The commentary of al-Nayrizi on Books I–VI was probably based on Ḥajjāj's translation, but was heavily edited.[87] A second (not independent) translation of the *Elements* was made in the last third of the ninth century by IsḤāq Ibn

[81] Section 4 above.

[82] Glasner (2009, Chap. 8).

[83] See commentary on I.1.1, I.1.3.

[84] See Appendix below.

[85] There are no allusions at all to the added 14th and 15th book.

[86] Djebbar (1996, 96–8); Brentjes (2001b, 39; 2006, 171–4; 2001a, 24).

[87] On the relation between Hajjaj's original translation and al-Nairizi's commentary see Engroff (1980, 19); Brentjes (2001, 24–8). See also the section on al-Nayrizi below.

Ḥunayn and was revised by Thābit Ibn Qurra. The validity of this description has been questioned recently.[88] Hebrew translations from the Arabic were made in Provence in the thirteenth century by Moshe Ibn Tibbon, his nephew Ya'aqov ben Makhir, and R. Ya'aqov (perhaps Ya'aqov Anatoli). A few more mixed versions, anonymous compilations, and paraphrases are known.[89] Sarfatti referred to the mathematical jargon of Moshe Ibn Tibbon and Ya'aqov ben Makhir as "the language of the translators."

Alfonso was not familiar with the language of the translators. Perhaps he was not even familiar with the title of the Hebrew *Elements*. Whereas the translators refer to the *Elements* as ספר היסודות or ספר השרשים, Alfonso refers to it as ספר העקרים.[90] As he himself testifies (II.3.0), he used al-Ḥajjāj's Arabic translation and numbered the propositions following al-Ḥajjāj. He quotes verbatim two passages from al-Nayrizi's commentary on the first book, so we may assume that at least the beginning of this commentary was on his desk. Did he have in his possession besides al-Nayrizi's commentary also an original version of Ḥajjāj's translation? Explicit references are found in *SMA* to propositions of *Elements* Book I[91] and one reference to Book V[92]; in Part III propositions of *Elements* V and VI are assumed without references.[93] In two of the references to Book I and in the one reference to book V the proposition numbers are not identical to those known to us through Heiberg's edition, but it is doubtful that we can learn much from these differences.[94]

The intriguing question is whether Alfonso had direct access to Ḥajjāj's translation or only to al-Nayrizi's commentary that includes Books I–VI. Alfonso refers once to Book X, but no specific proposition is mentioned; the reference is general and cannot confirm acquaintance with *the text* of Book X.[95] Alfonso ascribes to Euclid statements that Euclid did not make.[96] The most serious reason to doubt that Alfonso had access to a complete copy of the *Elements* is the glaring absence of any mention of Book XII. Alfonso is highly interested in the method of exhaustion, but when discussing this method, he refers to Bryson and Archimedes, never to Euclid.[97] It is thus not certain that Alfonso had first-hand acquaintance with the complete text of the *Elements*.[98]

---

[88] Brentjes (2018).

[89] See Lévy (1997b); Elior (forthcoming).

[90] Sections III.0, III.11.

[91] Explicit references: to I.11; to I.13 = Heiberg I.17; to I.12 = Heiberg I.13 (all in II.3.2); to I.34 (I.2.2), to I.16, I.26 (II.3.7), to I.33 (II.4.1), to I.47 (III.11).

[92] To V.13 = Heiberg V.12 (in I.5.4).

[93] The propositions used are: V.8, 9, 10, 12, 14, 16, 19, 24; VI. 1, 2, 4, 8, 12, 16, 20, 22, 23, 25, 31.

[94] It is not clear whether these are author's or copyist's errors, or whether they carry some information on the version that Alfonso used. On the numbering of the proposition in the Arabic versions of the *Elements* see Jaouiche (1986, 138–144); Lévy (1997a, 223, n. 59).

[95] See II.5.5.

[96] In I.1.2 he ascribes to Euclid the three circles example, perhaps assuming that it is included in *Elements* XII. In I.3.4 he ascribes to Euclid the statement that any solid is divisible to pyramids, perhaps assuming that it is included in *Elements* XI.

[97] I.1.3–4.

[98] On the secondary transmission of Elements in Arabic see Brentjes (2001, 39–40).

Archimedes (ארשמידס)[99]
Sections I.1.2, I.1.4, I.2.4, II.5.5 (mentioned by name), III.18 (not mentioned by name)

In Parts I–II several propositions from the first book of *The Sphere and the Cylinder* are alluded to, but neither proposition numbers nor quotations of the announcements are given. Archimedes' book was translated into Arabic twice,[100] and then into Hebrew by Qalonymos ben Qalonymos (Provence, ca. 1317). Property 18 of Part III is based on Archimedes' *On the Sphere and Cylinder* I.21, but significantly modified. There is no way to know whether Alfonso had direct access to Archimedes' book or whether he used a secondary source. It seems that he did not use the Hebrew translation of the *Sphere and Cylinder.*

Alfonso refers several times to Archimedes' *Measurement of the Circle*, which is highly relevant to the project of *SMA*. Roshdi Rashed established that there were two Arabic translations of the *Measurement,*[101] and Tony Lévy has shown that there were two Hebrew translations, to which he refers as *HA* and *HB*.[102] While *HB* was made by someone who was familiar with the mathematical terminology of the Provençal translators (perhaps Qalonimus ben Qalonimus), the translator of *HA* was not familiar with that language and used the Sephardic mathematical vocabulary.[103] Ruth Glasner suggested that translation *HA* was made by Alfonso himself, perhaps for his private use.[104]

Ptolemy (בטלמיוס)
Sections III.12, I.2.1 (implicitly)

Alfonso presumably draws on Ptolemy's table of chords (*Almagest* I.11) when he adduces his first example of non-isometric superposition. Moreover, it is possible that his rudimentary concept of function was inspired by Ptolemy.[105] In Part III he quotes and proves Ptolemy's famous theorem (*Almagest* I.10),[106] explicitly ascribing it to Ptolemy. It is thus possible that he used the *Almagest* itself rather than a popular presentation of it.[107]

---

[99] Three of the four manuscripts of Qalonymos' Hebrew translation of *The Sphere and the Cylinder* offer an unusual spelling of Archimedes' name: ארשמידאת or ארשמידאס. In *SMA* as well as in one of the two Hebrew translation of *The Measurement of the Circle*, the name is spelled ארשמידס, which is the correct transcription of the common medieval Arabic spelling ارشميدس.

[100] On the Arabic translations see Rashed (1997, 3–4). See also Rashed (1993).

[101] Rashed (1997, 4–5).

[102] Lévy (2011, 129).

[103] Lévy (2011, 134–7).

[104] Glasner (2013).

[105] See Pedersen and Jones (2010, 49, 79–81).

[106] "In a quadrilateral inscribed in a circle, the sum of the products of opposite sides with one another is equal to product of its two diagonals with one another" (Toomer 50–1).

[107] The quotation is neither a literal translation from the Arabic nor a verbatim quotation from the Hebrew translation by Ya'aqov Anatoly (Cf. Wien ms. 194/40, IMHM 1317, fol. 20b20–12a11). Alfonso "translates" the theorem to his own language of *ribui*. See Part III, introduction and commentary on Property III.12.

Simplicius (סנביליקיוס)
Section II.2.3

Simplicius wrote a commentary on the definitions, postulates, and common notions of the first book of the *Elements* that was known to al-Nayrizi,[108] and Alfonso mentions Simplicius's statement about the proof of the parallels postulate after al-Nayrizi.

The construction of the quadrature of the lunes that Alfonso recites (III.23) was preserved in Simplicius' commentary on the *Physics* I.[109] This commentary was not translated into Arabic, but was translated into Latin in the thirteenth century. We do not know what Alfonso's direct source was.

Greek Authors Not Mentioned By Name

Nicomedes
Sections III.29–32

Alfonso was acquainted with the construction of the conchoid and some of its uses. The study of this curve is attributed to the Greek mathematician Nicomedes, who lived in the third century BC. Nicomedes' work is known today through some remarks of Pappus and Eutocius. We do not know how Alfonso learned about the conchoid.

Arabic Authors

SMA provides important information about the circulation of Eastern Arabic texts in the Iberian Peninsula. With the exception of Ibn Rushd, all the Arabic authors referred to explicitly by Alfonso lived in the east. It is likely that in addition to the authors listed below, Alfonso relied on one or more Arabic geometrical compilations that we have not identified.

Al Nayrizi (אלניריזי)
Sections II.2.3–4

A Persian mathematician and astronomer (865–922), who wrote a commentary on Euclid's *Elements*, which he accessed in al-Hajjāj's translation.[110] Al-Nayrizi's commentary circulated in the Iberian Peninsula, and in the twelfth century was translated into Latin by Gerard de Cremona. It was also one of the sources of Yehuda

[108] Sabra (1969); Jaouiche (1986, 18–19).

[109] Simplicius *In Phys*. 61–69; Langermann (1995, 34).

[110] Two extant copies of the commentary are preserved: Codex Leidenesis 399 (edited by Besthorn and Heiberg 1897–1905) and Qom ms. See Brentjes (1996, 203, note 13, 206); Lo Bello (2003, pp. xv–xvi, xxviii).

ben Shlomo in the geometrical part of his *Midrash ha-Ḥokhma*.[111] Alfonso copied verbatim two lemmas from the first book of al-Nayrizi's commentary,[112] so we can assume that he had it on his desk.

Ibn al Haytham (בן הייתם)
Section II.2.1

Ḥassan Ibn al-Haytham (965–1039) was a leading mathematician and scientist, born in Basra and died in Egypt.[113] Alfonso was acquainted with his *Commentary on the premises in Euclid's Elements*.[114] Ibn al-Haytham wrote also a treatise on the quadrature of the circle and two treatises on the quadrature of the lunes.[115] There is no indication that Alfonso was acquainted with either one of these treatises. See commentary to Sections I.1.1 and III.23.

Ibn Sīnā (בן סינא)
Sections I.1.1 (mentioned by name) II.1.1 (not mentioned by name)

A physician and philosopher of Persian origin (980–1037). Alfonso refers to Ibn Sīnā as "The Sage." Szpiech who studied thoroughly Alfonso's book *Mostrador de justicia* found a few references to Ibn Sīnā, and has also shown that Alfonso was familiar with Ibn Sīnā's now lost *Oriental Philosophy*.[116] In *SMA* Alfonso refers briefly to Ibn Sīnā along with Ibn Rushd as representatives of the view that circular and rectilinear areas cannot be compared (I.1.1). In addition, without mentioning Ibn Sīnā's name, Alfonso recapitulates his argument against motion in the void (II.1.1). He could have found it in the *Physics* part of the *Shifā'* or, perhaps, in some later presentations of this argument.

Ibn Ridwān (אבו עאלי בן רצואן)
Section III.33

A physician and astronomer who lived in Egypt (988–c. 1061). Alfonso refers to Ibn Ridwān's argument against *quies media*. We have not located the reference.[117]

Ibn Rushd (בן רשד)
Sections I.1.1, I.2.1

The highly influential Andalusian philosopher and commentator (1126–1198) is mentioned by name twice as the representative of the strict conception of geometry that forbids comparison of nonhomogeneous magnitudes. It is possible that Ibn Rushd's commentaries were Alfonso's direct source when he refers to Aristotle.[118]

[111] See Lévy (2000, 302, 309).

[112] See Section II.2.3.

[113] On Ibn al-Haytham's contribution to mathematics see Rashed (2013a, b).

[114] This text was translated into Hebrew by Moshe Ibn Tibbon.

[115] Rashed (2013a, Chap. 1).

[116] Szpiech (2010).

[117] A manuscript which is partly in Hebrew letters testifies that Ibn Ridwan's commentary on Ptolemy's *Tetrabiblos* circulated among Jews in the Iberian Peninsula. See Koningsveld, p. 102, item 90.

[118] See commentary to sections I.1.1, I.1.3, I.2.3, I.3.2–4.

Arabic Authors Not Mentioned by Name who Could have Inspired Alfonso

Thābit Ibn Qurra

Sections II.3.6, III.28 (not mentioned by name)

A Sabian mathematician, astronomer, and translator who lived in Baghdad (836–901). He attempted two proofs of the parallels postulate. Alfonso was probably acquainted with the first. Also, a theorem in spherical trigonometry by Thābit is reproduced in III.28.

Al-Ghazali

Section II.1.3 (not mentioned by name)

A Persian philosopher and religious thinker (1058–111). Al-Ghazali's *Maqāṣid al-falāsifah* (*Intentions of the Philosophers*), which draws heavily on Ibn Sīnā's Persian encyclopedia *Danesh Name* (*The Book of Science*), includes several arguments against atomism.[119] Alfonso was probably acquainted with Al-Ghazali's *Maqāṣid,* a book that was influential among Spanish Jews. This text was translated into Hebrew, perhaps three times, but Alfonso presumably read it in Arabic.[120] He recapitulates Al-Ghazali's arguments against atomism and against infinity.

Al-Tūsī and the school of Marāgha

Sections I.1.2, I.2.1, II.5.1, III.33 (not mentioned by name)

A Persian mathematician and scientist (1201–1274), who was in charge of the Marāgha observatory. Alfonso seems to have been aware of several mathematical ideas that originated in the Marāgha schools: [i] the analysis of superposition (I.1.1, I.1.2, I.2.2); [ii] the three-circles example, namely the equality of a circle and the sum of two small circles (I.1.2); [iii] the analysis of the rolling movement of a cylinder on a plane (I.2.1, II.5); [iv] most notably, the construction of the Tūsī couple (III.33). Alfonso describes the construction in detail and proves that the resulting composite motion is linear. The proof is not identical to Tūsī's. A possible explanation is that Alfonso had some second-hand knowledge of the work done by al-Tūsī's and his students, adopted their ideas, and completed the details himself.

---

[119] In his dissertation Murdoch (1957) argues that Duns Scotus, who was the main Latin source for geometrical arguments against indivisibilism, drew on the Latin translation of al-Ghazali. Podkonski (2006) contends that this thesis should not be accepted without reservation, and in a later presentation Murdoch put the thesis more cautiously: one source of the geometrical argument against indivisibilism was al-Ghazali's *Metaphysics* (Murdoch 2009, 19).

[120] See Appendix.

Jewish Authors

The only medieval Jewish scholar named in *SMA* is R. Moshe ha-Levi of Seville. Notable in his absence is Maimonides, with whose writing Alfonso was well acquainted.[121] Maimonides raised the question of the two lines that approach one another but do not meet (the asymptotes) and, following Apollonius, he offers a solution using a hyperbola.[122] Alfonso raises the same question (III.29), does not mention the Maimonides' solution, and offers another using a conchoid.

Moshe ha-Levi of Seville (ר׳ משה הלוי ממדינת אשבלייא)
Sections II.1.5–II.1.7

Alfonso dismisses the attempt by "Moshe ha-Levi's and others" to prove the parallels postulate by means of motion. Freudenthal identifies Moshe as the thirteenth-century physician and philosopher Moshe ben Yossef, known in Arabic as Abu 'Imrān Musa' al-Lawī al-Ishbilī.[123] Moshe was described by Freudenthal and Zonta as a "committed Avicennian philosopher."[124]

We do not know in what year Moshe died, but of all scholars mentioned in *SMA*, Moshe of Seville was the closest to Alfonso in time, place, and cultural milieu. Seville was a flourishing city in the thirteenth century and its connections with North Africa on the one hand and with Italy on the other were intensive. It was one of six cities in Spain and the Maghreb that became cultural centers under the rule of the AlmoḤad dynasty.[125] Barnabas Hughes (who edited Fibonacci's *De Practica Geometrie*) conjectures that mathematical works circulated among the scholars of these six cities. Also, Italians merchants, especially from Genoa, settled in Seville.[126] Moshe was interested in mathematics and it is not impossible that in his youth Alfonso had some interaction with him and perhaps gained access to his notebooks or, with his help, to some mathematical texts that came from the east.

[121] Sadik (2016b, 35) writes that Alfonso refers to Maimonides more than seventy times in his writings, more than to any other post-Talmudic Jewish author.

[122] Maimonides, *Guide* I.73; see Freudenthal (1998). Alfonso does not refer to this passage, and offers a different, more complex solution using a conchoid (which apparently interested him more). Also, Maimonides asserts in his commentary on the *Mishna* (*'Eruvin* 1.5) that the ratio of the diameter diagonal to its circumference cannot be known, but only approximated. Alfonso does not mention this statement either.

[123] Freudenthal (2005, IX, 4); Hebrew (1991, 986). See also Lévy (1992, 40); *Encyclopedia Judaica*, vol. 14, pp. 551–2.

[124] Freudenthal and Zonta (2012, 232).

[125] Hughes (2008, xxi). Seville fell into the hands of the Christians in 1248.

[126] Verlinden (1953, 200–2).

Christian Authors

Only one Christian author, the thirteenth-century Italian Campaus, is mentioned by name in *SMA*. The question naturally arises whether, or to what extent was Alfonso familiar with Scholastic philosophy of his time, and was interested in it.

Johannes Campanus (קנפאנו)
Section II.5.6

Johannes Campanus of Novara was an Italian mathematician and astronomer (1220–1296), who published a Latin edition of Euclid's *Elements* (based on the Arabic) in 15 books.[127] Albert of Saxsony distinguished several senses of squaring the circle, and attributed to Campanus a "squaring" of the circumference, rather than of the area.[128] Campanus' text is problematic and Marshal Clagett, who published the text *De quadratura circuli*,[129] remarked that he was "somewhat reluctant to concede that this elementary piece, exhibiting such immaturity, could be the work of the same man who composed the commentary on the *Elements* of Euclid."[130] Alfonso, too, refers to this text in derogatory terms.[131]

Contemporary Scholastic Scientists Who Could Have Inspired Alfonso

There is no conclusive evidence that Alfonso had direct acquaintance with Scholastic texts. He shared, however, some interests with the schoolmen of his time and was probably aware, to some extent, of the discussions that took place in the Christian universities. The following examples are notable:

[i] The incommensurability of the planetary periods (II.5.5)
In the *Mostrador* Alfonso adduces an argument against eternal return from the incommensurability of the planetary periods:

> If the relations between the planetary periods were rational and commensurable, it could be that all the individuals of the world will return at some point to the same way and the same order exactly as they have been at a different time. But if those relations were irrational, not rational and commensurable, as it is proved in Chapter 10 of the Book of Euclid, it would never happen in any way that all the individuals of the world return in such a way and in such order that they seem exactly as they were at another time.[132]

---

[127] Busard (2005). Books XIV and XV are late additions to the *Elements*.

[128] Grant (1974, 172. Col. 1).

[129] Clagett (1964, I, 581–605).

[130] Clagett (1864, I, 581).

[131] Section II.5.6.

[132] Quoted from Szpiech (2006, 524–5). Szpiech' translation.

Alfonso could, perhaps, have found a brief reference to this question in Ibn Rushd.[133] The issue, however, attracted much more attention in Christian circles. Since Augustine the Stoic conception of eternal return, associated with the astronomic notion of the great year, raised objections among Christians. It comes up in a comment by Henry Bate that is attached to his Latin translation of Ibn Ezra's *Sefer ha-'Olam*[134] from 1281. Bate defends the thesis that the celestial configurations would sometimes return, commenting that this might have been Ibn Ezra's intention.[135] Ibn Ezra, however, does not say this and could not have been Alfonso's source.[136] Belief in eternal return was one of the theses prohibited in the famous condemnation by the bishop of Paris in 1277. Duns Scotus, suggests Grant, may have been the first after the condemnation to use the incommensurability argument to reject the possibility of precise cyclical returns[137]: if some of the ratios between planetary periods are irrational, precise cyclical return is ruled out.[138] The argument appears in both versions of Scotus' commentary on Peter Lombard's *Sentences*, the Paris version and the Oxford version. In the Parisian version Scotus refers to the tenth book of Euclid's *Elements*:

> Some celestial motions or celestial bodies are perhaps improportional and incommensurable, since the magnitudes on which they are located can be incommensurable and improportional as is evident from the tenth book of Euclid on the side and diagonal of the square.[139]

Scotus turns to *Elements* X in order to make the point that uniform motions along incommensurable lines are incommensurable. The reference is dubious and Scotus omits it from the later Oxford version. Could there be any indirect connection between Scotus' and Alfonso's use of the argument from incommensurability against eternal return and their reference to *Elements* X?

What was Alfonso's meaning in the reference to *Elements* X? A reasonable explanation is that he refers here to the fact, established by Euclid, that there are infinitely many types of irrational lines.[140] This explanation of Alfonso's reference to

[133] Grant (1971, 107). We could not locate the passage in Ibn Rushd.

[134] *Sefer ha-'Olam* I #24.1–7. See Sela, forthcoming.

[135] Sela, forthcoming; see also Sela (2017).

[136] We are very grateful to Shlomo Sela for clarifying this point. Grant (1971, 111–6) led us to assume that Ibn Ezra could have been Alfonso's source, but Shlomo Sela ruled out this possibility.

[137] Grant (1994, 499).

[138] Szpiech (2006, 524–5).

[139] Grant (1971, 120, note 97).

[140] In *Elements* X 115 Euclid proves that "from medial straight line there arises irrational straight lines infinite in number, and none of them is the same as any of the preceding"; in the second part of X.111 he proves that "the apotome and the irrational straight lines following it are neither the same with the medial straight line nor with one another".

Elements X in the *Mostrador* is supported by his reference to *Elements* X in *SMA*: "Every two unequal lines can be cut into very small parts, so that they are perceived as having a rational ratio to each other. Yet, presumably, there is no rational ratio between them, since there are infinitely many kinds of irrational ratios, as was proved in Euclid's tenth book."[141] *Elements* X is thus adduced in *SMA* and in the *Mostrador* for the same purpose: to argue that the probability of two lines (or of two planetary orbits) to be incommensurable is higher than their probability to be commensurable. Interestingly this statement is explicitly made in a surprising theorem by Nicole Oresme in his book *De proportionibus propotionum*:

> Proposition X. It is probable that two proposed unknown ratios are incommensurable because if many unknown ratios are proposed it is most probable that any [one] would be incommensurable to any [other].[142]

Nicole Oresme was occupied with this question and advanced the thesis of the incommensurability in two of his books *Ad pauca respicientes* and *Tractatus de commensurabilitate vel incommensurabilitate motum celi*.[143]

Alfonso wrote his book after Scotus and Bate, but before Oresme. He shares with the former the reference to *Elements* X and with the latter the deeper understanding of the relevance of *Elements* X to the possibility of cyclic return. It is possible that Alfonso was aware of some pre-Oresmeian stage in the scholastic discourse on planetary incommensurability. It can also be cautiously suggested that Oresme could have been informed on some ideas of Alfonso. Alfonso could, perhaps, have been the source of Oresme's somewhat vague allusion to the Ṭūsī couple too.[144]

[ii] The divisibilists-indivisibilists controversy

The fourteenth century saw an increased interest in the question of continuity in the Latin west. A lively controversy took place between the indivisibilists, who accepted some mode of existence of indivisible parts, and the divisibilists who defended Aristotle's theory of continuity.[145] The same dilemma underlies Parts I–II of *SMA*. No definitive conclusions can be drawn, but several examples that were adduced in the Scholastic arguments were also examined by Alfonso: [i] The example of the rolling cylinder that plays a major role in *SMA* was mentioned by Henry of Harclay, but was also noted before in Greek antiquity and in the school of al-Ṭūsī.[146] [ii] The view that inequality between infinite magnitudes is possible was shared by Alfonso, Harclay and some other scholastic scholars.[147]

---

141 Section II.5.5.

142 Grant (1966, 247).

143 Grant (1966; 1971).

144 See commentary to III.33.

145 Murdoch (1957, 15–16); Grellard and Robert (2009, 2).

146 I.2.1, II.5.1–4.

[iii] The example of the separation of two touching surfaces, distinguishing between two possible accounts of the way that air filters in.[148]

[iv] The example of the moving of the ray[149] was mentioned by Henry of Harclay, but also in several Arabic and Hebrew texts.

None of these examples proves that Alfonso was informed about the divisibilist-indivisibilist controversy, but it may not be accidental that he was so interested in the issue of divisibility at the time that it was a subject of controversy at the universities of Paris and Oxford.

Geometrical Compendia

Several encyclopedias, textbooks, and geometrical compendia circulated in the west in Alfonso's time,[150] some of which are no longer extant. The third part of SMA may include some contributions by Alfonso himself, but is mostly a selection from secondary and perhaps primary sources.[151] The closest that we found is a Hebrew geometrical compendium (hereafter *HGC*),[152] that was presumably composed either in the fourteenth-century Provence or in the fifteenth-century Italy. *SMA* shares with *HGC* four propositions: III.12 (Ptolemy's theorem), III.18 (Archimedes' SC I.21), III.23 (Hippocrates' quadrature of the lunes), and III.28 (a lemma of Thābit Ibn Qurra that is copied in the margin of book XI of *HGC*).[153] It was suggested by Tzvi Langermann that in III.23 (the squaring of the lunes) Alfonso used *HGC* and abridged it.[154] If Alfonso had access to *HGC* and used it, *he did not adopt the vocabulary* but remained faithful to the Iberian-Hebrew mathematical language that was based on the vocabulary of Abraham Bar-Ḥiyya. Afonso and the compiler of *HGC* could have relied on a common Arabic source.[155]

---

147 Murdoch (2009, 23 esp. note 26). Alfonso adopts this view in I.2.1 and II.5.4.

148 I.4.4.

149 Section II.1.3.

150 Berggren in Katz et al. (2016, 383–4); Rashed (2012, 727); Hogendijk (1994); de Young (2002).

151 The closest to verbatim quotations from first-hand sources ae al-Nayrizi's lemmas in Part II and Ptolemy's theorem in Part III.

152 *HGC* is a long text, preserved in the library of the Jewish community of Mantua and carries the title *The book of Euclid* Langermann (1984, 74; 2004); Glasner.

153 Two of the theorems common to *SMA* and *HGC* appear also in Ibn Hud's *Istikmal*: Ptolemy's theorem and Archimedes' *Sphere and the Cylinder* I.21.

154 Langermann (1996a, 41).

155 Langermann (1996a, 35) suggests that in addition to his *Recension of Euclid's Elements*, al-Maghribī wrote also a geometrical compendium titled *The Book of Euclid*.

7 Genre and Language

SMA is a mixture of two literary genres: philosophical discussion and formal, Euclidean-type geometrical writing. Philosophical language is used in Parts I and II (except Section II.3); mathematical language in Section II.3[156] and Part III (except for the last paragraph of the III.33) and the extant beginning of Part IV. Obviously, the philosophical vocabulary differs from the mathematical.[157] Some interesting differences are due to the dissimilar structure of the philosophical and geometrical arguments.[158] Less obviously, the mathematical language somewhat differs from part to part. Notable differences between the parts are:

[i] Different terms used to denote "proposition"

In Part II Section 3 the term used for "proposition" (شكل) is *tavnit*.[159] This is also Alfonso's standard word when he refers to a proposition in Euclid's *Elements*. *Tavnit* is the translation of *shakl* in both senses of the Arabic term, namely "proposition" and "figure."[160]

In Part III and in Part V (as we can infer from the outline), the term for proposition is the less common word *segula* (corresponding to the Arabic خاصة, Latin *proprium*). In Aristotelian philosophy *segula* is a property that is inseparable from the subject, but is not its defining property.[161] This word is used in the sense of "proposition" in several Arabic mathematical texts that were written in the Iberian Peninsula.[162] Ibn Hūd (d. 1085) opens many sections in

---

[156] Alfonso's attempt to prove the parallels postulate.

[157] A few examples: the word ראוי (appropriate) appears 25 times in the general discussion and is absent in the mathematical parts; the root חיב (to be necessary or obligatory) occurs 15 times in the general discussion and not once in the mathematical discussion. The root גבל (to limit, also to construct) appears 26 times in Parts I–II (24 times as verb twice as noun), not once in Part III. The verb נדמה (first person plural of דמה) is used in Part II (II.1.5, II.2.1, II.4.1) in the sense of "imagine", usually in the context of imagining moving of lines; In III.24 and III.25 it is used in the simpler sense of "assume".

[158] For instance, the word כי occurs 91 times in Parts I–II, only once in Part III, whereas the combination אם כן is very frequent in Part III and in mathematical passages of part II (86 times) and occurs only once in the philosophical Part I. The word וכי seven more times in Parts I–II, and once in the philosophical paragraph of Part III, i.e., the end of Property 33.

[159] In one context (II.2.3), when citing two lemmas of al-Nayrizi, Alfonso translates *shakl* as *haqdama* rather than *tavnit*.

[160] The Provençal translators usually translate *shakl* (in both senses) as *temuna* (Sarfatti 1968, 193–4). Ofer Elior remarks that in manuscripts Firenze 137 and Mantua 1 (that is based on Firenze 137) of the Hebrew translation of the *Elements* the term "tavnit" is used.

[161] Aristotle *Post. An.* I.13 79a6–9. Langermann (1992, 273–4) shows that this term was used in medicine to refer to the special properties of medicines that cannot be accounted for in terms of the basic qualities.

[162] E.g., Katz et al. (2007, 667).

the geometrical part of his *Istikmāl* with the introductory sentence "on the properties of" followed by lists of geometrical objects.[163] Abū al-Ṣalt (1068–1134) uses it frequently in the arithmetical section of a scientific encyclopedia.[164] Among the Hebrew mathematical works that were written in Alfonso's intellectual milieu we can find the term *segula* in his contemporary IṣḤāq Isra'eli's *Yesod Ólam*.[165]

In Part IV the term for "proposition" is *yesod*, meaning basic principle or element. This accords with the Iberian tradition, e.g., Bar-Ḥiyya's *Yesodei ha-Tevuna*,[166] and IṣḤāq Isra'eli's *Yesod Ólam*.[167]

Do the three terms—*tavnit*, *segula*, and *yesod*—reflect the language of the different texts on which Alfonso relied when he was writing Parts II, III, and IV, respectively? Or did he use these three terms because he ascribed significance to the differences in meaning between them? On these questions we can only speculate. It may be that he used the term *tavnit* for the six propositions of II.3, because he consulted Ibn al-Haytham who used the term *shakl* in exactly the same context.[168] The term *segula* was used in Part III and was meant to be used in Part V, two parts that are purely geometrical. Geometrical proposition can be perceived as the unfolding of the properties of a geometrical object. *Yesod* is used in Part IV that is also geometrical but has a deep philosophical significance for Alfonso: the propositions of Part IV were supposed to provide a geometrical infrastructure to Plato's cosmology and thus to establish universal basic principles.

[ii] Terminological innovations in Part III

Part III opens with a set of definitions. Some of the terms defined are incompatible with those used in Parts I–II.[169] For instance, in Part III Alfonso coins the term מיושר הקוים for polygon, whereas in Parts I–II he used שטח ישר הקוים. Also, there are in Part III a few single occurrences of words that do not belong to the Sephardic Hebrew typical of SMA, but rather to the Provençal Hebrew:

אצטוונה. In Parts I–II Alfonso uses 11 times the word אסטונא for cylinder (the apparently mistaken spelling אצטולא occurs once in Part I); in Parts III–IV the

163 E.g., Hojendick (1991, 232, 235, 241, 245, 247, 252, 254, 256).

164 This text was later translated into Hebrew, probably by Don Benveniste ben Lavi in Saragossa. See Aradi (2015, 11). We are grateful to Naomi Aradi for letting us consult this translation.

165 He uses the expressions משפטים וסגולות, סגלה נפלאה and דינים וסגולות (*Sha'ar* C, *Sha'ar* D). Isra'eli uses the word *Segula* to refer to a mathematical proposition that characterizes a mathematical object or concept, e.g., the *segula* of parallels (*Limud* 18), of triangles (*Limud* 20), of similar triangles or of ratios.

166 *Yesodei ha-Tevuna*, 51.

167 *Sha'ar* D is systematically divided into propositions. The term for proposition is *limud*. In *Sha'ar* B and *Sha'ar* C several terms are used to divide the text: *yesod musad*, *'iqar efshar*, or simply *'iqar*. The text is from 1310.

168 See *Commentary on the Premises of Euclid's Elements,* Hooper-Sude's edition, Arabic text pp. 46, 50, 55. Al-Tūsī also uses *Shakl* in his suggested proof.

169 See commentary on the definitions of Part III.

spelling אצטוונה is used three times. אסטונא is the simple transcription from Arabic; אצטוונה was used by the Provençal translators.[170]

מוגשם. The term מוגשם is a straightforward loan translation of the Arabic مجسم and was commonly used by the Provençal translators.[171] Alfonso adopts the unusual (perhaps idiosyncratic) word מושגם that occurs 13 times in Parts II and III of *SMA*.[172] The single occurrence of מוגשם is in III.0, definition [g] (see comment on terms there).

יחס. Alfonso's word for ratio is ערך. The noun is used 171 times in all parts of *SMA* and the verb מתערכים (meaning being in proportion) three times. The Provençal word for ratio is יחס, and this word appears once in Part III of *SMA*, in a sentence that is actually out of context.[173]

The arguments listed above point out some differences in the usage of mathematical terms between Part III and the mathematical passages in Parts I–II. It is likely that Part III was written later than Parts I–II relying on new source (or sources).

8 Appendix: Translation Comparisons[174]

Ibn Rushd

Tahafut al-Tahafut

This book was translated into Hebrew by Qalonimos ben David under the title הפלת ההפלה. Alfonso refers to it as הכחשת ההכחשה.[175]

Epistle on the Possibility of Conjunction

An anonymous Hebrew translation is titled אגרת אפשרות ההדבקות. Alfonso refers to this text as ספר הדיבוק האלהי.[176] He quotes one sentence from the book.

---

[170] Shmuel Ibn Tibbbon in his Hebrew lexicon used the spelling אצטוונה (Sarfati 177), and this form was adopted by the other Tibbonides in their translations of Euclid, by the compiler of the Hebrew Geometrical Compendium, by Qalonimus in his translation of Archimedes and by other Provencal translators and writers. The word is found also in Provencal philosophical translations, e.g., in the anonymous translation of Ibn Rushd's long commentary on the Physics, Paris ms. 154b.

[171] See Sarfatti (1968, 195).

[172] The Mishnaic word שגם does not seem to be relevant.

[173] III.25. See note to the text there.

[174] Since most of the manuscripts mentioned here are relevant only to the Appendix they are not listed in the bibliography, but instead are referred to directly here.

[175] *Teshuvot*, see Hecht (1993, 367–8).

[176] *Teshuvot*, see Hecht (1993, 372); Gershenzon (1984, 12); Szpisch (2006, 634).

Alfonso	Anonymous Hebrew translation
שאין בכל המציאות לבד האלוה וכי זה הוא הייחוד שלו. וכי לכך נאמר במשיח שהיה מן אלוהים וכי האלוהות התלבש באנושות ולקח בשר בבטן האשה מרים כשנתעברה מן המשיח.[177]	ואין בנמצא רק השם ישתבח כל הנמצאות וזה הוא הייחוד. וזה הוא כנגד לאשר הם טועים[178] שאומרים במשיח שהוא בן אלהים והאלוהות יצה (כך!) מאשה ושהאל נכנס בבטן מרים ושנשאה משיח.[179]

Middle Commentary on De Generatione et Corruptione

Alfonso	Qalonimos ben Qalonimos' translation	Falaquera, *De'ot ha-Filosofim*
ושאלו היו כל הדברים נקראים בשמו היה בתכלית האמת, כמו שאלו היו הענקים והטבעות מהזהב נקראים בשם זהב היה בתכלית האמת.[180]	ויאמר כי ההוה ממנו, אם נקרא בשם הדבר אשר ממנו היה, הוא מאמר בתכלית האמת, כמו שאם נקראת האצדעה זהב היה אמת.[181]	ואמ' כי ההווה ממנו אם נקרא בשם דבר שמתהווה ממנו היה נאמר בתכלית הצדק כמו שאם נקרא התלי זהב היה זה אמת.[182]

Al-Ghazali

Alfonso was acquainted with both *Muqāṣid al-falāsifah* and *Tahāfut al-Falāsifa.*[180]

The *Tahāfut* was translated into Hebrew by ZeraḤya ben IṣḤāq ha-Levi under the title הפלת הפילוסופים but referred to by Alfonso as ספר ההכחשות.[181]

[177] I.3.4.

[178] Ibn Rushd *MC De gen. et corr.*, Kurlad edition, 59. The underlying text is: "The account given in Timaeus… says that it is a sort of substratum prior to the so-called elements as gold to the artefacts made of gold… but in the case of generation and corruption it is not possible to call something by the name of the thing from which it has come to be" (*De gen.* II.1329a17–24, Joachim's translation).

[179] Parma ms. 69a5 6.

[180] Much has been written on the study of al-Ghazali among the Jews. See, for instance, Harvey (2001); Freudenthal and Zonta (2012, 244, 247–248).

[181] Szpiech (2006, 620).

The *Maqāṣid* was translated into Hebrew perhaps three times, the first translation was made by IṣḤāq Albalag at the end of the thirteenth century and was completed by IṣḤāq Pulgar (who was Alfonso's student and opponent after his conversion). The second was made by Yehuda ben Shlomo Nathan, in Provence in the fourteenth century. The third anonymous translation was the one which Moshe Narboni used when writing his commentary on this text.[182] Steven Harvey notes (note 18) the similarity between Yehuda's translation and the anonymous translation. The comparison below supports Harvey's suggestion that these two translations are not independent.

Maqāṣid al-falāsifah

Alfonso	Albalag's translation	Yehuda ben Shlomo Natan's translation	Anonymous translation
וזה שאמרו שאם היינו מניחים קו ג״ד בלי תכלית מצד ג׳ והיינו מניעין את קו א״ב בעגלתו לצד ג׳ עד שיהיה נכחי לו, יהיה זה אפשר בלי ספק. ואז נניעהו מן הנכחות כדי שיתקרב אליו ב יהיה תחלה על יושר נקדה בקו ג״ד ואחר כך יהיה על יושר שאר הנקודות, נקדה אחר נקדה, עד שיחזור להיות נכחי בצד האחר וזה שקר.[183]	האחת מהם שאנחנו לו הנחנו קו ג״ד בלי תכלית בצד ג׳ והנענו קו א״ב בעגולה אל צד ג מקו ג עד שב נכחו היה זה הנעה אפשרית בהכרח. ולו הנענוהו מהנכחות אל צד הקורבה ממנו הנה אי אפשר מבלתי שירמוז ממנו נקודה היא ראשית הנקודות הרמוזות עד אחר זה הרמוז שאר הנקודות עד שישוב מקדמו בתכלית אל הנכח מהצד האחר לו זה שקר.[184]	אחת מהן שאם הנחנו קו ג״ד באין תכלית בצד ג׳ ונענו קו א׳ בעגול אל צד ג׳ מצד ג״ד עד ששב אל מקבילו, הייתה זו ההנעה אפשרית בהכרח ולו הנענוהו מצד הנגוד אל צד הקורבה ממנו, הנה אין ספק שיהיה נוכח נקודה ממנו אשר היא ראשית הנגוד. ואחר זה יהיה לנוכח שאר הנקודות עד שישוב מן הנוכחות כשיגיע אל ההקבלה מן הצד האחר וזה שקר.[185]	אחת מהם שאנחנו לו הנחנו קו ג״ד בלי תכלית בצד ג׳ והנענו קו א״ב בעגולה אל צד ג׳ מקו ג״ד עד שב בנכחו היה זה הנעה אפשרית בהכרח. ולו הנענוהו מהנכח אל צד הקורבה רומז הנה אי אפשר מבלתי שירמוז ממנו נקודה היא ראשית הנקודות הרמוזות. עוד אחר זה ירמוז שאר הנקודות עד שישוב מהרמיזות בתכלית על הנכח מהצד האחר וזה שקר.[186]

[182] See Harvey (2001, 364).

[183] II.1.1.

[184] Paris ms. 901, IMHM 26857 fol 60b24–61a6.

[185] Paris ms. 904, IMHM 26860 fol 42b9–20.

[186] Oxford Bodl. ms. 1346, *Add.* 4048, IMHM 22372 fol 31b31.

Euclid

Euclid's *Elements* (*στοιχεία*, اصول) was translated into Hebrew under the title *Sefer ha-shorashim* both by Moshe Ibn Tibbon and by his nephew Yaʿaqov ben Makhir.[187] Alfonso referred to it as *Sefer ha-ʿIqarim.*

Elements Postulate 5

Alfonso	Moshe Ibn Tibbon's translation	Yaʿaqov ben Makhir's translation	Geometrical compendium
כשיפול קו ישר על שני קוים ישרים והיו שתי הזויות יחד אשר בצד אחד פחות משתי נצבות, כשימשכו שני קוים אלו באותו צד יפגשו בהכרח.[188]	ואם נפל קו ישר על שני קוים ישרים ושם באחד משני הצדדין שתי הזויות הפנימיות קטנות מנצבות, הנה שתי הקוים הישרים כאשר יוצאו בצד [189]ההוא יופגשו.[189]	ואם נפל קו ישר על שני קוים ישרים ושם באחד הצדדים הזויות הפנימיות קטנות משתי נצבות, הנה שני הקוים הישרים אשר יוצאו בזה הצד יפגשו.[190]	ואם נפל קו ישר על שני קוים ישרים ושם באחד הצדדים הזויות הפנימיות קטנות משתי נצבות, הנה שני הקוים ההם יפגשו אם יוצאו בצד ההוא.[191]

[187] In the translation of R. Yaʿaqov and in Firenze manuscript 137 the book is called ספר אוקלידס without any additional title. I am greatful to Ofer Elior for this comment.

[188] SMA I.5.3.

[189] Paris, Private collection Ms. IMHM 39116, fol. 1b18–20.

[190] Oxford Ms. Hunt. 16, IMHM 19290, fol. 2a5–6.

[191] Mantua com. 2, Ms. IMHM 783 fol. 2.b13–15.

Ptolemy

Ptolemy's Theorem, *Almagest* I.10

Alfonso	Ya'aqov Anatoli's translation
כשיהיה תבנית בעל ד׳ צלעות ישרות בעגול יהיה רבוי שתי צלעותיו המתנגדות זו בזו יחד כרבוי שני אלכסוניו זה בזה. **כיצד?** נשים תבנית אבג״ד בתוך עגול א"ב ג״ד ונוציא א״ג ב״ד. אומר שרבוי א״ב בג״ד וא״ד בב״ג יחד כרבוי א״ג בב״ד. וכבר הזכיר בטלמיוס	ויהיה עגול שבו ד צלעות עליהם אבג״ד ונוציא קוי א״ג ב״ד ונבאר שמרובע א״ג בב״ד שוה למרובע א״ב בד״ג וא״ד בב״ג.
המופת בזה. והוא שנשים זוית הג״ד [שוה לזוית אג״ב וזוית בא״ג] שוה לזוית בד״ג. ישאר זוית גה״ד שוה לזוית אב״ג. אם כן רבוי ג״ד ביתר זוית דג״ה שהוא א״ב כרבוי ד״ה ביתר זוית ה׳ שהוא א״ג. ולפי שזוית אג״ד שוה לזוית הג״ב וכן זוית גא״ד שוה לזוית הב״ג ישאר זוית גד״א שוה לזוית בה״ג. אם כן רבוי ב״ג ביתר זוית בג״ה שהוא א״ד כרבוי ב״ה ביתר זוית ה׳ שהוא א״ג וכבר היה רבוי א״ב בג״ד כרבוי א״ג בה״ד אם כן רבוי א״ד בג״ב וא״ב בג״ד יחד כרבוי א״ג בב״ד. [192]	**המופת בו** שנשים זוית אב״ה בזוית דב״ג ולפי שזוית דב״ג שוה לזוית אב״ה אם אנחנו שתפנו [זוית] הב״ד והוספנוה על כל אחת מהן תהיה זוית אב״ד שוה לזוית הב״ג וזוית בד״א שוה לזוית בג״ה, לפי שמיתרן קשת אחת. ואם כן זויות משלש אב״ד שוות לזויות משלש בג״ד. ולזה יהיה יחס ב״ג אל ג״ה כיחס ב״ד אל ד״א. ואם כן מרובע ב״ג בד״א שוה למרובע ב״ד וג״ה. ועוד לפי שזוית אב״ה שוה לזוית דב״ג וזוית בא״ה שוה לזוית בד״ג תהייו זויות משלש אב״ה שוות לזויות משלש בג״ד. ואם כן לזה יהיה יחס א״ב אל א״ה כיחס ב״ד אל ד״ג. ואם כן מרובע ב״א בד״ג שוה למרובע ב״ד בה״א וכבר היה מבואר שמרובע ב״ג בא״ד שוה למרובע ב״ד בג״ה. אם כן כל מרובע א״ג בב״ד שוה למרובע א״ב בד״ג וא״ד בב״ג יחד. [193]

Anatoli uses the preposition ב (which in this context means "by") to denote product. Alfonso's uses his favorite term רבוי which means half product.[194] The factor 1/2 introduced by the use of this term is irrelevant because in this context it appears in an equation and the factor can be cancelled.

[192] Section III.12.

[193] Wien Ms. 194/40, IMHM 1317 fol. 11b–12a. Also Munich Ms. 182, IMHM 1638.

[194] See discussion of this term in the introductory note to Part III.

Edition and Translation Conventions

The Edition

The text of *SMA* was edited by Gita Gluskina and published with a Russian translation and comments in 1983. We consulted her edition as well as her introduction, notes, and two articles (that were translated for us from the Russian). In order that our edition be compatible with hers, we followed Gluskina's Latin transliteration of the letters of the diagrams:

a	א	g	ז	m	מ	s	ק
b	ב	h	ח	n	נ	t	ר
c	ג	i	ט	o	ס	u	ש
d	ד	j	י	p	ע	v	ת
e	ה	k	כ	q	פ		
f	ו	l	ל	r	צ		

The editing methods are somewhat different. Gluskina's method was to provide the reader with a precise transcription of the manuscript, adding her own corrections in the apparatus. Our policy was to offer the reader a readable text, so where we believed the text to be mistaken we put what we took to be the correct reading in the text, citing the manuscript reading in the apparatus. The justification for this policy is that the manuscript is often misleading in the rendering of the letters of the diagrams, and must be corrected in order to make the text comprehensible. Most of the corrections were already made by Gluskina, and we usually adopted her corrections. In a few places our interpretation of the text and consequently our corrections were different from hers.

We did not comment in the apparatus when we corrected the readings of graphically similar letters as ב כ or ח ה or sometimes ח א. We did not reproduce some single letters scattered in the text and apparently irrelevant, and we also skipped repeated words. We completed truncated words without noting it, and only in cases that it was not obvious we added brackets.

The division of the text into four parts (she'arim) is Alfonso's; the division into sections and the numbering of the paragraphs are editorial additions. Added numbers and titles are in square brackets.

The Translation

Our translation policy was to offer a literal translation if possible in order to preserve some of the flavor of Alfonso's style. In this vein we tried to translate words consistently. For instance, we rendered רכב always as "superpose" even if in some cases “be congruent” sounds better; we use the more literal translation “we consider triangle *ABC*…” of the Hebrew נשים משלש אבג', rather than the more elegant expression “let *ABC* be a triangle….” In Part III, however, we preferred mathematical clarity, so the translation is somewhat freer. For instance, we use “rectangle” instead of “four-sided right-angled figure.”

In those cases where more than one translation of a word was used, the different translations are listed in the glossary.

We do not insist on strictly parallel punctuation of the Hebrew and English versions, and a few times divide in English a long sentence that could not be divided in the Hebrew text.

Notations

[]	Editors’ additions. What appears within brackets is not a part of the text.
[…]	Missing text.
[?]	Unreadable text (no. of question marks = no. of unreadable words).
<>	Parts of the text that seem to be irrelevant or out of place.

Chapter 2
English Translation and Commentary

0 [Outline]

[93b] This book includes five parts. The first [serves as] an introduction to the book and inquires whether the imagining [ציור] of equally divided motion [תנועה שוה בחלקיה] can qualify as one of the first principles of geometry. The second is on the accidents associated with motion, which prevent its imagining from being a first principle of geometry, and how several earlier scholars became misled by them. The third is about the *propria* of rectilinear magnitudes and areas that are useful in this science. The fourth explains how a body is divided into surfaces, the surface into lines, and the line into points, as Plato taught, and the way of evaluating and measuring them by one another. The fifth is about some of the *propria* of divisions of bodies into surfaces, of surfaces into lines, and of lines into points, which are needed for the rectifying of the circle, and for the measurement of the sphere and the cylinder and the cone and the spherical annulus and the square annulus and the circular annulus and their sections, and their enclosing lines and surfaces, and on transforming them, one into the likeness of another, and on equating them to rectilinear [bodies], and on the division of angles into any number of equal parts.

R. Glasner, A. Baraness, *Alfonso's Rectifying the Curved*,
Sources and Studies in the History of Mathematics and Physical Sciences,
https://doi.org/10.1007/978-3-319-77303-2_2

Terms

Alfonso's basic vocabulary is that of Abraham Bar-Ḥiyya.[1] The terms that are introduced in the outline are discussed in their respective contexts below.[2]

Commentary: The Title of the Book

The title of the book *Meyasher 'Aqov* is not mentioned at the beginning, but at the end of Part I (I.5.4 below). It is taken from Isaiah 40:4 והיה העקוב למישור, commonly translated "and the crooked shall be made straight." "Rectifying the Curved" is a literal translation that alludes to the geometrical meaning.

I Part I, Being an Introduction to the Book

[I.0. Poetical Preface]

Alfonso the author said: it is me speaking here. Blessed be God forever, amen and amen, who tuned and summoned my heart.[3] The sages' saying about believing: I have exerted myself and found, believe; I have exerted myself and have not found, do not believe.[4] I pursued learning consistently without distraction, and much study is the exertion of the flesh.[5] One should not exert oneself for nothing in vain,[6] [but] seek to find the right words[7] to

[1] Bar-Ḥiyya's vocabulary is notable throughout *SMA*. Here we find מוצק for pyramid, and the unusual spelling כידור side by side with the common spelling כדור for a sphere. See Sarfatti (1968, 86, 122). See also Section I.1.4 below.

[2] The word סגולה (*proprium*, property) was discussed in the introduction. The key terms to the understanding of Alfonso's agenda, שער, מדד, ציר and שוה בחלקיה (evaluate, measure, imagine, equally divided) are discussed in Sections I.3.4, I.4.2 and I.5.3 below; דמות (form) in I.5.5, שברים (sections) in I.5.4 and רבוע (square) in III.0. The word טבעת (spherical, square or circular), translated here as "annulus", appears only in the outline and we have no information on its use. It may be a translation of the Arabic *ḥalaqa*.

[3] We understand the word זִמֵּן in this sentence as a verb.

[4] Allusion to Bab. Talmud, *Megila* ch. 1 6b, tr. M.L. Rodkinson.

[5] Ecc. 12:12. Most English translations use "weariness." We used "exertion" to be consistent with the translation of the previous sentence from the Talmud.

[6] Isa. 40:17.

[7] Ecc. 12:10, following here the New International Version.

acquire a clear knowledge, [94a] emanated from the holy spirit.[8] To show me his glory and his greatness,[9] [to] behold the beauty of the Lord and to inquire in his temple[10]; to glorify his name and greatness by the mensuration of the circle, which from my youth to my old age was the one thing I desired from the Lord,[11] who created everything for his glory,[12] everything for his honor and Majesty,[13] and the whole earth is full of his glory.[14]

Commentary: Poetical Language in SMA

Poetical introductions are common in Hebrew philosophical books. Alfonso uses poetical language here and in paragraph I.3.1. In both cases the issue is his quest for wisdom. Biblical verses are also quoted in II.4.2. Here as well as in several more paragraphs in the sequel,[15] Alfonso enjoys playing with several meanings of words derived from the root מצא and also on combinations of the roots מצא and מצע (e.g., מציאות אמצעית, מציאת שטח אמצעי). He also sometimes uses anagrams such as פותחין - פוחתין.[16] It is not clear whether מושגם- מוגשם is another intended anagram or an error.

[I.1. Presentation of the Problem: Can Equality Between Rectilinear and Circular Areas Be Established]

Introductory note: concepts of equality (between magnitudes)
Alfonso's aim, as he announces it in the paragraph below, is to "inquire whether the existence of a rectilinear area equal to a circular area can be *truly* established." In order to achieve this goal one must have an adequate concept of equality between straight and curved lines and between areas bound by such lines. Two concepts of equality were known in Greek geometry and underlie Alfonso's discussion:

[i] Equality defined by superposition (*Elements* I–IV)
Euclid's basic notion of equality is derived from that of superposition: "things which are superposed on one another are equal (*isos*) to one another" (*Elements* I,

---

[8] The Biblical source is Numbers 11:17 or 11:25, ויאצל מן הרוח ,אצלתי מן הרוח. The expression ברכי אצולה מרוח הקדש appears in Yehuda ha-Levi's poem *Barkhi Atzula*, and the exact expression מרוח הקדש אצולה in Yehuda al-Ḥarizi's Taḥkemoni, p. 24.

[9] Dt. 5:24.

[10] Ps. 27:4.

[11] Ps. 27:4.

[12] Allusion to Isa 43:7.

[13] Ps. 96:6.

[14] Isa. 6:3.

[15] Sections I.1.3–4, I.3.3, II.1.6, I.4.1, I.5.5, II.1.6–7, II.5.6.

[16] Sections III.26–7.

common notion 4).[17] Using common notions 2 and 3[18] and Proposition II.14,[19] the concept of equality can be easily extended to all rectilinear areas. As stated by Hilbert: "Two polygons are said to be of equal area when they can be decomposed into a finite number of triangles which are respectively congruent to one another in pairs."[20]

The simple concept of equality based on that of congruence works well for areas bounded by straight lines, but is not useful when areas bounded by curved lines are involved. Euclid's concept of equality between curved areas is limited to simple congruence: two circles are said to be equal if their diameters are equal.[21] This definition can serve neither as a basis for arithmetic of circular areas nor for comparing circular and rectilinear areas.

Whereas Euclid obscures the role of motion in defining superposition, Ibn al-Haytham openly and assertively admits that motion can be used in geometry and describes basic geometrical concepts such as angle or circle by "imagining motion."[22] Rashed sees in Ibn al-Haytham's analysis of the concept of angle in terms of motion a historical and epistemological turning point: a new conception of the geometrical object and a new ontology are made possible by the introduction of motion as a primitive notion.[23] Alfonso follows Ibn al-Haytham and advances his idea.

[ii] Equality defined by inequalities (*Elements* V, XII).
The idea of defining equality by conditions stated in terms of inequalities comes up first in Euclid's theory of proportions (*Elements* Definition V.5).[24] The norm underlying this theory is the homogeneity of the magnitudes involved. Euclid's requirement of homogeneity is stated for ratios between magnitudes, but applies as well to equality between magnitudes. Ratio is restricted to magnitudes "of the same kind" or homogeneous magnitudes (Definition V.3).[25] Definition V.4 as well as the use of ratios throughout *Elements* testify that ratios neither between magnitudes of different dimensions nor between straight and curved lines or areas were allowed. However, observes Lachterman, "Euclid nonetheless opens the door a crack to transgression of homogeneity…"[26]; "the crack in the doorway left open by a certain

[17] The term "superposition" is discussed in the introductory note to Section I.2 below.

[18] Asserting the preservation of equality under addition and subtraction.

[19] "To construct a square equal to a given rectilinear figure."

[20] Hilbert *Grundlage der Geometrie* # 18. In the first 34 propositions of *Elements* I equality is used in the simple sense of congruence; from Proposition I.35 the term is used in a wider sense of equality of rectilinear areas. Proclus comments: "it means that the areas, not the sides, are equal, for the statement is about included spaces, the areas" (*Commentary* 398 Heiberg; Morrow's translation 314).

[21] *Elements* III, definition 1. Euclid defines also similarity between sections of a circle.

[22] Hooper-Sude 41, 43.

[23] Rashed (2015, 91, 440).

[24] On equalities between ratios see introductory note and commentary to III.24–27.

[25] See Heath (1908/1956, II, 116–120); Mueller (1971, 2; 1981, 132); Lachterman (1989, 33–4). Mueller suggests that the requirement of homogeneity was "reworked" by Euclid and that it cannot be ascribed to Eudoxus the way Euclid presents it.

[26] Lachterman (1989, 35).

ambivalence about homogeneity becomes a breach wide enough to admit a sequence of mathematical conceptions and techniques, increasingly antagonistic to the pre-understanding that Euclidean teaching continued to address."[27] The most important among such rebellious "conceptions and technique" was the double indirect method[28] (the method of exhaustion): equality can be rigorously established by ruling out all possible inequalities using indirect argument. For nonhomogeneous magnitudes A, B, it can be proved that $A = B$ by ruling out the possibilities that $A < B$ and $A > B$. The method is used by Euclid,[29] and more freely by Archimedes who establishes equalities between a triangle and a circle, the surface of a sphere and a plane circle, or between a sphere and a cone.[30]

Alfonso is aware of the great advantages of the double indirect proofs, but his aim is to revise the concept of superposition and thus to make equalities between nonhomogeneous magnitudes more intuitive and explanatory.

[I.1.1. Presenting the Two Positions]

My intention is to inquire whether the existence of a rectilinear area equal to a circular area can be truly established, [that is] neither by way of approximation as earlier scholars established, nor by way of skeptical reasoning that some [scholars] espoused,[31] as we learn from the writings of Aristotle and his followers; rather, by way of true decisive demonstration, free of any approximation or sophistry [זיוף]. For many years it was my ambition to find it, because I disagreed with what Ibn Rushd wrote about it, that a genuine demonstration is impossible, and that a straight line cannot be equal to a circular line because the parts of one line cannot be superposed [ירכבו] on the parts of the other, and that a rectilinear area cannot be equal to a circular area for the same reason itself, i.e., the parts of the rectilinear area cannot be superposed on all parts of the circular one. The sage Ibn Sīnā was also of the opinion that an arched line or area can never be transformed into straight ones.

Terms

רכב[32]

The verb רכב, "to ride," is used in *SMA* in the sense of "to superpose."[33] The root *rkb* occurs about 80 times in the first two parts of *SMA* and is the main key term in Alfonso's mathematical "ideology." He inquires into its mathematical meaning and

---

[27] Lachterman (1989, 37).

[28] We are grateful to Nathan Sidoli for suggesting this term.

[29] Elements XII.2, XII.10, XI.11.

[30] Archimedes, *Measurement of Circle* 1; *Sphere and Cylinder* I.33, 34.

[31] Literally: the thought of those who have doubts.

[32] We are grateful to Donna Shalev for reading this paragraph and for her comments.

[33] In the discussion of atomism (Section I.3) מורכב and הורכב are used in the sense of "composed of."

the possibility of extending it beyond its original Euclidean denotation. Alfonso does not quote biblical verses and his interpretation of *rkb* is not influenced by Maimonides' famous chapter dedicated to this verb.[34] Still, we shall see that his mathematical interpretation has a spiritual dimension and is deeply associated with his Christian theology.[35]

Euclid introduces the term *epharmozein* in *Elements* I, common notion 4. The verb basically means to fit on or put on, and in Common Notion 4 it is used in the active voice "*ta epharmozonta ep allēla isa allēlois estin.*"[36] The meaning is symmetrical and Heath uses in his translation the symmetrical verb "coincide": "things which coincide with one another are equal to one another."[37] In Proposition I.4 (in which Common Notion 4 is applied) the verb is used both in the active and in the passive voices, the latter in the phrase "*epharmozomenou tou ABG epi to DEZ.*" The meaning is asymmetrical and Heath uses the asymmetrical verb "be applied to."[38]

The Arabic root used in al-Nairizi's commentary (that Alfonso probably had on his desk) for the active *epharmozein* (in common notion 4[39]) is *ṭbq* (طبق),[40] meaning "cover, make coincident or congruent, cause to coincide, superpose."[41] In Proposition I.4, where Euclid uses the passive voice *epharmozomenou*, the Arabic translations use the asymmetrical Arabic verb *rkb* (ركب), in the sense of to ride: ركبنا مسلس ابج على مسلس دهز.[42] Thus, two forms of one Greek verb were rendered by two different verbs in Arabic. In Leiden manuscript where the verb *ṭbq* is used in the text, the verb *rkb* is added in supra scriptum.[43] It is, thus, possible that a variant of common notion 4 in which the verb *rkb* was used circulated in the Iberian Peninsula and could have been Alfonso's source.

How were the active and passive forms of the word *epharmozein* transmitted into Hebrew? In Yisḥāq-Ṭhābit's Arabic translation of the *Elements* the Arabic verb *ṭbq* is used in Common notion 4 and *rkb* in Proposition I.4 as in Leiden manuscript of al-Nayrizi's commentary. In the Provençal Hebrew tradition (based on Yisḥāq-Ṭhābit's version) the Arabic verb *rkb* (in Proposition I.4 passive sense) is translated into its Hebrew equivalent *rkb* by all translators.[44] The Arabic verb *ṭbq* (in Common

[34] *Guide* I.70.

[35] See Sections II.1.7; II.4.2; Introduction Section 3 argument ii.

[36] Common Notion no. 4 in Heath' list (1908/1956 I, 224); no. 7 in Heiberg's list (1883, I, 10).

[37] *Elements* Heath' translation I, 224–5, 249.

[38] *Elements* Heath translation I, 247, 249.

[39] According to Heath' numbering.

[40] Besthorn's edition 28.12, also in the manuscripts of Yisḥāq-Ṭhābit's version. I am grateful to Ofer Elior for this information. Elior also noted that Ibn Sina uses انطبق in his rendering of common notion 4 and Proposition I.4.

[41] Wehr's dictionary, 645.

[42] Besthorn's edition 52.19. Also in the manuscripts of version.

[43] Besthorn's edition, Introduction, p. 28, lines 12–3, p. 29, footnote 5.

[44] Moshe Ibn Tibbon (Paris ms. fol. 2b24–5; Oxford ms. fol. 2b9), Yaaqov ben Makhir (Oxford ms. fol. 2b6) and the compiler of the Hebrew Geometrical Compendium (Mantova ms. 2, fol. 3b13, 14), all use הורכב.

Notion 4 active sense) required the coining of another Hebrew term. Moshe Ibn Tibbon uses the phonetic loan translation *dbq* (دبق, דבק). The Arabic root *ṭbq* is phonetically (and also semantically) similar to the more common root *dbq*. In Hebrew and in Arabic *dbq* means to stick, adhere. In medieval philosophical Hebrew *dbq* had the meaning of being continuous with, and *mitdabeq* was the standard term for "continuous." In this sense *dbq* was used by Moshe's father Shmuel Ibn Tibbon. Moshe Ibn Tibbon translates the verb in common notion 4 as התדבק בשיווי,[45] namely fit on exactly without excess or defect. This translation was widely adopted.[46] Ya'aqov ben Makhir used a simpler and apparently better translation הושם על,[47] and the compiler of the *Hebrew Geometrical Compendium* offers a compromise: כשיונח דבר על דבר ויתדבק עליו.[48]

Alfonso is the only Hebrew writer who uses *consistently and systematically* the Hebrew root *rkb* for both Arabic roots *ṭbq* and *rkb*. The concept of *harkava* is a major issue in his agenda, and there is no doubt that his insistent use of the root *rkb* throughout the book reflects his geometrical "ideology."[49] Consequently, we translate it always by "superpose," even in the few cases where "coincide" fits in better.

מופת, מופת חותך, מופת עצמי, ראיה, באור All these words are used by Alfonso to refer to proofs in geometry. When the focus is on the logical structure of the proof, Alfonso uses the term הקש.[50] *Mofet* is the standard Hebrew term for برهان (ἀπόδειξις, *demonstratio*). It is used in *SMA* first and foremost as a general term for a mathematical proof, and also as a title (מופתו, המופת) to mark the proof part of a proposition after the enunciation and the setting out. In II.3 he claims to offer a demonstration beyond any doubts (מופת נקי מכל ספקות) of the Parallels Postulate. It does not seem that Alfonso uses this expression to emphasize the deductive structure of the proof but, rather, the geometrical standards, notably allowing motion, but excluding actual infinity. To an argument that do not stand in these criteria, notably earlier attempts to prove the Parallels Postulate, Alfonso refers as באור (I.5.2, II.2.1, II.2.4, II.3.0).

מופת חותך, מופת נחתך are probably translations of برهان قطعي.[51] Here (I.1.1) Alfonso endeavors to provide a true decisive demonstration (מופת נחתך אמתי) of the squaring of the circle, probably meaning an explanatory constructive proof that does not rely

[45] Oxford ms. fol. 1b20; Paris ms. fol. 2a1. In Proposition I.4 where both the active and the passive senses are used Moshe Ibn Tibbon and the compiler of *the Hebrew Geometrical Compendium* use both רכב and **התדבק בשווי** (Paris ms. 3a2. Mantova ms. 3b16, 17).

[46] For example, in the Hebrew translations of Ibn Rushd by Qalonimus ben Qalonimus. See *MC Phys*. Hamburg ms. fol. 3b6; *MC Soph. Ref.* Paris ms. fol. 111a5.

[47] Oxford ms. fol. 2a10.

[48] Mantova ms. fol. 2b5–6.

[49] See Section I.2 below.

[50] See discussion of terms in I.2.3 below.

[51] חותך occurs in II.1.1, and is also the term used by Shmuel Ibn Tibbon in his translation of the *Guide*.

on indirect argument. The issue of constructive vs. indirect proof is important for Alfonso.

In I.2.3 Alfonso opposes מופת עצמי, essential demonstration, to ראיה (دليل, *demonstratio quia*), namely indirect demonstrations based on *per impossibile* argument. Alfonso, however, is not always consistent in his use of מופת and ראיה.[52]

We translate מופת as "demonstration", ראיה and באור a "proof" quoting the Hebrew word in brackets.

זיוף This word means falsity in general and sometimes sophistry (Arabic *al safsaṭa*). It was used in this sense in texts from thirteenth century Spain,[53] and also here. The Provençal translators Ya'aqov ben Makhir and Qalonimos ben Qalonimos in their translations of Ibn Rushd's epitome and middle commentary on the *Sophistical Refutations* use the term הטעאה for sophistical argument.[54] Qalonimos uses also the word זיוף several times, but not always consistently. In Chap. 11, Aristotle distinguished between eristic and peirastic argument,[55] and Qalonimos uses הטעאי and מזויף respectively.[56] In his translation of the middle commentary on the *Physics* Qalonimos uses זיוף for *eristikos sullogizontai*.[57]

Alfonso's use of the word זיוף (rather than הטעאה) adds to our impression that he was not acquainted with the Provençal vocabulary.

Commentary

Alfonso distinguishes between [i] early thinkers who denied the possibility of equality between areas bounded by straight lines and areas bounded by curved lines because such areas cannot be superposed on one another (or cannot be decomposed into finite number of parts that can be superposed on one another), and [ii] early thinkers who were willing to consider such equalities. The question underlying this controversy was raised in two contexts: in Aristotle's physics and in his logic.

The first context is the discussion in *Physics* VII.4 of comparability of entities, which Wardy describes as "one of Aristotle's dialectic exercises without the

---

[52] For example, in I.1.4 he refers twice to Archimedes' proofs, which should be ראיות, according to his standard, as מופתים.

[53] Klatzkin's *Thesaurus*, vol. I, 257.

[54] Aristotle's book *Sophistical Refutations*, named in Arabic *Kitāb al Safsaṭa* was translated into Hebrew by Qalonimos ben Qalonimos as ספר ההטעאה. This term was used earlier in Shmuel Ibn Tibbon's translation of the *Guide of the perplexed* and in other Hebrew texts. See Klatzkin, *Thesaurus* I.168.

[55] *Soph. Ref.* 11, 171b8. On this distinction and the notion of peirastic argument in this context see Schreiber (2003, 191, note 6).

[56] Ibn Rushd *MC Soph. Ref.* Arabic Jéhamy, vol. 2, 694; Hebrew 110b11–13.

[57] See Harvey's edition, text 186.4, translation 250, Hamburg ms. 3b16. The Arabic text of the middle commentary is not extant. Zeraḥya ben Shealtiel uses the word גנות.

complementary resolution of difficulties and exposition of refined coherent theory."[58] Aristotle contends that linear and circular motions are incomparable, because their comparison would imply the commensurability of the circle and the straight line.[59] Their incomparability is further supported by Aristotle's physics: the former is the natural motion in the sublunar realm, the latter in the celestial realm. The argument of *Physics* VII.4 is, however, problematic,[60] and the issue became a subject of debate between Arabic philosophers and mathematicians.[61]

The second context is Aristotle's criticism of the attempts that were made in the fifth century BC to solve the problem of the quadrature of the circle.[62] With the development of Aristotelian logic and of a strict axiomatic approach in the fourth century BC it was recognized that the solutions that were suggested in the fifth century were not rigorous. Aristotle portrayed these early attempts as examples of inadequate reasoning.

Alfonso takes as the representative thinkers of group [i] Aristotle and his followers Ibn Sīnā and Ibn Rushd. The reference to Ibn Sīnā may be to the *Shifā'*, the fourth book of the *Physics*: "when one thing and its limits do not coincide with another and its limits, it is not actually equal to it," or to a similar statement.[63] The reference to Ibn Rushd may be to several of his commentaries.[64] There are reasons to assume that when he wrote this section Alfonso used Ibn Rushd's long commentary on the *Posterior Analytics*:

> "The greater" and "the smaller" do not have the same meaning when said of straight and circular boundaries. I mean that it should not be said that the circular is greater than the [rectilinear] area. Rather, there is no true relation between the straight and the circular line.[65]

Similarly, in the middle commentary on the same text Ibn Rushd mentions that "a straight line cannot be superposed on a circular line,"[66] and in the middle commentary on *De caelo* he remarks that ratio is not defined between two arcs (that fall on two points) because ratio exists only between lines that can be superposed on one another.[67]

---

[58] Wardy (1990, 263).

[59] *Phys*. VII.4, 248a18–b7.

[60] Wardy (1990, 266–73).

[61] Rashed (2015, 441–9).

[62] See references in I.1.1 and I.1.3.

[63] *Shifā'*, *Physics* IV.5, McGinnis' edition and translation, vol. 2, 425–6. See discussion in Rashed (2015, 446–9). Also in IV.3. "So we say that in truth neither the straight line nor the circular line undergoes alteration as to become the other" (McGinnis' edition and translation, vol. 2, 414). It is also possible that Alfonso found a similar statement in Ibn Sīnā's *Oriental Philosophy*, a text with which he was familiar and that is no longer extant. See Szpiech (2010).

[64] See *MC Phys*. VII.5; *MC*. Hamburg ms. fol. 89b16–90a15; *Post. An,* Arabic Jéhamy, vol. 2, 396; Hebrew 10a; *MC Soph. Ref.* Arabic Jéhamy, vol. 2, 694; Hebrew Paris ms. fols. 110b-111a. The issue is not raised in the *Epitome on Logic*.

[65] *LC Post. An.* I comment 67, Badawi edition, p. 293. The complete passage is quoted in I.1.3 below.

[66] *MC Soph. Ref.* Arabic Jéhamy, vol. 2, 694; Hebrew Paris ms. fol. 111a3–5.

[67] Ibn Rushd, *MC De cael.* Arabic 92, Herbrew 21b8.

Alfonso lists in group [ii] early scholars who took up the challenge and tried to solve the problem of the quadrature of the circle, by using either approximation or what he considered to be sophistical arguments. When referring to the first subgroup, Alfonso most probably had in mind Archimedes. The second subgroup probably includes Antiphon, Bryson, and Hippocrates of Chios. The reference to "skeptical reasoning" may allude to the "aporias" about nature that Aristotle mentions immediately after the discussion of the false reasoning of Hippocrates and Antiphon.[68] Aristotle criticizes the arguments of these thinkers as eristic.[69]

The distinction between the two approaches, of the philosophers and of the mathematicians, was made already three centuries earlier by Ibn al-Haytham, who, in his treatise on the quadrature of the circle, associated approach [i] with the philosophers and the commentators (not mentioning names), whereas he, a mathematician, advocated approach [ii]:

> Many philosophers have believed that the area of the circle cannot be equal to the area of a square limited by straight lines. This notion has been revisited many times in their dialogues and commentaries… We have therefore [after considering the quadrature of the lunes] given our utmost attention to this notion, until it was made apparent to us through a proof that it is possible and that there is no ambiguity as to this possibility.[70]

There is no additional evidence that Alfonso was familiar with this treatise of Ibn al-Haytham.[71] The question of the equality between rectilinear and circular magnitudes, however, came up again in the school of Marāgha in the thirteenth century, and al-Tūsī also points out the distinction between the two approaches.[72] We will come across several more indications that Alfonso was somehow informed about some mathematical ideas that were raised in this school.[73]

[I.1.2. Drawing on Archimedes and Euclid]

We claim that there is no sense in what they [Ibn Rushd and Ibn Sīnā] say, because we find one circle equal to two smaller circles, as Euclid established, although the two circles together cannot be superposed on the one circle as a whole, nor all their parts on all its parts. Similarly, the circumference of a

---

68 *Physics* I.2185a18. In his long commentary on this passage Ibn Rushd uses the terms ספקות, ספקות טבעיות, ספקות על צד הניצוח (doubts, natural doubts, eristic doubts). See *LC Phys*. Not extant in Arabic; Anon. Hebrew translation Paris ms. fol. 9b.

69 For example, *Pr. An.* II.26, 69a27–36; *Post. An* I.9, 75b37–76a3; *Soph. Ref.* 11171b16–18, 171b34–172a8; *Physics* I.1185a14–17. More information about these arguments comes from the Greek commentators.

70 Rashed (2013a, 99).

71 Alfonso's two proofs of the quadrature of the lunes (III.23) do not follow Ibn al-Hytham's and nowhere does he allude to the last step in his predecessor's argument about the squaring of the complete circle.

72 Rashed (2015), text and translation 480–1; analysis 450–51.

73 See Introduction Section 6 and I.1.2, I.2.1, II.5, III.33.

sphere is equal to the area of a [plane] circle, and to the area of a round cylinder [94b], and to the area of a cone, as Archimedes has established,[74] [although] no part of the one can be superposed on a part of the other. Therefore, I do not give up the investigation, and think that there may be a possibility of squaring the circle although it is not the case that all the parts of the square are superposed on all the parts of the circle.

Commentary

In order to refute group [i] thinkers, Alfonso adduces several examples illustrating that even if no construction can be provided, equality of rectilinear and curved areas can be rigorously established. All the examples except the first are taken from Archimedes' uses of the double indirect method in *The Sphere and the Cylinder*, Propositions I.33 and I.34. The first, paradigmatic example to which Alfonso comes back several more times throughout *SMA* is **the three-circles-example,** i.e., the equality of a circle to the sum of two smaller circles. This example is mentioned precisely in the same context by al-Ṭūsī as well as by his student Shīrāzī,[75] thus adding some support to our impression that Alfonso was to a certain extent informed about the mathematical work that was carried out in the school of Marāgha.[76]

Alfonso ascribes the three-circles example to Euclid, and it should be asked why. Euclid nowhere states this example, and nowhere in the *Elements* does he allow arithmetic of circular areas. Moreover, in *Elements* VI.31—a generalization of the Pythagorean theorem—he refers only to similar rectilinear figures constructed on the sides of the right-angled triangle, avoiding the case of semicircles. The three-circles example goes back to Hippocrates of Chios, who began his famous argument on the lunes with the statement that in a right-angled triangle the semicircle on the hypotenuse equals the sum of the semicircles on the sides.[77] Alfonso was familiar with the work of Hippocrates.[78] So why did he ascribe it to Euclid?

The first, straightforward explanation is that Alfonso understood that Hippocrates' proof was incomplete, and knew that the missing step—that circles are to one another as the squares on their diameters—is given in Euclid's *Elements* XII.2. In *SMA* III.23, however, Alfonso presents two of Hippocrates' quadratures without mentioning the required reference to Euclid.

A second plausible explanation is "political": Alfonso wished to have Euclid, the ultimate authority on geometry, on "his side" of the debate. He wished to present the strict requirement of homogeneity as a pedantry of Aristotelian philosophers, rather than as a basic tenet of Euclidean geometry.

---

[74] Archimedes, *Sphere and Cylinder* I.33, 34.

[75] Rashed (2015, 484 (Arabic 485); 522 (Arabic 523)).

[76] See commentary to previous section and note with references there.

[77] Heath (1921, vol. I, 185).

[78] See commentary to I.2.3. In I.2.3 and I.2.5 he mentions Hippocrates by name.

A third possible explanation is that Alfonso did not have direct access to a complete copy of Euclid's *Elements*. There are several indications that he was not acquainted with Book XII.[79] It is possible that he used some secondary transmission version of the *Elements* included additions.[80]

[I.1.3. Drawing on Bryson]

We claim that this is why the sage Bryson, the geometer, said that since the circle is greater than any inscribed rectilinear area [שטח ישר הקוים עשוי בתוכו] and smaller than any circumscribing rectilinear area [שטח ישר הקוים עשוי חוצה לו], and [everything] that is enclosed between these two is smaller than the circumscribing [rectilinear area] and greater than the inscribed [rectilinear area], then the circle must undoubtedly be equal to this intermediate area. This sage indeed understood that the circle and the rectilinear area are both of the genus "area" and differ only in their enclosing bounding surfaces, and therefore the rectilinear area must undoubtedly be either greater or smaller than the circle. It is impossible to compare in terms of great and small two magnitudes that are not of the same close genus, as it cannot be said that the mountain is greater or smaller than the day. This is why any area that is neither greater nor smaller than the circle will necessarily be equal to it, and this is the figure to which the sage Bryson referred as an intermediate between the area outside and the one inside. On this he postulated a conjunctive conditional premise:

> **[Bryson's premise]** If there exists a rectilinear area intermediate between any inscribed area and any circumscribing area—it must undoubtedly be equal to the circle.

This is a true premise which can qualify as a first principle of geometry, side by side with the other valid first principles of geometry. Particularly it can qualify as a first principle and a major premise in [any argument] for the quadrature of the circle. [95a] It remains for the geometers to inquire whether the minor premise of this disjunctive conditional statement—which [asserts] the existence of the antecedent—is valid, namely to find an intermediate rectilinear area, less than any circumscribing rectilinear area and greater than any inscribed rectilinear area. Then [when the minor premise is established], the conclusion of the deduction would result, namely the existence of the consequent of the conjunctive conditional premise [i.e., that the intermediate rectilinear area is equal to the circle].

---

[79] See Introduction, Section 6.

[80] The construction of the squaring of the lunes, for instance, is found in a few Hebrew manuscripts It is appended to the Paris Private collection manuscript of the *Elements* fol. 169b; it is also included in the *Hebrew Geometrical Compendium* Mantova ms. fol. 70b. On this text see Introduction Section 6.

Terms

עשוי בו, עשוי חוצה לו, עשוי בין Euclid defined the terms "inscribed" and "circumscribing" (*eggraphesthai eis* and *perigraphesthai peri*) in book IV of the *Elements*. Euclid uses consistently the same two verbs in the construction propositions of Book IV: *eggrapsai* ("to inscribe" e.g., in IV.2), *perigrapsai* ("to circumscribe", e.g., in IV.3). In the process of the transmission of the text into Arabic and Hebrew some variation was introduced. The Arabic text preserved in ms. Leiden 399 (presumably al-Hajjāj's translation, revised by al-Nayrizi) follows Euclid faithfully translating the two verbs in the definitions *marsum dakhilan* and *marsum kharidj*.[81] The same verb رسم is used also in Proposition IV.2 and other construction propositions. In Iṣḥaq-Thābit version (at least in some manuscripts), however, we find the verb رسم in the definitions of *Elements* IV, but the verb عمل in Proposition IV.2: "we wish *to make* in a given circle a triangle…,"[82] and in similar propositions. Moshe Ibn Tibbon, Ya'aqov ben Makhir, and the anonymous compiler of the *Hebrew Geometrical Compendium* follow the Arabic text. They use the verb רשם in the definitions, הנרשמת בתוך, הנרשמת סביב,[83] and the verb עשה in Proposition IV.2 and similar propositions: לעשות בעגולה, לעשות על העגולה.[84]

Alfonso does not use the verb רשם in this context,[85] but only the verb עשה: עשוי בו, עשוי חוצה לו.[86] He also coins the term לעשות בין for enclosing a figure between two polygons. It thus seems that in this context he did not consult one of the Arabic versions of *Elements* IV, but perhaps the passage of Ibn Rushd (quoted below), in which he could have found the expressions سطح يعمل داخل, سطح يعمل خارج.

הקש Alfonso uses the terms מופת, ראיה, ביאור when he discusses the geometrical aspects of a proof; he uses the term הקש (قياس, syllogism) when he considers the logical structure of a proof. He refers specifically to:

הקש התנאי conditional syllogism. It can be conjunctive (based on הקדמה תנאית מדובקת) as in the case of Bryson's attempt to square the circle, or disjunctive (based on הקש תנאי מתפרק) as in the case of Archimedes' proofs using the double indirect method.

הקש הסיבוב circular argument. Alfonso uses this tern when referring to al-Nayrizi's attempt to prove the Parallels Postulate.

---

[81] Besthorn edition part II fasciculus 1, p. 8 in the online file, fol. 51r of the Arabic text.

[82] Rabat ms. fol. 142–143; Paris ms. fol. 41a. Euclid uses consistently the verb *eggraphesthai.*

[83] The compiler of the *Hebrew Geometrical Compendium* used both expressions לעשות בעגלה and לרשום בעגלה.

[84] For example, Moshe Ibn Tibbon's translation Paris ms. 95b; Compendium, Mantova ms. 29b.

[85] He uses this verb רשם in other contexts in the sense of "describe" or "mark." See glossary.

[86] See also Moshe Ibn Tibbon's translation of Elements I.46.

Historical Commentary

Alfonso credits Bryson with the understanding that the circle and the square both belong to the genus "area," and even if they cannot be known to be equal, they are comparable in terms of great and small. This understanding was naturally taken for granted before Aristotle argued to the contrary in *Physics* VII.4.

Bryson lived in the fifth century BC, and the information about his attempt to square the circle comes from brief critical remarks in Aristotle's *Posterior Analytics* and *Sophistical Refutations* and, mainly, from the more detailed reports of Themistius on the former text and of pseudo-Alexander on the latter text.[87] Ibn Rushd mentions Bryson's construction in his middle commentary on the *Sophistical Refutations*, and in his middle and long commentaries on the *Posterior Analytics*.[88] The passage in the long commentary seems to be the best candidate:

> And the premise that Brusen (بروسن, ברוסן) postulated is that things that are greater than one thing and smaller than [the same] thing are equal. This premise, used mainly by geometers, is true,[89] and is not specific to magnitudes.[90] "The greater" and "the smaller" do not have the same meaning when said of straight and circular boundaries. I mean that it cannot be said that the circular is greater than the [rectilinear] area. Rather, there is no true relation between the straight and the circular line. The geometers use this premise when they argue that the measure of the circle results from multiplying [ضرب] half its diameter by half its circumference by establishing that this entity is smaller than any area made outside (سطح يعمل خارج) the circle and larger than any area made inside (سطح يعمل داخل) the circle. It follows that the area that results from multiplying half the diameter by half the circumference is equal to the circle, because [both] it and the circle are smaller than one thing itself, i.e., any area that is made outside the circle, and are larger than one thing itself, i.e., any area made inside the circle[91]

[87] *Post. An* I.9, 75b39–76a2; *Soph. Ref.* 11, 172a2–7; on the commentators see Heath (1921, I, 223).

[88] *MC. Soph. Ref.* Arabic Jéhamy, vol. 2, 695; Hebrew Paris ms. fol. 111a, 20–24; *MC Post. An.* Jéhamy, vol. 2, 396.5; The passage from the LC is sited below. Bryson is not mentioned in Alfarabi's commentary on the *Sophistical Refutations.*

[89] In Arabic صادقة (right). Among the Hebrew manuscripts, Munich and Parma manuscript read צודקת (right), Wien manuscript reads בלתי צודקת (not right). Comparison with the middle commentary confirms the first reading: Bryson's argument "is not demonstrative although he made right premises" (*MC. Post. An.* Paris ms. fol. 10a. 15–6).

[90] In the middle commentary "because they are general and comprehensive."

[91] *LC Post. An.* I, Badawi's Arabic edition, pp. 292–3 (there is no numbering of the texts-comments), text-comment 67 according to the numbering in the margin of the Hebrew translation: Munich ms. fol. 116b, Oxford ms. fol. 44a–b, Parma Palatina ms. 3022/2 De Rosi 285, IMHM 13751 (there is no page numbering, but texts-comments are numbered in the margin).

There are several reasons to suspect that Alfonso's source was Ibn Rushd's long commentary on the *Posterior Analytics*:

(a) This passage could have also been Alfonso's source in I.1.1 above, where he ascribed to Ibn Rushd the statement that the straight and curved lines are not comparable.
(b) The expressions used by Alfonso for inscribed and circumscribing polygons accord with Ibn Rushd's Long commentary: سطح يعمل داخل and سطح يعمل خارج. In the middle commentary Ibn Rushd uses سطح يقع في and سطح يقع خارج.[92]
(c) Ibn Rushd's text implies that Archimedes, in his demonstrations using the method of exhaustion, relied directly on Bryson, disregarding Euclid's *Elements* XII. Alfonso also links Archimedes' work directly to Bryson (I.1.4 below).
(d) The phrasing of the theorem by Archimedes in I.1.4 below is similar to Ibn Rushd's and different from Archimedes' original statement (see commentary on I.1.4 below).

One argument against the assumption that Alfonso's source was Ibn Rushd (whether from the long or the middle commentary on the *Posterior Analytics*) is the different spelling of Bryson's name. The name بروسن, following the Greek βρυσων is used in Ibn Rushd's middle and long commentaries on the *Posterior Analytics*. The Provençal translators render it in Hebrew as ברוסן, and in the middle commentary on the *Sophistical Refutations* as ברושם.[93] Alfonso uses the altogether different spelling פרוס.

Mathematical Commentary

According to Alfonso's analysis, Bryson's quadrature of the circle should be proved using the scheme of *modus ponens*[94]:

Major premise ($P \rightarrow Q$): if such a rectilinear area exists, then it is equal to the circle.
Minor premise (P): there exists a rectilinear area larger than any rectilinear area inscribed in the circle and smaller than any rectilinear area circumscribing the circle.
The required conclusion (Q): the rectilinear area is equal to the circle.

The major premise, to which Alfonso refers as "conjunctive conditional,"[95] was provided by Bryson. Establishing the minor premise is the main challenge that had not been achieved.

---

[92] Badawi edition, p. 69.
[93] *MC. Soph. Ref.* Paris ms. fol. 111a8.
[94] The rule of inference: if P (minor premise) and $P \rightarrow Q$ (major premise), then Q.
[95] See for instance Maimonides, *Introduction to Logic*, p. 37. See also Bobzien (2002, 360–1).

[I.1.4. Drawing Further on Archimedes]

On this [Bryson's] premise Archimedes based his demonstrations that the circle equals the product of half its diameter by half its circumference. This means that if there exists a rectilinear area that is neither greater nor smaller than the product of half its diameter by half its circumference, it would be the rectilinear area that equals the area of the circle. However, he [Archimedes] had never found a straight line that equals half the circumference of the circle, only approximately. And Archimedes added in his proof of Bryson's statement that if there were an intermediate area equal to the circle, as Bryson said, then there would have also been a straight line equal to the circumference; and vice versa, if there were a straight line equal to a circumference of the circle, there would have also been the intermediate rectilinear area whose existence Bryson conjectured. But he [Archimedes] had never found the minor premise [needed to complete Bryson's argument] because he found neither such an intermediate area nor a straight line equal to the circumference of a circle.

Also, on a similar major premise Archimedes based additional demonstrations—of constructing a plane circle equal to the surface of a sphere, or equal to the surface of a cylinder, or equal to the surface of a cone—namely, that if there exists an intermediate plane circle that is neither greater not smaller than any of these bounding surfaces, then it will be equal to it. [95b] But he established the minor premise of the conditional deduction, because he found the intermediate circle.

Terms

רבוע קו בקו Translated here as product. See Section III.0, commentary on terms.

Commentary

According to Ibn Rushd's presentation the area of the circle is the product of half its diameter by half its circumference, whereas Archimedes' original statement is purely geometrical: the area of the circle equals that of the right-angled triangle whose sides about the right angle are half the diameter and the circumference of the circle. Alfonso seems to follow Ibn Rushd, but is slightly more "geometrically correct": whereas Ibn Rushd uses the arithmetical term for product ضرب, (that was later translated into Hebrew הכאה), Alfonso uses the geometrical term "square."

It should be noted that Alfonso proceeds directly from Bryson to Archimedes, without mentioning (neither here nor anywhere else in the extant part of *SMA*) Euclid's use of the method of exhaustion in *Elements* XII.[96]

[96] Heath (1921, vol. I, 224).

[I.2. Introducing Alfonso's Concept of Mental Superposition]

Introductory note: the concept of superposition

Following Ibn al-Haytham, Alfonso does not try to obscure the fact that superposition is motion, and that motion is a basic principle of geometry. He distinguishes two concepts of superposition: Euclidean superposition and a more complex notion of superposition that he names "mental superposition." The former underlies the notion of equality between homogeneous rectilinear magnitudes; on the latter he hopes to be able to base a concept of equality between nonhomogeneous magnitudes. Such a distinction was made before by al-Ṭūsī,[97] so this is another indication that Alfonso had some, probably indirect information about the mathematical and astronomical activity that was carried out in al-Ṭūsī's school.

[i] Euclidean superposition

The notion of geometrical superposition is introduced by Euclid in common notion 4: "things which are superposed on one another are equal to one another." This common notion was considered by many to be one of the weak links in the Euclidean edifice.[98] Superposing is motion, and according to both Plato and Aristotle motion belongs to the realm of physics, not of geometry.[99] Motion is a change of place or position and neither place nor position is a part of the conceptual basis of Euclid's *Elements*.[100] This problem was noted by Euclid himself, who tried to minimize the use of superposition.[101]

The question whether superposition can be guaranteed to be isometric was not asked by Euclid, but was raised by the Arabic geometers who assumed motion to be a basic geometrical principle. Thābit Ibn Qurra acknowledges the use of superposition for measuring but is aware of the problem that movement may introduce distortion:

> The superposition (*intibāq*) of something with something else equal to it, when we imagine [the former] to be moved to [the latter] *while maintaining its shape*, and be placed upon [the latter] so as to measure it.[102]

[97] Rashed (2015, 454) refers to the two types of superposition in al-Ṭūsī as "superposition géométrique" and "superposition cinématique."

[98] Bertrand Russel commented (1903, 161 # 152 β) that defining equality in terms of superposition is circular because superposition assumes the rigidity of the superposed body, but the notion of rigidity depends on this of equality.

[99] Plato *Tim.* 37c–38c. Aristotle *Meta.* E.1.

[100] Euclid deals with the notion of "position" in his *Data*.

[101] "It seems clear that common notion 4, as here formulated, is intended to assert that superposition is a legitimate way of proving the equality of two figures… At the same time it is clear that Euclid disliked the method and avoided it wherever he could." Heath (1908/1956 I, 225).

[102] Sabra (1968, 19–20), Sabra's translation with a minor modification; Jaouiche (1986, 151); Rashed and Houzel 43–46.

Arab geometers offered several requirements that guarantee isometry, namely guarantee that the points of the moved magnitude describe equal and parallel lines and thus distances and angles are preserved under superposition.[103] Thābit Ibn Qurra suggested that the imagined moved line "should be placed in a solid which moves while [the line] remains fixed in it."[104] The rigidity of the three-dimensional body is taken for granted. Ibn al-Haytham suggested that "the motion should be one and simple, neither composed of motions nor composed of motion and rest, but one motion, unmixed with another and at the same time continuous.[105] Al Ṭūsī suggested that the superposition should be carried out at once.[106] Alfonso (I.2.2 below) adopts al-Ṭūsī's requirement.

Adopting Ibn al-Haytham approach,[107] Alfonso uses superposition several times, even when simple absolute geometry[108] would have sufficed, obviously to emphasize its legitimacy as a geometrical method.[109] Sometimes he uses superposition to show how geometrical lines actually coincide.[110]

[ii] Mental superposition

It could have been his delving into the subject of non-isometric mappings (I.2.1, II.5.3) that led Alfonso to a more sophisticated conception of superposition.[111] Whereas Euclidean superposition is induced by instantaneous motion, mental superposition is induced by continuous uniform motion and continuous motion is not necessarily isometric.

The concept of function actually had to wait until the seventeenth century, when a new approach to the study of motion evolved.[112] Yet, it was suggested that a concept of function tacitly played a role in the development of science long before the new era,[113] especially in astronomy, the mathematical science of motion. Pedersen argues that

> Spherical astronomy and planetary theory abound in mathematical relations which are best described as functions of one, two, or three variables. Such relations a mathematical astronomer like Ptolemy had to deal with, and one of the most fascinating but least

---

[103] Jaouiche (1986, 162–4).

[104] Sabra (1968, 21), Sabra's translation.

[105] *Commentary on the Premises of Euclid's Elements,* see Jaouiche (1986, 162).

[106] Rashed (2015, 454).

[107] See introductory note to Section I.1 above.

[108] Namely not relying on the parallels postulate.

[109] In II.3.2, II.3.4, II.3.5.

[110] For example, in III.33 when he proves that the planet always moves on the diameter of the deferent.

[111] Alfonso's notion has little or nothing to do with the methods of the Oxford calculators and with the controversial notion of "Bradwardine's function." See Maier (1982).

[112] Kleiner (1989, 283).

[113] Ponte (1992, 2–5).

investigated mathematical aspects of the *Almagest* is the rudimentary theory of function which can be extracted from its pages.[114]

Alfonso indeed draws on Ptolemy and also on al-Ṭūsī when he adduces his examples of non-isometric mapping. Three instances of mental superposition are studied in *SMA*: [i] the projection of the motion of a planet (that moves on the ecliptic) on the meridian (I.1.2); [ii] the rolling of a circle along a tangent line or of a cylinder on a tangent plane (I.1.2; II.5); and [iii] the projection of the motion of a planet carried on an epicycle on the diameter of the deferent circle (III.33).

[I.2.1. Two Examples of Non-isometric Superposition]

It follows that not any superposition of magnitudes on one another in any way can be the essential superposition that defines their equality with one another, as Ibn Rushd thought. This is because there are [on the one hand] magnitudes which are equal to one another but cannot be superposed on one another, [on the other hand] magnitudes unequal in quantity that can be superposed on one another.

[First example] For instance, half the zodiac, which is 180°, can be superposed in its entirety on the arc of the meridian which is 48°, so that each point of the big arc meets a corresponding point of the small arc, with no excess of the big [arc] over the small.

[Second example] In a similar way by the motion of rolling it becomes possible, as we shall show,[115] [i] that the area of a cylinder in its entirety would touch a plane area much bigger than itself, as well as a plane area much smaller than itself, and [ii] that the circumference of a circle would cover a very long straight line as well as a very short straight line, and [iii] an acute angle would be superposed on an obtuse angle without excess or defect, so that many of the natural philosophers and, a fortiori, other people would be stunned and would not believe it until they see it with their own eyes. And before they see it they would be very confused and would ask: how is it possible that each part of the big magnitude touches equally the corresponding part of the small magnitude without exceeding it, as it [the big magnitude] exceeds [the small] in the case of static superposition. They do not understand how motion causes this to happen.

Terms

זוית נרוחת Alfonso uses Bar-Ḥiyya's term זוית נרוחת for obtuse angle. Among the Provençal translators Moshe Ibn Tibbon uses זוית נרוחת, whereas Ya'aqov ben Makhir uses the term זוית נרחבת.

[114] Pedersen Jones (2010, 49, 80–81), See also North (1987).

[115] II.5.1.

Commentary

Alfonso adduces two examples of non-isometric superpositions, i.e., superpositions that map unequal magnitudes on one another.

The first example presumably draws on Ptolemy's astronomy, but Ptolemy is not explicitly referred to. The motion of a point on the ecliptic (for instance, the center of the sun) defines a one-to-one correspondence between two unequal arcs on the surface of a sphere. While the point traces an arc of 180° on the ecliptic (e.g., from the summer solstice to the winter solstice) the component of the motion in the north-south coordinate, namely the distance along a meridian will be 48° (from the tropic of the Cancer to the tropic of the Capricorn). In Ptolemy's table of inclinations an arc of 90° on the ecliptic corresponds to an arc of 23° 51′ 20″ on the meridian.[116]

The second example is the mapping defined by rolling: when a cylinder rolls on a floor, it can describe different rectangles on the floor. Alfonso could have found a discussion of this motion in al-Tūsī and in his student Shīrāzī, one more argument to support our hypothesis that he was to some extent familiar with the work of the school of Marāgha. Alfonso comes back to this example in Section II.5, and it will be explained in detail there.

[I.2.2. Modifying Euclid's Common Notion of Superposition]

Therefore I say that it is necessary to add a reservation to the [axiom of] superposition that Euclid introduces in the beginning of his book[117]: that superposing should not be carried out through motion or part after part, but rather all [96a] at once and instantaneously. Undoubtedly, those magnitudes the imagining of whose superposing on one another is instantaneous (not part after part as it is in motion) would equal one another; and vice versa,[118] equal magnitudes, their superposing on one another cannot be imagined unless […].[119] Since the equal magnitudes that we mentioned above cannot be imagined to be superposed on one another all at once in this superposition that we used to imagine, it necessarily follows that there must be another [kind of] superposition, superior in rank to the superposition as we usually imagine it, which is the essential cause of equality of magnitudes.

Commentary

In order to guarantee the isometry of superposition, al-Tūsī requires that the superposition be carried out at once.[120] Alfonso amends Euclid's common notion 4 by adding the restriction that the motion be instantaneous, not part after part. This is

[116] *Almagest* I.15, Toomer's translation, p. 72. This value is somewhat larger than the modern.

[117] *Elements* I, Common Notion 4.

[118] Literally "similarly."

[119] The sentence seems to be interrupted.

[120] Rashed (2015, 454). See introductory note to Section I.2.

one more indication that Alfsonso was to some extent familiar with the work of the school of Marāgha. With this added requirement Euclidean superposition can be safely used, and Alfonso indeed uses it freely several times.[121] He refers to Euclidean superposition as the "known," the "ordinary," "the superposition that we are used to imagine," and also as the "static" superposition. Alfonso knows, however, that this kind of superposition cannot support an extended geometry in which nonhomogeneous magnitudes can be compared. He seeks to define a new kind of superposition—to which he refers as "superior" or "mental" superposition—which can serve as "an essential cause of" a notion of equality between nonhomogeneous magnitudes. Aristotle requires that a demonstration be explanative and present the "cause" of the validity of the statement to be proved. The method of exhaustion provides an efficient way to prove equalities between nonhomogeneous magnitudes, but it cannot provide the "essential cause" of these equalities, because it is based on *per impossibile* arguments. It is Alfonso's vision that mental superposition should provide a more direct method of establishing such equalities. This point will be further elaborated in the next two paragraphs.

[I.2.3. Mental Superposition in Hippocrates]

And I surmise that Hippocrates the geometer alluded to this superior superposition when he claimed that if a circle could be divided[122] into lunes which are equal to rectilinear areas, then the circle would be equal to these rectilinear areas together. Undoubtedly, if Hippocrates was an expert in geometry as he was reputed to be, he would not have said that the circle is divisible into lunes which can together be superposed on the circle (assuming the ordinary superposition that we are used to imagine), so that they would be equal to it. Rather, he meant to say that there can be a ratio between an area whose boundaries are circular lines and an area whose boundaries are straight lines, drawing on [the example of] one circle being equal to two or three smaller circles. This is made possible only by the mental and superior superposition.

We infer this equality [of the circle and the smaller circles] by way of proof [ראיה, *demonstratio quia*] called in Arabic *dalil*, not by way of essential demonstration [מופת עצמי, *demonstratio propter quid*] that is called in Arabic *burhān*. Only by means of mental superposition can the circle be divided into figures that are equal to two [96b] or three small circles. From this [i.e., mental superposition] follows the equality of a lunar figure to a rectilinear area. One can think that this may imply also the equality of the circle to a rectilinear area. Therefore Hippocrates stated it as a conditional premise and hesitatingly. In a conditional premise it is not a requisite that the antecedent be necessarily true. If it were [true] it would not have been necessary to recapitulate it a second time as a minor premise of the deduction.

[121] II.3.2, II.3.4, II.3.5, III.33.

[122] Reading תחתך, rather than תחתוך.

Commentary

Hippocrates was the first known mathematician to compare areas bounded by straight lines to areas bounded by circular lines. The arguments that were ascribed to Hippocrates can be divided roughly into two: [i] his quadratures of specific lunes and [ii] his alleged suggestion to proceed from the quadrature of these specific lunes to the quadrature of the whole circle.[123] Of these two arguments [i] is correct but incomplete,[124] [ii] is obviously false,[125] and Aristotle mentions it as an example of false reasoning.[126]

The Aristotelian tradition (notably commentaries on the *Physics* and the *Sophistical Refutations*) focuses on Argument [ii]. Hippocrates is mentioned by name and criticized. The mathematical tradition (that goes back to Eudemus and came down to us through Simplicius' commentary on the *Physics* and references to Alexander's lost commentary)[127] focuses on Argument [i]. In this tradition, as Clagett and Langermann note, Hippocrates is not mentioned by name.[128] This is true also of Alfonso: in III.23 he does not mention Hippocrates by name, but presents in detail two variants of his mathematical argument [i].[129]

Alfonso was familiar with both the Aristotelian and the mathematical traditions. In the present passage he refers to Hippocrates by name and following the Aristotelian tradition, alludes to Argument [ii].[130] He could have relied on Ibn Rushd's long commentary on the *Physics*,[131] or on his middle commentary on the *Sophistical Refutations,* where Ibn Rushd ascribes to Hippocrates the false Argument [ii]:

> As one of the early geometers named Abukrat did. Since he made a square equal to a lunar figure he thought that he also found a square equal to a circle, because he thought that the circle can be divided completely into lunar figures.[132]

---

[123] Heath (1921 I, 184–200).

[124] See comment to I.1.2 above.

[125] The question of Hippocrates' "fallacy" is studied in detail in Lloyd (1987).

[126] Notably *Soph. ref.* 11, 171b13–15; *Phys.* I.2, 185a14–17, *Pr. An.* II.25, 69a30–35. In the *Physics* Hippocrates' name is not mentioned and there is an argument among modern commentators about it. See Lloyd (1987, 104, note 3); Heath (1921 I, 184–200).

[127] See commentary to III.23 below.

[128] See Langermann (1996, 36–7).

[129] Two of the four quadratures that Simplicius quotes from Eudemus. See Heath (1921, 187, 191–195); Netz (2004, 249–252).

[130] Here and in paragraph I.2.5.

[131] For example, Ibn Rushd, *LC Phys*. Comment I.10, Hebrew Paris ms. 9a20–26.

[132] *MC Soph. ref.* Arabic Jéhamy, vol. 2, 694; Hebrew Paris ms. fol. 110b20–24.

Alfonso dismisses Argument [ii] and considers it highly unlikely that a mathematician like Hippocrates suggested it. His remark—that Hippocrates was a reputed geometer—goes back to Greek sources.[133] According to Alfonso Hippocrates had in mind only Argument [i].

In his quadratures of lunes, as transmitted by Simplicius, he skips a crucial step, namely the proof that the ratio of two semicircles is as the ratio of the squares on their diameters. Hippocrates, as it seems, took it for granted. The proof was completed in the fourth century BC by Eudoxus using the double indirect method (the method of exhaustion). None of the quadratures is included in Euclid's *Elements* but Eudoxus' proof of the missing step is presented in *Elements* XII.2.

Alfonso was well acquainted with Hippocrates' work. The allusion to division of the circle either into two or into three circles is to two variants of Hippocrates' argument [i].[134] In the mathematical Part of *SMA* (III.23) he describes in detail two of Hippocrates' quadratures, as if unaware of the missing step, but here he definitely understands that this step is the crux of the proof. He thus ascribes to Hippocrates the use of indirect argument that is not a *demonstratio propter quid,* but rather a *demonstratio quia*. It cannot provide an explanation and the reader cannot truly understand the theorem. This is a major issue for Alfonso and he raises it again in the next section.

[I.2.4. Mental Superposition in Archimedes]

We suggest that it was on the imagining [ציור] of mental superposition that Archimedes based his demonstration of the equality of the circle to the product of half its diameter and half its circumference, because he established it on a *per impossibile* demonstration called in Arabic *burhān al-khalaf* with a disjunctive conditional premise: that the straight line is either bigger than, or smaller than, or equal to the arc of the circumference of the circle. If it is neither bigger nor smaller, it must be equal to the circle. But an equality between a straight line and an arc on the circumference of the circle must draw on mental superposition, as we mentioned above. Similarly, along the same lines Archimedes based on mental superposition his demonstrations of the equality of the area of the circle and that of a round cylinder, or of a cone or of the sphere. He could have arrived at this understanding drawing on [the example of] the lunar figure which implies the existence of a mental superposition that determines the ratio between magnitudes of different figures.

---

[133] Heath (1921, 186); Lloyd (1987, 107). It is mentioned by Simplicius and Proclus who praised Hippocrates as "a man of genius, when it came to constructions, if there ever was one." See, *Commentary* 213 Friedline, Morrow's translation 167.

[134] Heath (1921, 185–6).

Commentary

Eudoxus and Archimedes proved equalities between nonhomogeneous magnitudes using the double indirect method (the method of exhaustion),[135] but Alfonso ascribes to Hippocrates (to whom he ascribes Eudoxus' proof) and to Archimedes a premonition of his idea of mental superposition. Perhaps he felt that an indirect proof is philosophically insufficient and the geometer who uses it has to be convinced that a more explanative proof exists.

[I.2.5. Methodological Remark]

On such [conditional arguments] Hippocrates established that equality between a circle and a rectilinear area is natural and intelligible, although it is doubtful whether the geometer can [actually] construct this rectilinear area. They [the geometers] should pursue the inquiry about it because general knowledge achieved by reason is not sufficient according to their standards, although [97a] it is sufficient in physics[136] and metaphysics. For instance, the mind can see how to place a straight line, equal to another straight line so that its extremity is at a given point, but this [knowledge] is not sufficient for the geometer, who has to find the construction of this line and its actual existence, as Euclid did in the beginning of his book.[137] To this end the geometers have instruments such as the compass [מחוגה] and the drawing device [משרטטת] or other instruments. The less instruments needed [to construct] the magnitude and the easier to find, the superior would be the knowledge of this magnitude, and the closer to [Euclidean] construction.

Terms

מוגבל, להגביל Verbs and adjectives derived from this root גבל occur 24 times in Parts I–II of SMA, but not in Part III. In Parts I–II גבל has two meanings that are obviously related. Usually להגביל means "to limit" or "to bound." Here and in I.5.4 להגביל means "to construct," and מוגבל can be translated as "constructible."

משרטטת, מחוגה The Hebrew word מחוגה is biblical (Isa. 44:13). The root שרטט is Talmudic.[138] Similar instruments are mentioned in some Arabic texts.[139] The two

[135] Eudoxus' proof is presented in Euclid, *Elements* XI.2. Archimedes was the most sophisticated user of the method. For example, *Measurement of the circle* Proposition 1, *Sphere and Cylinder* I. 13, I.14, I.33, I.34.

[136] In the text "geometry" (השעור). Should probably be "physics" (השמע).

[137] *Elements* I.2.

[138] *Gitin* 7. The word was used also by Abraham Ibn Ezra (Sefer ha-'Ibur 4, 90).

[139] See for instance Brentjes (1996, 127).

words were not in common use in medieval mathematical texts. The suggested translation of מחוגה as compass is likely. The word משרטטת is used here and in III.28 below alluding to the instrument used to draw the conchoid.

מעשה The Greek kataskeuē, i.e., the construction part of the Euclidean proof, was translated into Arabic as عمل and to Hebrew by Moshe Ibn Tibbon as מעשה.[140] The word occurs four times in SMA,[141] and in all four cases the meaning accords with this interpretation.

Commentary

The subject of SMA is equalities between nonhomogeneous magnitudes. Such equalities have been proved rigorously, but indirectly, using the method of exhaustion. Alfonso contends that non-constructive *per impossibile* proofs, like those referred to in the previous two sections should not be sufficient for the Euclidean geometer. Geometers have to construct their objects with geometrical instruments. Euclid provided a formal construction even for an operation as simple as "to place at a given point a straight line equal to given straight line" (*Elements* I.2).

[I.3. Turning to Plato]

Introductory note

The introduction so far Alfonso aimed to convince his reader that a concept of mental superposition is essential in geometry. Before proceeding to the promised definition of this concept in Section I.4 Alfonso offers a second introduction (in Section I.3), in which he tries to justify his deviations from the Aristotelian-Euclidean main stream. To this end he turns to Plato.

Section I.3 is difficult and not as well-presented as Sections I.1–2. It looks like notes that Alfonso wrote when he tried to learn and understand what Plato said in the *Timaeus*. Gradually it becomes clear that only part of Section I.3 is relevant to the main stream of *SMA*, as Alfonso himself admits in I.3.4. His acquaintance with Plato was indirect. He learned about Plato mainly from th

e misleading presentation of Aristotle, probably through the commentaries of Ibn Rushd. Section I.3 was revised and two layers are still present in the text. Part IV of *SMA*, in which Alfonso intended to pursue his interpretation of Plato's is lost (except for the first 32 lines). Nonetheless, the little we know about Part IV helps us to understand the difficult Section I.3 and vice versa.

---

140 See Sarfatti 193 # 247.

141 I.2.5, II.1.6, II.3.2, II.4.1.

The outline (at the beginning of the book) summarizes in a few words the agenda of Part IV: to explain "how a body is divided into surfaces, and the surface into lines and the line into points, as Plato taught, and the way of *evaluating and measuring* them by one another." This scheme that Alfonso ascribes to Plato conflicts sharply with two basic assumption of Greek geometry and of Aristotelian philosophy:

[i] The dimensional-barriers assumption
In Greek geometry as well as in Aristotelian philosophy lines, surfaces, and solids are considered as entities of different kinds. Definitions I.3, I.6, and XI.2 of Euclid's *Elements* clarify that points are the *extremities* of a line, lines of surfaces, and surfaces of solids. They are not parts or components. In this vein Aristotle defines divisibility separately for each dimension: "of magnitude that which is continuous in one dimension is length, in two breadth, in three depth."[142] According to both Euclid and Aristotle, a surface is divisible into surfaces, for Plato surface is divisible into lines.

[ii] The impossibility of measurement assumption
The word *geometria* means land-measurement, but Euclidean geometry was not a science of measuring. It is the science that studies magnitudes, which *by definition* are infinitely divisible.[143] A natural geometrical measuring unit is nonexistent and, due to the problem of incommensurability of geometrical magnitudes, no measuring unit can be defined. Alfonso ascribed to Plato the conception of geometry as a science of "evaluating and measuring" magnitudes by one another. It was actually Ibn al-Haytham who boldly considers measuring as a principle of geometry,[144] and suggests to define magnitude in terms of measurement, rather than of divisibility: "This [the mathematical] body can be measured in that its dimensions can be measured, *and since it is susceptible to measurement, it is a magnitude.*"[145]
Examining the possibility of measurement in geometry Alfonso claims: "Unless assuming the division of magnitudes into indivisible parts: how can it be imagined?… If the infinitely many parts existed only in potentially, then they would not be able to measure distances at all."[146] He sought to extend the science of geometry, but by no means to deconstruct it. He knows well that the hypothesis of indivisible parts of magnitudes (which he ascribes to Antiphon), conflicts with both Euclidean geometry and Aristotelian philosophy,[147] and he rejects it. Plato was said to have supported a theory of indivisible parts,[148] but Alfonso

---

142 *Metaphysics* V.13, 1020a12–14.

143 Aristotle, *Meta.* V.13.

144 Huppert-Sude 21.12–14.

145 Huppert-Sude 22.13–14. For surface see 24.25.

146 I.3.5 below.

147 Euclid *Elements* I.10; Aristotle, *Meta.* V.13.

148 In the *Metaphysics*, referring to Plato, Aristotle says: "He gave the name of principle to the line—and this he often posited—to the indivisible lines" (*Meta.* I.9, 992a20–22).

expressly denies such an interpretation of Plato (I.4.3 below). Without assuming indivisible parts, Alfonso wishes to challenge Aristotle on both [i] and [ii] and he turns to the authority of Plato for support.

In I.3.2 Alfonso ascribes to Plato three theses. In I.3.3 he claims to have achieved a new understanding and in I.3.4 he presents again the three theses. To facilitate the comparison of the first and second presentations of the three theses we add in brackets numbering and abbreviated titles.

[I.3.1. Alfonso's "Intellectual and Moral" Principle]

[Translation: Gad Freudenthal[149]]

And I was thinking for a long time on this mental superposition, trying to understand what it is and how can I imagine it. I wandered through the alleys of inquiry [namely] the books of Aristotle and his followers. I saw the various rebukes and charges that he addressed to Plato and Empedocles and other Ancients, and they did not seem right to me. This follows from one principle, which is both intellectually and morally right, that God, blessed be he implanted in my heart, and which I have not seen Aristotle usually employ. For me it was a great revelation and a first step to [the understanding of] some puzzles. This principle is that in my view it is inappropriate to ascribe to scholars, who spent all their days inquiring into the sciences, such errors as are committed by those who have never occupied themselves with science. A fortiori one should not ascribe to scholars who spent all their days with science and inquiry and composing books about them errors that are not committed even by irrational animals. Rather, when we find [in a scholar's writing] anything [97b] that is seemingly erroneous, it is without doubt appropriate to search for a way to do justice to it, so as to render [his affirmations] true and [show why] they seem acceptable to his intellect.

Commentary

As he himself testifies, Alfonso learned about Plato's cosmology from Aristotle. He was aware of the bias that characterized Aristotle's presentation of his predecessors. He offers as a rule of conduct to consider seriously the teaching of scholars known for their wisdom, even if at first glance it seems false. Alfonso was highly interested in the work of early thinkers, notably Antiphon, Bryson, Hippocrates, and Plato, and blames Aristotle for criticizing these thinkers too readily. Gad Freudenthal showed that in his *Theshuva la-Meḥaref,* Abner of Burgos pronounced a similar principle, thus providing additional support to the identification of Alfonso as Abner.[150]

---

[149] The translation of this section is taken from Freudenthal (2005, IX, 2–3).

[150] Freudenthal (2005, IX, 3).

[I.3.2. Three Theses Ascribed to Plato: First Presentation]

[Three Theses Ascribed to Plato]

And I gather from Aristotle's charges against Plato answering [Plato's] statements: [i. on *hiuli*] there is an all-receiving substratum and this is the *hiuli;* and [ii. on dimensional reduction] that bodies are divisible into triangles and lines, and [iii. on weight] that the cause of heaviness of the heavy bodies is the large amount of elementary *geramim* of which the body is made, and the cause of levity is the small amount of these *geramim.*

[Five Questions Ascribed to Aristotle]

And Aristotle was puzzled by it and said: [a] how is it possible that the *hiuli* be surfaces? [b] And if they[151] were surfaces, how could they be pyramids, as pyramids are bodies sharpened above and wide below and are not surfaces? [c] Furthermore, if the *hiuli* be surfaces, how could surfaces be divisible into triangles? For if they could, it would be more appropriate to call the triangles *hiuli,* than the surfaces. [d] Furthermore, if the *hiuli* be triangles, how can it be lines? [e] Also if in his opinion the *geramim* of body were surfaces, how is it possible that their amount is the cause of heaviness, as it is impossible for one surface to be heavier than another, even if the other is bigger?

Terms

היולי, גרמים The two terms *Hiuli* and *geramim* (in the plural) appear seven times on *SMA*, *always* when referring to Plato.[152]

The word היולי (Arabic الهيولى) stands for the Greek húlē, usually used to denote the Aristotelian matter. Alfonso borrows the Aristotelian term *hiuli* to refer to Plato's receptacle (*hupodokhē*),[153] an entity whose essence is difficult to understand.[154]

The term *geramim* (Arabic إجرام, singular جرم) is difficult. Usually both גוף and גרם stand in medieval Hebrew for "body."[155] In *SMA*, however, *guf* and *gerem* are not

[151] The *hiuli* is referred to sometimes as singular, sometimes as plural.

[152] *Gerem* in the singular appears once in I.4.3.

[153] *Tim.* 48E–51B.

[154] Plato had a difficulty to explain the receptacle's mode of existence (*Tim.* 48E–52C), and used several metaphors to explain it. Gold persists, whereas the things that come to be in it are temporary. In Aristotle's presentation: "the truest thing to say that each of them is gold." *De gen. et corr.* II.1329a21. An associated concept is *chora*, i.e., place (*Tim.* 52A, 52D).

[155] The word *gerem* was common in astronomy but less in geometry. Sarfatti (1968, 202–3) considers it as a synonym of *geshem*, i.e., body.

synonyms. They appear in one and the same sentence, having different meanings: *guf* means "body", *gerem* has a more specific meaning: It is used (in the plural) to describe the Platonic components of a body, but the exact meaning of this term is less clear.[156] Three possible interpretations should be considered:

[i] *Geramim* are three-dimensional geometrical entities, perhaps Plato's regular solids. *Prima facie,* this interpretation seems to be the most straightforward as it accords best with the literal meaning of the word גרם in Hebrew or جرم in Arabic. It should be noted, however, that of the four regular solids Alfonso mentions only the pyramid. It could in principle be argued against this interpretation that *hiuli* and *germaim* cannot coexist as two *archai* of physical reality (because of the Aristotelian axiom that two bodies cannot coexist at the same place),[157] but Alfonso rejects this axiom.[158]

[ii] *Geramim* are two-dimensional entities. This interpretation agrees well with Aristotle's testimony, on which Alfonso relies. In *De generatione et corruptione* Aristotle says that whereas for Leucippus and Democritus the "primary existences" were bodies, for Plato in the *Timaeus* they were planes,"[159] and adds that "for the Platonists the triangle is one."[160] Criticizing Plato, Aristotle ascribes to him the association of weight with planes,[161] and Alfonso accordingly refers to *geramim* as the cause of weight. Interpretation [ii] accords better than Interpretation [i] with everything that Alfonso says in Section I.3. The problem with this interpretation is that it conflicts with the lexical meaning of the word *gerem.* A possible justification can be found: the verb جرم in medieval Arabic had also the meaning of "to cut off,"[162] and the (very unusual) verb להגרים in Hebrew has a similar meaning.[163] It is, therefore, possible that Alfonso thought of *geramim* as thin layers or sections.

---

[156] In I.3.3 below the word *geramim* is not mentioned, and Alfonso says that the cause of heaviness is "bodies triangles." The two words appear one after the other with a bigger than usual space between them. Apparently, the copyist had a confused or corrected text before him and could not decide between the two words. Perhaps this sentence belongs to an earlier stratum of the text, before Alfonso settled on the term *geramim.*

[157] *De caelo* III.6, 305a18–19.

[158] See Sadik (2007, 5).

[159] *De gen. et corr.* I.2315b26–31.

[160] *De gen. et corr.* I.2316a11–12. *MC de gen. et corr.* I.3.2 Kurland 8.35–6, 46–7. Ibn Rushd writes "Plato" where Aristotle "Platonists."

[161] Ibn Rushd (in both his short and middle commentaries on *De caelo*) explicitly says that Plato associated weight with the planes. Ibn Rushd, however, does not use the word اجرام, but the word السطوح. *Epitome on De caelo*, Horesh edition 67. 3,5; *MC De caelo* III.3.3.2.1, Arabic, p. 286; Hebrew Berlin ms. fol. 61a.

[162] Lane's Arabic Lexicon, vol. I, 412–4.

[163] Ben Yehuda dictionary, vol. 2, 844.

[iii] *Geramim* are the Aristotelian minimal parts.[164] In paragraph I.4.3 Alfonso suggests this with the obvious intention of clearing Plato of the charge of atomism. There is no indication, however, that when he wrote Section I.3.2–4 he thought of the *geramim* in this sense.

Due to these uncertainties, we leave the two terms, *hiluli* and *geramim*, untranslated. The discussion of these two terms is continued in I.3.4.

מוצק This term was coined by Abraham Bar-Ḥiyya, as the equivalent of the Arabic مخروط. It was used for both pyramid and cone.[165] Alfonso uses מוצק or מוצק ישר שטחים for pyramid and מוצק עגול or מוצק האצטונה for cone.

Commentary

Gershenzon argues that when Abner/Alfonso mentions Plato by name, he usually refers to Plotinus.[166] Several references to the fifth *Ennead* have been identified.[167] The discussion of Plato's cosmology in *SMA*, however, does not suggest that he was acquainted with the second *Ennead*, in which Plotinus deals with Plato's cosmology.[168] Alfonso admits right at the beginning that he learned about Plato's cosmology from Aristotle's critical presentation. His direct sources could have been Ibn Rushd's middle commentaries on *De generatione et corruptione* and on *De caelo*. In I.3.2 Alfonso presents three theses that he ascribes to Plato and five doubts that he ascribes to Aristotle. Let us follow the rather tortuous transmission of these theses and questions from Plato through Aristotle to Alfonso.

Three Theses Ascribed to Plato (Marked [i] to [iii] in the Text)

Thesis [i]: On hiuli

The first thesis asserts the existence of a material substratum. Alfonso's brief first statement of the first thesis could well rely on Ibn Rushd's middle commentary on *De generatione et corruptione*, where he could find the term that he uses "the receptacle of everything."[169]

---

164 See commentary on I.4.3 below.

165 Sarfatti (1968, 86–7, 124).

166 Gershenzon (1984, 10).

167 Szpiech (2006, 662); Hecht, note 240; Sadik, notes 93, 262, 287, 391.

168 Sadik refers once in his dissertation (note 353) to *Ennead* II. The reference is not sufficient to confirm direct familiarity with Plotinus' text.

169 *MC De gen. et corr* II.1.1, قابل للكل (http://dare.uni-koeln.de/sourceviewer?type=text&textid=26, Chunk 37), translated by Moshe Ibn Tibbon into Hebrew also מקבל הכל, Kurland edition 59.52.

Thesis [ii]: On Dimensional Reduction

Plato associates the four elements—fire, air, water, and earth—with four (of the five) regular solids—pyramid, octahedron, icosahedron, and cube, respectively. The sides of the four solids can be decomposed into two basic triangles.[170] At this point Plato's reduction of physical bodies into geometrical elements terminates.[171] Not carrying on the reduction beyond surfaces to lines and points was an essential issue for Plato.[172]

Aristotle nowhere offers a detailed description of Plato's geometrical reduction. In his *De generatione et corruption* he ascribes to Plato the statement that the elements are decomposable *into planes*,[173] and to "Platonists" the view that the triangle is the indivisible part of plane areas.[174] These points are well presented and elaborated by Ibn Rushd.[175] Aristotle's description in *De caelo* III is more confusing.[176] He refers to "those who perceive fire as pyramids,"[177] but does not associate the other regular solids with the other elements. More problematically, Aristotle muddles up his criticism of Plato's geometrical reduction with his criticism of the Pythagorean additive cosmology: "The same theory which composes solids of planes clearly composes planes of lines and lines of points."[178] Alfonso was apparently misled by Aristotle: whereas Plato offers a two-stage reduction (from

---

[170] *Tim.* 48E–49A; 52A–B.

[171] Plato cautiously remarks (Tim. 53D–E): "the principles which are prior to these God only knows, and he of men who is the friend of God."

[172] The Pythagorean agenda to build up solids of points (or numbers) collapsed after the discovery of the incommensurability of the side and the diameter of the square (or the pentagon). Plato, apparently wished to "save" the Pythagorean idea of mathematization of natural bodies, but to avoid the pitfall of transition from surfaces to lines (that was equivalent to the extraction of square root).

[173] Aristotle, *De gen. et corr.* II.1329a23. This was also well mentioned by Falaquera: "it was stated by Plato in his book *Timaeus* that bodies are made of planes" (*De'ot ha-filosofim*, Parma ms. 65b5–6. It does not seem, however, that Falaquera was Alfonso's source.

[174] *De gen. et corr* 316a12.

[175] Ibn Rushd, *MC De gen. et corr.* Kurland, 8.43–9.49; 60.1. According to Ibn Rushd Plato bases his theory of indivisible areas on the triangle, because all areas are divisible into triangles, and the triangles are further divisible into smaller triangles but not into other areas. The triangle is, thus the beginning of area.

[176] The *Timaeus* is referred to explicitly three times, but Aristotle mentions also some close variants on the theme, referring to "those who construct bodies of planes" or to "Some who subject all bodies to generation by means of composition and separation of planes" (e.g., *De caelo* III.1, 299a1–4).

[177] "*Those* who start from fire as a single element.... *some* of them give fire a particular shape, like those who make it a pyramid..." (*De caelo* III.3, 304a9–10; III.5, 304a20, 304b4).

[178] Aristotle, *De caelo* III.1299a7–9; Ibn Rushd's middle commentary on *De caelo* III.3.1 Arabic 284.7–14; Hebrew Berlin ms. fol. 60b3–10. According to Aristotle's testimony "they [the Pythagoreans] say that a moving line generated a surface and a moving point a line, the movements of unites will also be lines, for the point is a unit having position" (De an. 409a3). Much has been written on this subject. See for instance Guthrie (1962, vol. I, 256–265); Philip (1966).

three-dimensional bodies into three-dimensional regular polyhedra, into two-dimensional planes), Alfonso ascribes to him a four-stage reduction from bodies to points (in the outline of *SMA*), or a three-stages reduction (in paragraph I.3.2).

Thesis [iii]: On Weight

Plato offers two accounts of weight.[179] According to the first, weight is due to the attraction of like to like.[180] According to the second account, which is "alluded to almost in passing,"[181] the weight of a body is determined by the number of triangles.[182] Considerations of weight provided Aristotle with his strongest argument against Plato: "If it is the number of planes in a body that makes one heavier than another, as the *Timaeus* explains, clearly the line and the points will have weight."[183] Alfonso's account could have been based on Ibn Rushd's middle commentary on *De caelo* III.1 or IV.2.

Five Questions Ascribed to Aristotle (Marked [a] to [e] in the Text)

> Question [a] is asked by Aristotle in *De generatione et corruptione* II.1, is rendered by Ibn Rushd in the middle commentary on this text, and accurately quoted by Alfonso. Aristotle (with good reason) criticizes Plato of not explaining the relation between the receptacle and the geometrical components[184]: Plato "takes the analysis of the elements, which are solid, as far as surfaces, though it is impossible for these surfaces to be the 'nurse' and the primary matter."[185]
> Question [b] is less accurately quoted. Presumably it relies on *De caelo* III.5, where Aristotle argues that fire cannot be pyramids because not every part of a pyramid is a pyramid.[186] In this chapter, Aristotle also ascribes to Plato the consideration that the pyramid is the sharpest among figures and therefore corresponds to fire.[187]

---

[179] O'Brien (1984, Chap. 5) tries to reconstruct Plato's "geometrical theory of weight," mainly from *Timaeus* 55D–56C. See also Cornford 222–3.

[180] *Timaeus* 62c–64a. See also Cornford (1937, 264–6).

[181] O'Brien (1984, 79).

[182] *Timaeus* 56B. Fire is the lightest "being composed of the smallest number of similar parts."

[183] *De caelo* III.1299b24–300a4; IV.2308b15–19; see comment on I.3.4 below. This argument is mentioned by Ibn Rushd in his epitome on *De caelo* (Horesh edition 66. 3–8), and in detail in his middle commentary (III.3.3.2.1, Arabic, p. 286; Hebrew Berlin ms. fol. 61a).

[184] The relation between them is not clear in Plato either. Plato introduces the receptacle after announcing a "new beginning" (*Tim.* 48E) and does not explain how the new analysis accords with the previous one.

[185] *De gen. et corr.* II.1329a16–24, Quotation from II.1329a22–3, Williams' translation; Ibn Rushd, *MC De gen. et corr.* II.1.1, Kurland edition, pp. 58–60.

[186] *De caelo* III.5, 304b3–6; III.7, 305b34–306a1, 306a30–34: Ibn Rushd, *MC De caelo* III.5.3, Arabic 319.6–12, Hebrew Berlin ms. fol. 70a21–27).

[187] *De caelo* III.5, 304a11.

Questions [c] and [d] cast doubts on the possibility of dimensional reduction. Question [e] recapitulates the arguments against Plato's theory of weight that Aristotle develops in *De caelo*.[188]

[I.3.3. Correction of the First Presentation]

I thus said, in the spirit of the principle that I mentioned above [I.3.1], that it is inappropriate to ascribe to a sage like Plato obvious errors like those that Aristotle ascribed to him. Rather one should seek an interpretation of what he said that makes sense. I say that the stronger the queries that Aristotle's raised against Plato are, and the more they [seem to] indicate Plato's senselessness, the more they indicate to me [98a] the truth of his [Plato's] teaching and their high value.

When I further learned that it was written on the entrance to his academy "Those who are not erudite in logic and geometry should not enter here," it occurred to me that Plato's words could be viewed from two perspectives: from the perspective of natural objects, which is the major part of his teaching, and from the perspective of geometry, in which he was interested. I was further encouraged when I found that Plato had not said that the surfaces and the solids are *hiuli* for the bodies, as Aristotle understood him,[189] nor [did he say] that the amount of the bodies[190] triangles is the cause of the weight. He did say that the bodies are divisible into triangles and into lines, and that the heavy body is made up of more parts and the light body of less.

Commentary

Alfonso claims to have made two discoveries: first, he learned about the famous inscription engraved over the door of Plato's academy[191]; second, he finds that theses [ii] and [iii] that stated in I.3.2 do not render faithfully what Plato said. Alfonso claims to gain a new understanding of theses [ii] and [iii], but it is not certain that the new understanding was based on the discovery of a new source. Gluskina suggests that Alfonso may have relied on a late Platonic text,[192] but there is no evidence to support this conjecture. The correction of theses [ii] and [iii] could have been based on a more attentive reading of Ibn Rushd's middle commentary on *De caelo*.

---

[188] *De caelo* III.1, IV.2; Ibn Rushd *MC De cael*. Chapters III.3.2.1; [III.5. 3]; IV.1.3.

[189] In I.3.2 above we do not find this statement of thesis [ii] but the revised one. Perhaps I.3.2 was corrected after the reconsideration reported in I.3.3.

[190] The two words *gufim meshulashim* appear one after the other with a large space between them. Perhaps the original manuscript was confused or corrected and the copyist hesitated which word to choose. See commentary on the word *geramim* after I.3.2 above.

[191] The tradition about this script goes back to late antiquity. See www.math.ucdenver.edu/~wcherowi/courses/m4010/historians.pdf.

[192] Gluskina's edition, p. 93, note 53.

The core of the new insight that Alfonso claims to have gained is that Plato's theory can be subject to two interpretations, "natural" and "mathematical." Alfonso could have been inspired by Ibn Rushd's distinction between "natural inquiry" and "mathematical inquiry" in this commentary.[193] Alfonso's revised interpretation of Plato's theses is presented in the next section.

[I.3.4. Three Theses Ascribed to Plato: Second Presentation]

[The Three Theses Ascribed to Plato (Order of Presentation Changed)]

[i. On *hiuli*] I say that according to Plato, there are two types of division of body into parts. **The first** in terms of natural science [into] what he calls *hiuli* and *geramim*.[194] On it[195] he said that it is a substratum of everything, whose outside is actuality and whose inside[196] is potency, and that if all things were named after it, then it would have been the truest [entity], in the same way that if all necklaces and rings made of gold were called gold, then [gold] would have been the truest [entity].[197]
[iii. On weight][198] And he said that when there are in one body more *geramim* than in another body equal to it in its size and in the quantity of its magnitude [i.e., of the same volume], then the form of this body naturally seeks the place that is appropriate to its quantity, namely the naturally small place. On the other hand, the form of the body that has less parts, because its parts are sparse and need a bigger wider place, would naturally require the big place appropriate to it. And the natural places are assigned to[199] the bodies according to their short or long distance [98b] from the circumference of the sphere. Therefore, the smaller among them would be nearer to the center of the world and the bigger among them would be closer to the circumference of the sphere. And since the place of the whole and of the part is the same in the natural bodies, as Aristotle claimed, the part seeks the proper place of the whole. This is the cause of levity and gravity according to Plato and Democritus and Leucippus and the other ancients.
[ii. On dimensional reductions] It is not my intention here to deal with natural science, neither to confirm nor to deny the existence of these natural *geramim*. I mentioned it here in the passing, in order to [arrive at] the **second**

---

[193] *MC De caelo* III.5.3. The Arabic text is defective; Hebrew Berlin ms. fol. 70a8.

[194] See commentary on I.3.2.

[195] On this reference see commentary on thesis [i] below.

[196] The opposition of חיצוני vs. תיכוני was used by Abraham Bar-Ḥiyya (Klatzkin, *Thesaurus*, vol. IV, 191).

[197] Translating בתכלית האמת as "truest" follows Joachim's and Williams' English translations of Aristotle.

[198] The order of the discussion of the three theses is different from the order in I.3.2 above.

[199] Literally: limited to.

type of bodily parts according to Plato, namely those parts that are expressed in terms of evaluating and measuring, which I intend to discuss here. It was established by Euclid that the area limited [by straight lines] is divided into triangles, and that the body [limited by plane surfaces] is divided into pyramids. Plato said it too. It remains for me to inquire how the body is divided into areas and the area into lines, as Plato said. And this [inquiry] is in terms of evaluating and measuring.

Terms

שעור ומדידה Alfonso uses the combination שעור ומדידה seven times in *SMA*. Both verbs שער and מדד are often followed by the specification "by one another" and refer to the act of measuring. It is not clear at this stage why Alfonso uses insistently the two terms together. The closest that Alfonso comes to clarifying this issue is in I.5.3, and we discuss it in the commentary on this paragraph. We translate שער as "evaluate" and מדד as "measure" to distinguish between them.

Commentary

The distinction between the natural and the mathematical analysis is a main issue in Alfonso's second presentation of Plato's three theses. Alfonso addresses first the "natural" theses [i] and [iii], and then turns to the "mathematical" thesis [ii].

Thesis [i]: On hiuli

Grammatically the pronoun הוא (the third sentence, translated "it") can refer either to "body" (in the first sentence) or to "hiuli" (in the second). The expression "substratum of everything (מונח אל הכל)" supports the second interpretation. Characterizing *hiuli* as a compound entity—potentiality inside and actuality outside[200]—seems to conflict with the prevalent notion of *hiuli* as pure potentiality, but Alfonso's notion of matter, notably of prime matter, was different from Aristotle's. Whereas for the latter prime matter is pure potentiality, for the former it is a substance.[201]

Thesis [iii]: On Weight

The revised presentation of thesis [iii] most probably relies on Ibn Rushd's middle commentary on *De caelo* III.5. According to Plato fire particles are the finest.[202] Aristotle, seeking to refute the assumption that the four elements are made of

---

[200] This description calls to mind Aristotle's descriptions of *body* as form outside, matter inside. *De cael.* II.13293b13–16; *Phys.* III.7207a35–b1.

[201] See Sadik (2007, 2); see also Introduction Section 3, argument ii. Perhaps Alfonso concept of matter is somewhat similar to that of Duns Scotus. See Gordon (2016, Section 1).

[202] *Tim.* 56A–B.

indivisible components, argues against Plato that the fire particles should be the largest.[203] Guthrie suggests (following Stocks) that Aristotle's argument is based on the observation that when water evaporates the volume increases.[204] Assuming (with Aristotle) the impossibility of void, this implies that the air particles are larger than the water particles and consequently need more room.[205] The place that the element of fire occupies is the largest because its particles are the largest.

Ibn Rushd examines more closely the hypothesis that the four elements are composed of indivisible units and the possible assumptions about similarity in number of particles, in particles' size and in total volume. Eventually he accepts Aristotle's view that the particles of the four elements differ in size and that the fire particles are the largest.[206]

Alfonso, presumably ignorant of Plato's original statement on the size of the fire particles, follows Aristotle and Ibn Rushd.[207] The element whose particles are smallest occupies the smallest natural place around the center whereas the element whose particles are largest occupies the largest place at the circumference of the sublunar region. In order to explain the movements of the elements, he turns to the theory of attraction of like to like that he ascribes to Plato, Leucippus, and Democritus. His source was presumably Ibn Rushd's middle commentary on *De gen. et corr.* I.7.[208]

Thesis [ii]: On Dimensional Reduction

It now becomes clear to Alfonso that he is interested mainly in Plato's thesis [ii], notably in its mathematical interpretation. It will become relevant in Part IV (of which only the beginning survived). Euclidean geometry provides the means to evaluate a complex figure by simpler ones *within each dimension*, e.g., two-dimensional polygons by triangles, or three-dimensional polyhedra by pyramids. Alfonso ascribes these two statements to Euclid, but they are not stated as propositions in the *Elements*. The first of the two statements was made by Plato,[209] the second by Aristotle.[210] Alfonso's goal was to provide the steps in which the *dimensional barriers are crossed*: "how the body is divided into areas and the area into lines, as Plato said."

[203] *De caelo* III.5, 304a23–34.

[204] *De caelo* Loeb Classic edition 300–301.

[205] "The element of water will be smaller than that of air, but the least quantity is contained in the greater. Therefore, the air element is divisible" (*De caelo* III.5, 304a33–b1).

[206] *MC De caelo* III.5.3, Arabic 318, Hebrew Berlin ms. 70a9–18.

[207] Tomas Bradwardine, Alfonso's English contemporary in his *Tractatus de Proportionibus* Written in 1328, suggests that "The four elements are linked in continuous proportionality." It seems that Alfonso was not familiar with his work.

[208] Aristotle *De gen.* I.7. 323b11–16. Ibn Rushd *MC. De gen. et corr.* Is available online via the DARE project. Both Aristotle and Ibn Rushd mention by name only Democritus in this context.

[209] *Tim.* 53C.

[I.3.5. The Confusion About Indivisible Parts]

And I wondered whether Plato thought that body is composed of areas and area of lines and line of points, as in Antiphon's argument about the squaring of the circle. Because when he [Antiphon] inscribed a polygon in a circle, and then bisected the arc on each of the sides of the polygon, and joined the straight cords of all those arcs, and carried on this [procedure] for every arc, he thought that eventually the division would terminate in the [ultimate] parts of which the straight line is made up and of which the circumference of the circle is made up. But this, as Aristotle notes, contradicts with [99a] what the geometers say, namely that the line is not composed of points and that magnitudes are infinitely divisible.[211]

At this point I became very perplexed about the conclusion of my inquiry described above, namely that there must be a mental superposition, superior to the superposition that we are used to imagine,[212] because unless by division of the circular lines [סובבים][213] into indivisible parts, how can it be imagined? Then I turned to re-examine Antiphon's argument, being intrigued by the question how he endeavored to contradict the first principles of the science in which he was an expert. And it occurred to me that perhaps he meant to say that a continuous magnitude is composed of indivisible only in the sense that Plato understood it, namely in terms of evaluating and measuring, rather than in terms of natural composition. He [Antiphon] had not claimed that there are infinitely many divisions in act, thus contradicting the first principles of geometry. On the other hand, I know that if the infinite parts existed only in potentially, then they would not evaluate distances at all.

Historical Commentary

Antiphon's suggestion to square the circle by inscribed polygons was very briefly alluded to by Aristotle,[214] but described in more detail by the Greek commentators.[215] Alfonso could have found the information in one of the commentaries of Ibn Rushd.[216] Solomon Luria's claim to find in Alfonso new evidence on the existence

---

[210] *De caelo* I.4, 303b1.

[211] *Physics* VI.1231a24–26, 231b15–16.

[212] I.2.2 above.

[213] Gluskina corrects סובבים to שעורים.

[214] *Soph. ref.* 11, 172a7; *Phys.* I.2185a17.

[215] Heath (1921, 221–3).

[216] For example, *MC Soph. Ref.* Arabic Jéhamy, vol. 2, 694; Hebrew Paris ms. fol. 110b26–111a4; *MC Phys.* Hamburg ms. fol. 3b3–6. *LC Phys.* I.11 Paris ms. fol. 9a26–b6. In the long commentary Antiphon's name is mentioned but the spelling is different—אנטיקן.

of a mathematical theory of infinitesimals among ancient atomists has already been dismissed by Vlastos.[217]

Mathematical Commentary

Alfonso is by no means willing to give up Euclid's geometry and Aristotle's theory of continuity. The thesis of indivisible parts is too radical as it undermines the very basis of the science of geometry. It is, however, appealing as it offers natural measuring units and makes the definition of mental superposition possible. His solution is that Antiphon adopted the hypothesis of indivisible parts "in terms of evaluating and measuring, not in terms of natural composition."[218] Perhaps Alfonso has in mind Ibn al-Haytham definition of magnitudes in terms of measurement.[219] The bottom line is that "evaluation and measurement" can be achieved neither by actual division nor by potential division. In the next section, *Alfonsi* will offer a way out.

[I.4. Turning to Aristotle]

The suggested solution is that indivisible parts can exist in an intermediate existence between potentiality and actuality that is defined by motion (I.4.2 below). It is important to emphasize that although Alfonso was considered to be a Platonist and as a critic of Aristotle, he relies most heavily on the latter. In Section I.4 he endorses Aristotle's definition of motion (*Physics* III), his theory of continuity (*Physics* VI), and his rejection of void (*Physics* IV).

[I.4.1. An Intermediate Existence Between Potentiality and Actuality]

Then it occurred to me that there may be another existence, neither actual nor potential, in which distances are divided into infinitely many parts. It is not this existence, but an existence that is in motion, which is intermediate between potentiality and actuality. When I saw also that the superposition by motion is intermediate, as it gives rise to infinitely many divisions while in motion, and it is the intermediate existence between potentiality and actuality.[220] Since there are infinitely many divisions of the small line as of the big line, it is possible to superpose on it by motion any line bigger [99b] than it

[217] Vlastos (1995, 294–5).

[218] In I.5.3 below Alfonso will try to solve the problem of "natural composition" by identifying the Platonic *geramim* with the Aristotelian minimal parts.

[219] Quoted above. See introductory note to Section I.3.

[220] The Hebrew sentence is not syntactically correct, so the translation is not certain,

by any amount (as I have illustrated in the case of the semi-zodiac and the latitude arc of the meridian).[221] And all these infinite divisions occur in a finite time in this intermediate existence between potentiality and actuality.

Terms

מרחק The term מרחק, translated here as "distance," was used in *SMA* to denote [i] distance in philosophical discussions of divisibility[222]; [ii] distance between two lines,[223] a concept that was not defined by Euclid, but introduced by some of his followers; [iii] radius of a circle.[224] The term מרחק in the geometrical senses [ii] and [iii] was used by Abraham Bar-Ḥiyya.[225]

Commentary

The conception of motion as mediating between potentiality and actuality goes back to Aristotle, who defines it as incomplete actuality.[226] In Ibn Rushd's words: "a prerequisite of motion is the existence of potentiality, joined to its actuality."[227] Following the examples of non-isometric superpositions that he analyzed above (I.2.1), Alfonso understands that motion can be a powerful tool in geometry. The issue is further explained in the remaining lines of Part IV. According to Euclid's conceptualization only the two extremities of a line are actual points and only the extremities of an area are actual lines. In the first proposition of Part IV the side of a rectangle is moved, maintaining a right angle with the base. Through this motion every height on the base become as actual as the side of the rectangle, and every point of the base becomes as actual as its endpoint.

[I.4.2. Defining Mental Superposition]

Since [I understood this] I had to acknowledge that the imagining [ציור] of equally divided motion [התנועה המשתוה בחלקיה] is the mental superposition mentioned above that evaluates distances, one by the magnitude of another. This gives rise to infinitely many divisions neither potential nor actual but

[221] I.2.1 above.

[222] For example, I.4.1–2, 4.4, I.5.4 and in III.33.

[223] For example, II.2.3–4, II.4.1, and III.29. The term was used by Bar-Ḥiyya in this sense, e.g., Ha-Meshiha Part I, #12, p. 9.

[224] The term was used by Bar-Ḥiyya in this sense, e.g., *Ha-Meshiha* Part I, #33, p. 17; #53, p. 34.

[225] See Klatzkin, *Thesaurus*.

[226] *Phys.* III.2, 201b30–32.

[227] הנה מתנאי מציאות התנועה מציאות הכח מחובר בשלמותה Ibn Rushd, *MC. Phys.* Hamburg ms. fol. 23b22–23 Hebrew translation by Qalonimus ben Qalonimus (this commentary is extant only in two Hebrew translations). A similar expression ממוצע בין הכח הגמור לפעל הגמור was used by Alfonso's contemporary Moshe ben Yehuda with respect to first matter. See Eisenmann (2009, 411).

in between them. These divisions follow the same order, because the motion is equally divided, and thus its parts evaluate the distance by the same quantity [שוה על ערך].[228] This agrees with what is established in the *Physics* that distance, time, and motion are all proportional.[229] This proportion would not be maintained for unequally divided motion, [but] for motion that is equally divided as time and distance are equally divided. Distance is analyzable into indivisible parts in the intermediate existence between potentiality and actuality, as the mind grasps it [the distance] as existing all at once, and evaluates one part by another, in the same way that it [the distance] grasps the "before" and the "after" when imagining time as existing all at once,[230] and evaluates the ratio of the relation between them.[231]

I know that it is the imagining of the superposition of distances on one another *all at once* that is the proper way of equating them with one another (whether similar or dissimilar in their shapes), namely by imagining the simple similarly divided motion, [100a] in the same way that the static distance along which the motion proceeds is simple and similarly divided. Since the parts of motion are similar to one another, the parts of the static [distance] are evaluated by one another, and since the motion is simple and is not made of motions that differ in their velocity, the parts of the moving [entity] are equal to the parts of the static [entity], not as in [the example of] the semi-zodiac and the latitude arc.

Terms

ציור/ צייר The word ציור occurs 42 times in *SMA* and the verb (צייר) 13 more times. Most frequently, *Ṣiyyur* applies to operations: to motion (27 times), to superposition (11 times), to division (twice). A few times it applies to magnitudes: to lines and points (9 times) and to time (once).

The word is used in philosophical texts to translate the Arabic تصوّر *(taṣawwur)* that was used by Ibn Rushd as a translation of the Greek *noēsis*.[232] *Taṣawwur* was translated into Latin (in different contexts) as "conceptio" or "conceptus," as "imaginatio" and as "formatio."[233]

We suggest that in his use of the term *ṣiyyur* Alfonso did not draw on the philosophical tradition, but rather on the mathematical. It is possible that *ṣiyyur* in *SMA* was not the translation of *Taṣawwur*, but rather of the Arabic تخيل (*takhyil*), that was associated with geometrical imagination. In his *Commentary on the First Book of*

---

[228] Or "in the same ratio." See Terms below on the translation.

[229] *Physics* VI.1231b18, VI.4.235a15.

[230] Alfonso alludes to Aristotle's definition "time is the number of motion with respect to before and after" (*Phys.* IV.11219b1).

[231] The sentence is difficult. Presumably its meaning is that distance and time are divisible into indivisible parts in the same way by motion.

[232] Wolfson (1973, 484–5).

[233] Wolfson (1973, 490). Shmuel Ibn Tibbon defines it in his glossary as "the actuality of the mental capacity in the human soul." The word *taṣawwur* appears in Maimonides *Guide* and is translated by Munk into French as "conception," and by Pines into English as "spontaneous perception."

Euclid's Elements Proclus claims that "*noēsis* in geometry occurs with the aid of imagination (*phantasia*)."[234] The Greek verb *noein* was used several times in Euclid's *Elements*, and attracted the attention of modern commentators. Lachterman associates *noesis* with "generative motion."[235] Netz interprets *noein* as the mental act of following a virtual diagram, and translates it as "imagine."[236] Ibn al-Haytham used *takhyil* many times in his *Commentary on the Premises of Euclid's Elements*.[237] Alfonso was familiar with this text,[238] and his use of *ṣiyyur* is similar to that of Ibn al-Haytham's *takhyil*.

Whatever the history of the word *ṣiyyur* was, it is important to understand that it is a mental activity or operation of the mind, often the operation of moving (translating or rotating) a geometrical object in the imagination. We thus translate *ṣiyyur* as "imagining," rather than as "conception."

שוה החלקים/ משתווה החלקים/ מתדמה החלקים These terms are used several times in I.4.2 and I.5.3. We translate שוה/משתוה החלקים as "equally divided," מתדמה החלקים as "similarly divided."[239] The meaning of these terms is explained in the commentary below.

ערך, יחס, הצטרפות ערך was the standard term for ratio in the mathematical vocabulary based on Bar-Ḥiyya and was widely used by Sephardic authors; **יחס** was the common term used by the Provençal translators. יחס is not a part of Alfonso's vocabulary and occurs only once in *SMA*.[240] The word **ערך** is used many times in *SMA* in the sense of "ratio." Here and in I.5.3 in the expression **ערך שוה** it can also mean quantity or "measured value." The word was used in this sense by Bar-Ḥiyya:[241] אם אתה שומע ממני מילת ערך הוי יודע כי אני אומר אותה על המדה או על המספר או המשקל. The two meanings of ערך do not differ significantly from each other because measuring magnitudes by one another is equivalent to determining the ratio between them. The next sentence[242] suggests that the translation "by the same quantity" may be more precise here.

הצטרפות is a more general term for relation.[243]

---

[234] Proclus, *Commentary* 55 Friedline, Morrow's translation 44.

[235] Lachterman (1989, 88–9).

[236] Netz (1999, 52).

[237] For example, with respect to dimensions, in pp. 3–5 of the Arabic text, see Hooper-Sude edition.

[238] See commentary on II.2.1.

[239] The fairly literal translation homeomerous belongs to the vocabulary of natural science rather than of geometry. מתדמה החלקים and sometimes also משתווה החלקים are the Hebrew terms for homeomerous (a term that was used several times in Aristotle's *De generatione et corruptione* to denote a substance in which the structure of the parts and of the whole are identical). This meaning is not relevant in the present context.

[240] See Introduction, Section 7, example iii.

[241] See Klatzkin, *Thesaurus*, III.172.

[242] "Distance is analyzable into indivisible parts."

[243] Cf. Klatzkin, *Thesaurus*.

Commentary

This difficult section was probably meant to be the climax of Part I of *SMA*: the promised concept "mental superposition" is at last defined. The definition is rather obscure: "the imagining of equally divided motion." Motion is the means to "evaluate and measure" magnitudes by one another, and mental superposition is the mental operation of imagining it. Alfonso relies here on Aristotle's analysis of continuity in *Physics* VI. Aristotle thesis is that distance, time, and motion are isomorphic: "either all of them are composed of indivisibles and are divisible into indivisibles or none."[244] On the one hand, the structure of uniform motion is the same as that of the simple *continua* time and distance. On the other hand, motion is endowed with a special capacity to divide. Uniform motion can divide time, distance, and other continua equally. Since all the entities involved are continuous, motion can be used to divide the other magnitudes into infinitely many parts, and by means of these divisions to map them on one another. Magnitudes x and y can be equally divided by motion into sequences of small parts Δx_1, Δx_2, Δx_3, ..., Δy_1, Δy_2, Δy_3, ... Given this division, Δx_1 can be mapped on Δy_1, Δx_2 on Δy_2, etc. Alfonso contends that this holds also when the division is carried out indefinitely.

[I.4.3. Identifying Plato's Elementary *geramim* with Aristotle's Minimal Parts]

This is the correct view that should be ascribed to Plato, Antiphon, and other who acknowledge [the existence of] infinitely many indivisible parts: the *geramim* about which they agree are not infinite in number, neither indivisible in their magnitude nor indivisible in their form. In the same way that it [the Aristotelian minimum] is the part that is the smallest in each kind of natural body, as Aristotle acknowledged in the *Physics*, so is the *gerem* for them [Plato, Antiphon and others] is the part which is the smallest in universal body of the whole world. In the same way that a body cannot get indefinitely large in the quantity of its magnitude, so it cannot get indefinitely small. Both smallness and largeness are limited and given by nature.

Although the majority of philosophers used to ascribe to Plato the view that there are in a limited body infinitely many indivisible parts with void in between them, this is an impossibility, and it is inappropriate to ascribe to any scholar unless contentiously. [Plato] should only profit from being disgraced,[245] in the manner of a respectable man towards loved ones who disagree with him.

---

[244] *Phys.* VI.1, 231b18; see also VI.4235a15.

[245] According to the Jewish tradition a man should not profit from the disgrace of another (המתכבד בקלון חברו אין לו חלק לעולם הבא). Gluskina noted that Abner of Burgos uses this expression in *Sefer Tshuvot la-Meharef.* The use of this somewhat unusual expression was one of the linguistic arguments that supported her identification of Alfonso with Abner. We are indebted to Shalom Sadik for letting us use his copy of *Sefer Tshuvot la-Meharef.*

Commentary

The intention of this paragraph is obviously to clear Plato (and even Antiphon) from the charge of atomism and to harmonize Plato's views with Aristotle's. The outcome of the long excursus on Plato's *Timaeus* was the distinction (in I.3.4) between physical parts ("natural *geramim*") and mathematical parts ("parts that are expressed in terms of evaluating and measuring"). The mathematical parts were "moved" (in I.4) to the intermediate existence defined by motion. Here Alfonso comes back to the physical parts, Plato's *geramim,* and identifies them with the Aristotelian *minima naturalia*, which are not building blocks from which bodies are composed, but rather the smallest parts into which a physical body can be divided while maintaining its form.[246] Whereas Aristotle's minimal parts are defined for each kind of body, Plato's *geramim*, according to Alfonso's interpretation, are universal minimal parts.

Alfonso strongly objects to the philosophers who ascribe to Plato a strong version of atomism—infinitely many indivisible parts in a finite body and void[247]—and refers to them as contentious.

[I.4.4. Excursus on Void][248]

I know about someone who held that body is divisible to infinitely many indivisible parts, each part moving over a limited distance instantaneously, [100b] in order to avoid the conclusion that void exists that is not [?]. This happens when a rigid body is separated from another rigid body that was previously in contact with it, because necessarily another soft body like water or air immediately comes in between them. He said that one of two alternatives must be true of the incoming air: Either [a] it includes a part or parts, each of which comes in together (not [b] part after part, in which case void would have existed in between the two bodies from the time they start to separate until the end of the entering of the part of air in between them). Or [b] that there would not be in the incoming air a part that enters together, but [rather the air enters] part after part. This [account a] is possible only if the body of air is divisible into and reconstructed from the planes of which it consists, so that each of them moves over the aforementioned distance all at once. He accepted this [account of] division and motion because of his denial of void. This is indeed absurd. But it is not the context here to discuss the existence of void, so let us return to our subject.

---

[246] There is a minimal amount of water that can be considered water. This theory was briefly alluded to by Aristotle, and developed in the Arabic Aristotelian tradition by Ibn Rushd. For a summary of the subject see Glasner (2009, 143–6).

[247] This charge could be based on Aristotle's complaint that "Plato in the *Timaeus* says that matter (*hule*) and place (*chora*) are the same." *Physics* IV.2, 209b11. See also Keyt (1961).

[248] The text is confused and we tried to clarify it by adding brackets and parentheses.

Commentary

Alfonso addresses the question whether a temporary void is formed when two smooth surfaces in contact are separated. The question was raised by Lucretius, whose book was discovered in the west in the fifteenth century, but the question infiltrated into scholastic discussions earlier.[249] Grant mentions two approaches to the question: [i] that a momentary void between the two surfaces forms, [ii] that air consists of plane layers which infiltrate instantaneously so that void does not form even temporarily. Most scholastic scholars in the fourteenth century adopted approach [i], and we do not know who was the proponent of approach [ii] to whom Alfonso refers. The best-known proponent of approach [ii] was Blasius of Parma, who lived after Alfonso.[250] Probably a less known scholar suggested this thesis before Blasius.[251]

Alfonso's Account [a] = Grant's Account [ii] and Alfonso's Account [b] = Grant's Account [i]. Alfonso dismisses Account [ii]: it is absurd to introduce a hypothesis of instantaneous motion just in order to avoid the situation of temporary void. Alfonso adds that it is not the right context to discuss this subject, the right context was, perhaps, his book *New Philosophy* that is now lost. Alfonso's stance on the issue of void is explained neither here nor in the *Mostrador*.[252] From the criticism of Iṣaaq Pulgar we learn that Alfonso did not deny the possibility of void.[253]

[I.5. Adding Two New Superposition Postulates to Geometry]

Introductory comment

The Euclidean first principles include definitions, postulates (that are specific to geometry), and common notions (that announce general truths).[254] The first three postulates of Euclid geometry are known as the constructional postulates[255]: [1] "to draw a straight line…," [2] "to extend a straight line…," and [3] "to describe a circle…." Alfonso suggests to add two new construction postulates to geometry: [*] "to

[249] There were several discussions of the continuity of air that took place in Alfonso's lifetime. For instance, Adam Woodland's argument that when a sphere is placed on a plane "some *continuum* would have been interposed, and consequently those objects would not touch." A similar argument was made by Buridan (Grellard 2011, 72–3).

[250] Grant (1981, 91–2).

[251] Grant (1981, 86–7).

[252] See Sadik (2007, 138–9).

[253] Gluskina (1974, 70); Sadiq (2007, 134).

[254] The first principles of a science, according to Aristotle are posits, suppositions, and definitions (*An. post* I.272a15–24). The use of these terms, remarks Barnes, was not consistent and "this terminology remained in a state of flux" *An. Post.* Barnes' edition (1993, 99).

[255] See Mueller (1981, 15–17).

imagine an equally divided motion" (I.5.1) and [**] "to divide a given magnitude by equally divided motion" (I.5.3). Alfonso's reference (in I.5.2) to [*] as התחלה להתחלות may mean that he thought of placing it as the first postulate of geometry. The two new postulates [*] and [**] are discussed after I.5.1 and I.5.3, respectively.

[I.5.1. First Addition: A Postulate of Imagining Motion]

[Postulate *] It is known that imagining motion is one of the first principles of geometry. Imagining the superposition of triangles on one another that Euclid used,[256] is not confined to imagining the moving of one triangle from its place to the place of the other triangle, nor is the imagining of rectilinear motion alone one of the basic principles of geometry, but also the imagining of circular motion. The equality of all the lines that connect the center [101a] of a circle to its circumference cannot be imagined unless by imagining their rotation about the center and their superposition on one another, or of one on all of them. Similarly, we cannot imagine the equality of all the lines that connect the center of a sphere to its surface unless through imagining the tracks described by the rotation of half of the big circle about a fixed diameter. Also, the circular cone is [imagined] by rotating a right-angled triangle about one of the sides that subtend the right angle which is kept fixed. Similarly, the straight-angled cylinder is imagined either by the rotation of a right-angled area [i.e., a rectangle] about one of its sides or by the linear moving of its basis in a right angle along its axis. Also, Euclid could not have truly established the equality of the two base angles in an isosceles triangle unless by imagining the rotation of the triangle until it is superposed on itself.

But Euclid covered it up artificially,[257] when he subordinated[258] it to other things, either unintentionally, not noticing how the imagining of motion is implied in what he said about the superposition of an area upon itself, or [intentionally] in order to conceal it from his students, so as to spare them the difficult idea that the imagining of motion is one of the first principles of geometry. Thus, he concealed it as far as he could. When he used superposition he did not mention motion, although [motion] is implied in it. To this end he established before [he used superposition in *Elements* I.4] the construction of a straight line at a given point equal to a straight line given at a different place [*Elements* I.2], in order that the superposition of the lines of one triangle on another will be thus imagined without mentioning motion. He also assumed the equality of the lines connecting the center of the circle to its circumference, although this too can be established only by motion.

---

[256] *Elements* I.4.

[257] Literally "by forcing."

[258] By not admitting that it is an axiom. The Hebrew word is not clear and the translation is uncertain.

Commentary

Postulate [*] introduces into geometry the operation of "imagining motion." The idea is not new. The first steps towards the modern geometry of transformations[259] were made in the Arab world. As Rashed observes, it started with Ibn al-Haytham in the tenth-eleventh century, and was revived by al Ṭūsī and his followers in the thirteenth century. Continuous translation and rotation of lines were used in geometrical contexts such as attempts to redefine parallel lines or to prove the parallels postulate.[260]

Alfonso sharply criticizes Euclid's "double standard" with respect to motion. He was not the first to criticize Euclid on this issue. Thābit Ibn Qurra notes Euclid's uses of motion in his treatise on the parallels postulate.[261] Alfonso's criticism is systematic. He summarizes the definitions and propositions, in which Euclid explicitly or tacitly uses them:

- Propositions I.4. Euclid uses superposition (common notion 4) in order to prove equality of triangles. Euclid, contends Alfonso, evades the natural approach to geometry, namely imagining motion, and contrives artificial solutions instead (such as proving Proposition I.2 before I.4 in order to justify the simple operation of moving a line segment). With the postulate of imagining motion, Proposition I.2 would not be needed.
- The definition of circle (I, def. 15). Euclid does not mention motion, but his definition is not constructive and does not prove the existence of a circle. Here also Alfonso most likely followed Thābit Ibn Qurra in criticizing it:

  From this we derive the understanding and knowledge of what we said regarding the equality of the lines extending from the center of the circle. For we have attained that knowledge only by continuously moving straight line in our imagination in such a way as to apply it to all those distances and extensions.[262]
- The definitions of sphere, cone, and cylinder (XI, defs. 14, 18, 21). Euclid explicitly defines these solids as generated by the rotation of plane figures. It should be noted that these definitions by motion were subject of controversy among Arabic mathematicians.[263]
- Proposition I.5 on the equality of the base angles in an isosceles triangle. Euclid uses an auxiliary construction to prove this theorem. Alfonso altogether ignores Euclid's proof and comments that "Euclid could not *truly* establish the equality of the two base angles in an isosceles triangle *unless* by imagining the rotation of the triangle until it is superposed on itself." He most probably alludes here to

[259] That is, geometry that allows translation, rotation, and reflection of geometrical magnitudes.

[260] Rashed (2015, 440).

[261] Sabra (1968, 20); Jauiche (1986, 151–2).

[262] Sabra (1968, 20).

[263] 'Umar al-Khayyām criticized Ibn al-Haytham for his "permissive" interpreting of Euclid's definitions of the circle and the sphere. Jaouiche (1986, 186–7); Rosenfeld and Youschkevitch (1997, 143).

Pappus' alternative proof: Pappus imagines the isosceles triangle as two congruent triangles *ABC* and *ACB*,[264] taking for granted that the isometry of rotation. Alfonso somehow learned about Pappus' proof,[265] and considered it to be *truer* than Euclid's, probably because in Pappus' proof follows directly from the essence of being an isosceles triangle, whereas Euclid's proof that uses an auxiliary construction is less explanative.[266] For Alfonso imagining motion is התחלה להתחלות,[267] namely the most basic postulate. In this vein he uses several times superposition where simple plane geometry is sufficient.[268]

[I.5.2. The Postulate of Imagining Motion Should Replace the Parallels Postulate]

We say that since Euclid did not want to disclose that motion is one of the first principles of geometry and concealed it as far as he could, he had to postulate a premise [101b] (on which a significant part of geometry relies) as self-evident, but which is not [self-evident] and should be established by demonstration, like the other propositions that are established by demonstration. It is [the Parallels Postulate] "if a straight line, falling on two straight lines make the interior angles on one side less than two right angles, the two straight lines, when extended in this side would necessarily meet."[269] It is not self-evident that these two lines necessarily meet, because their getting nearer to each other when extended does not imply that they meet. We know that there exist two lines, extended in the same direction, and as they are extended they get nearer to each other, but they never meet, even if produced to infinity. This premise [the Parallels Postulate] can be proved [תתבאר] according to the majority of recent scholars, by imagining motion. I, however, dismiss these proofs [באורים], not because of [the use of] motion itself but, rather, because of some accidents associated with it, as I shall show in Part II. I will then explain this premise by way of the superposition that they accept. It follows from what was said above that the imagining of motion is [not only] an indispensable first principle of geometry, but also that it should be *the first of the first principles* accepted by these early scholars.

---

264 Proclus, *commentary* 249–50 Friedline, Morrow's translation; 194–5; Heath (1908/1956 I, 254).

265 The proof is known to us via Proclus' commentary. No Arabic translation of Proclus' commentary is mentioned in the bibliographical literature but, comments Rashed (2015, 8, note 3), his ideas were not unknown to Arabic mathematicians. Rashed names al-Sijzī as an example. There is evidence that the author of the *Hebrew Geometrical Compendiun* (preserved in manuscript Mantova com. 2) was familiar with Proclus commentary (Glasner, 2019).

266 On the question of the incompatibility of mathematical techniques and Aristotle's notion of demonstration as explanatory see Mancosu (1996, Chap. 1). Harary (2008), Harari (2012).

267 I.5.2 below.

268 For example, II.3.2, II.3.4, III.33.

269 Euclid, *Elements* I, Postulate 5, known as the parallels postulate. Alfonso's translation differs from the three main Hebrew translations by Moshe Ibn Tibbon, Ya'aqov ben Makhir and R. Ya'aqov.

Commentary

Pursuing his criticism of Euclid, Alfonso contends that if the postulate of imagining motion were accepted, it would have been possible to prove the controversial Parallels Postulate. Here, however, Alfonso is wrong. As we know today, many attempts were made throughout the ages to prove the parallels postulate (usually by introducing equal or weaker premises) until it was proved in the nineteenth century that this postulate is independent of Euclid's other postulates.[270]

Alfonso argues that earlier thinkers who tried to prove the parallels postulate failed, not because they used motion, but because of associated errors. Part II is dedicated to the analysis of these errors and to Alfonso's own attempt to prove the parallels postulate.

The example of the asymptotes, two lines that approach one another but do not meet, was known to mathematicians since antiquity. The hyperbole was known to Apollonius,[271] and was mentioned by Proclus precisely in the same context of his discussion of the Parallels Postulate.[272] This example circulated in Arabic in the middle ages and Alfonso could have found it in Maimonides' *Guide*,[273] with which he was familiar. Alfonso, however, could have in mind another example of a curve that comes closer to a straight line but never meets it: the conchoid, which he studies in III.29–32.

[I.5.3. Second Addition: The Postulate of Measurement]

In the same way that the imagining of motion *qua* motion is a first principle of geometry, so can evaluating [by] equally divided motion be one of the principles of geometry; specifically [the evaluating] of the quantities of magnitudes [ערכי השעורים][274] and of their measuring by one another, as it is the task of the geometer to study the quantity of magnitudes[275] and their measuring by one another. He cannot achieve this unless by cutting continuous magnitudes into equal parts, since for him [the geometer] a continuous magnitude [102a] can be dissectible and equally divided and at the same time remain in itself continuous.[276] In this respect his discipline is called *Ḥokhmat ha-Tishboret*, because of the sections [*shevarim*] that he [the geometer] makes. It would be appropriate to derive this [the breaking down magnitudes into sections] from that [entity] for which it is natural, and this is found only in equally divided motion and in the time that counts and evaluates it.[277] We do

[270] For a brief summary and bibliography see https://en.wikipedia.org/wiki/Parallel_postulate.

[271] For a study of the transmission of this example see Freudenthal (1988).

[272] Proclus, *Commentary* 192, Friedline, Morrow's translation 151.

[273] *Guide* I.73 tenth premise. Pines' English translation I, 210. For a discussion of this example see Freudenthal (1988).

[274] Or "ratios of magnitudes." See Terms on I.4.2.

[275] Or "ratio of magnitudes." See Terms on I.4.2.

[276] Literally: "and of equal parts."

[277] Aristotle defines time as the number of motion. *Physics* IV.11, 219b1.

not know,[278] however, how the quantities[279] are determined and become known in terms of the distances measured by equally divided motion, where do we procure such motion, and how can the geometer use this imagining.

I say that the geometer can achieve this by dividing distance into parts equal to one another, because this way the distance would be potentially similar to the time and to the equally divided motion, and can be measured by them. When [the number of] magnitudes increasingly grows maintaining the same ratio, the ratio between them would always be the same even when [the process] is carried out to infinity. This agrees with what Euclid established in Proposition V.13,[280] and with what the natural philosophers say, namely that the ratio of the revolutions of the sun in its orb to that of the moon in its orb is always about 1–13, while they concede that the world is eternal. This is one of the merits of the intellect that it can know and comprehend infinite things according to [the pattern of] the finite things that preceded them, and how difficult all this would be for him who does not have this intellectual capacity, [fully possessed by] God alone.[281] **[Postulate** **] This is a first principle of this science, namely to imagine a number of equal divisions in the same ratio continuing indefinitely, in the same way that they [the equal divisions] exist in equally divided motion.

Terms

גלגל Here and in III.33 the term *galgal* refers to a spherical shell (of the sun, moon, Venus, and Mercury) and is translated "orb." In I.3.4 *galgal* refers to the outermost sphere and is translated there as "sphere."

שעור, מדידה, שברים The verb **שבר** is the Hebrew parallel of the Arabic كسر.[282] It occurs only twice in *SMA* (in the outline and here) in the plural form *Shevarim*, and is translated here as "sections."

The word **שעור** (*shi'ur*) is very frequent. As a noun (Arabic قدر) it is the standard word for "magnitude," as a nominal form of a verb (Arabic تقدير), it stands for the act of evaluating one magnitude in terms of another.[283] Alfonso uses several times the expression ומדידה that we translate "evaluating and measuring." The second word *medida* presumably stands for قياس or, perhaps مساحة.[284] When the expression שעור ומדידה first occurs (in I.3.4) it is not clear whether the two words are synonyms

[278] The word is not certain; the translation follows Gluskina's reading.

[279] Or "ratio of magnitudes." See Terms on I.4.2.

[280] According to Heiberg's numbering should be the twelfth.

[281] The sentence is difficult, and the translation is not certain.

[282] The term *tishboret* was used by Bar-Ḥiyya to denote area, and *shevarim* was used by Ibn Ezra with the same denotation. Sarfati (1968, 137, 142).

[283] It was used by Ibn al-Haytham in this sense. See for instance Hooper-Sude, Arabic text 2, 6, 37.

[284] Ibn al-Haytham mentions the term مشاحة for practical measurement. See Hooper-Sude Arabic text 37. This word has a Hebrew phonetic translation משיחה, that was used by Bar-Ḥiyya and by Alfonso himself for measurement and for area. Alfonso uses it in I.1.0 above and in the translation of Archimedes' Measurement of the circle that was made supposedly by him (Glasner 2013). For Bar-Ḥiyya see Sarfatti (1968, 124).

(and the repetition is a stylistic device[285]) or whether there is a noteworthy difference between them. At first glance one could get the impression that שעור is the task of בעל חכמת השעור (or משער),[286] namely the theoretical geometer, whereas **מדידה**, which is traditionally associated with land measuring, is the task of the practical geometer, namely בעל חכמת התשבורת. In II.5.6 Alfonso indeed confirms that *tishbotet* is practical geometry. This interpretation, however, is doubtful. Alfonso definitely seeks to revise *theoretical* geometry and intends the postulate of measurement to be a part of the axiomatic basis of Euclid's geometry. Perhaps Alfonso borrows the words שברים and מדידה from practical geometry and imports them into theoretical geometry.

Commentary

After the postulate of imagining motion Alfonso present his postulate of the measurement by motion. A magnitude can be divided by the motion of a lower-dimensional magnitude, which cuts equal sections (*shevarim*) one after the other, and still to maintain its continuity and unity.[287] This concept is stronger than potential divisibility, but does not involve the breaking down of the magnitude into separate magnitudes as in actual division. Given two magnitudes, each divided into equal sections, the ratio between the sections of the one magnitude to those of the other remains stable even when the process is continued indefinitely. The number of annual revolutions of the sun and the number of annual revolutions of the moon are two infinite sequences that the ratio between their respective terms remains constant. A similar idea was offered by Alfonso's contemporary Henry of Harclay,[288] who held that inequality between infinite entities is not an absurd and that a ratio between two infinite entities can be defined.[289] Alfonso uses the understanding achieved here in Section IV.1 below.

[I.5.4. Concluding Remark on the Aim of the Book]

Having understood the imagining of mental superposition, how it serves as a first principle of geometry [102b] *per se*, and how geometers can use it in actual construction of magnitudes, I now come back to my original intention in this book, namely

[i] to find a constructible straight line that equals the circumference of a circle, and each of the infinitely many curves of different species [מינים], and

---

[285] Like תכלית וקצה in II.1.1 below.

[286] I.1.3; I.2.3.

[287] See also Introduction, Section 4.

[288] "It follows therefore that if to one revolution of the sun are added twelve revolutions of the moon for an infinite time, there would now be a greater number of revolutions of the moon than of the sun, and similarly that there are potentially more numbers exceeding two than exceeding 100. This argument holds even if one proceeds to infinity" (Dales 1984, 299).

[289] Murdoch (2009, 22–3); Wood (2009, 40, note 8).

[ii] to find a constructible rectilinear area equal to the area of a circle, and to the area of the surface of a sphere, and to the area of the surface of a circular cylinder, and the area of the surface of a cone, and of the rest of the infinitely many of curved areas of different genera [סוגים], and
[iii] to find a constructible body with straight surfaces equal to the body of a sphere and to the body of a cylinder, and to a torus, or to the rest of the infinitely many bodies having curved surfaces of different species [מינים], and to transform all these species, each to the likeness of the other *by means of the plane that is common to all of them*,
[iv] to find a way to divide any rectilinear angle or an angle between arcs on a surface of a sphere to any number of equal parts that we wish, or in any ratio that we wish, rational or irrational, this being necessary for the transform of magnitudes to one another *by means of a plane* common to them.

Therefore, I call this book *Meyasher 'Aqov*,[290] as this is its main goal and it is very useful in mathematics.

Terms

מדובר, אלם See II.5.5 below.

מוגבל, להגביל See I.2.5 above.

Commentary

Alfonso recapitulates the agenda of the book that was given in the outline, with an important addition: the transformation of three-dimensional bodies into one another are carried out *by means of the plane that is common to all of them*. In paragraph I.3.4 above he summarizes his discussion of Plato's reduction: "It remains for me to inquire how the body is divided into areas and the area into lines, as Plato said. And this [inquiry] is in terms of *evaluating and measuring*." In IV.1 he takes the first step: he "covers" the rectangle by moving its side along its basis maintaining a right angle with it. Perhaps Alfonso alludes here to a similar method of moving a plane in parallel to itself through the three-dimensional solid. This idea can be considered as a prelude to Cavalieri's principle:

> Suppose two regions in three-space (solids) are included between two parallel planes. If every plane parallel to these two planes intersects both regions in cross-sections of equal area, then the two regions have equal volume.[291]

Alfonso meant to carry out this project of evaluating and measuring by motion in Part IV and we do not know how much of it was eventually achieved.[292]

---

[290] Following Isa. 40:4.

[291] https://en.wikipedia.org/wiki/Cavalieri's_principle.

[292] See introductory note to Section I.3 above.

II Part II, On Accidents Associated with Motion, Which Prevent its Imagining from Being a First Principle of Geometry, and how Several Early Sages were Confused by it

Introductory remark

Alfonso holds that there is nothing about motion itself that should prevent its being used in geometry, but it was often used inappropriately by past thinkers. The errors, he insists, should be pointed out, so that the geometer would be able to avoid them and to use motion freely in his demonstrations. Alfonso divides previous attempts to use motion in geometry into two groups: those that involve essential errors, notably those that admit actual infinity (listed in Section II.1), and those that are not a priori absurd, but are nonetheless erroneous (listed in Section II.2). The first two examples (II.1.1, II.1.3) originated in physical arguments against actual infinity. The other three arguments originated in mathematics, notably in attempts by previous geometers—Moshe of Seveille (II.1.5), Ibn al-Haytham (II.2.1), and al-Nayrizi (II.2.3)—to prove Euclid's fifth postulate, known as the parallels postulate.

Part II of *SMA* is a good example of the infiltration of methods and ideas between the two disciplines, physics and geometry. On the one hand, mathematical arguments are used in philosophical-physical discussions of infinity and continuity (following Aristotle, Ibn Sīnā, and al-Ghazali)[293]; on the other hand, philosophy affected the development of the science of geometry, for the good and for the bad. Since Aristotle's *Physics* VI, the study of physics deepened the understanding of continuous magnitudes, which are the subject matter of geometry. Yet, some tenets of Aristotelian physics—such as the denial of actual infinity, or the insistence that lower dimensional mathematical objects, i.e., planes and lines are not true "objects" and cannot be moved[294]—hindered the development of geometry. This kind of "interdisciplinarity" was typical of the fourteenth century in the west, but the majority of Alfonso's examples draw on Greek and Arabic sources.

[II.0. Foreword]

Alfonso said, some people associate with motion external accidents, which prevent its imagining from being a first principle of [103a] geometry. These [accidents] are of two types: first, accidents which essentially are impossible to imagine by the intellect and qua such cannot qualify as first principles of

[293] Mcginnis (2010, 5, 23) remarks that Arabic natural philosophers preferred "mathematical-style proofs" against the possibility of actual infinity.

[294] *Meta* VI.1; XI.2, 1060a13–15. Planes and lines can be moved accidentally as parts of a moving three-dimensional body.

any science; second, secondary accidents of geometry which, *qua* being secondary cannot qualify as first principles.

Terms

מקרה The subject of Part II is "accidents concomitant with motion," which is an unusual expression. The word מקרה (accident, Arabic عرض) is used in Aristotelian as well as in Kalām philosophy mainly to denote an attribute of substance,[295] not of motion. Here it is used in a loose sense of "something associated with."

[II.1. Motion in Geometry: Essential Errors]

[II.1.1. First Error: Rotating a Finite Line in an Infinite Plane]

To the first type belong those accidents which are essentially impossible to imagine by the intellect. An example is their accepting [the imagining of motion in geometry] when they imagine circular motion of any straight line bounded by an extremity and end. They say that if we posit a line *CD*, infinite in the direction of *C* and rotate a [finite] line *AB* in its circle in the direction of *C* until it becomes parallel to it [*CD*], this is possible without any doubt.

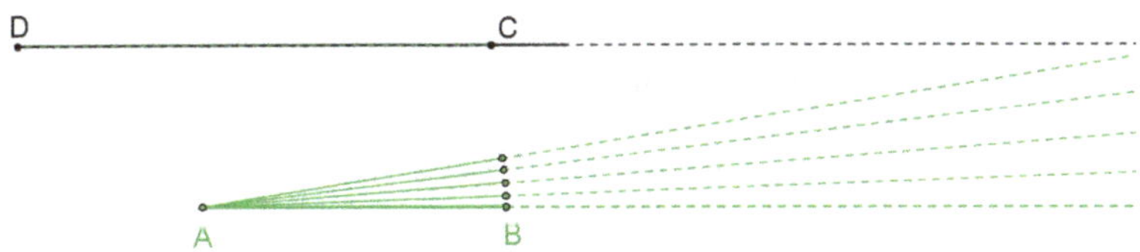

Then we continue to rotate it [*AB*] from the parallel position in order that *B* would approach it [*CD*], so that the straight extension [of AB] would first fall on a point on line *CD*, and then on the rest of the points, one after the other, until it [*AB*] will come to be parallel [to *CD*] on the other side. But this is impossible, because the [change of] inclination [of AB to CD] without its [*AB*′s] straight extension falling on [all] these points is impossible. And with its straight extension falling on these points is also impossible in the case of an infinite line, since its [*AB*′s] extension should first fall on one point [of *CD*] and then on the rest of the points, but in an infinite line there is no first point. For each point [of *CD*] on which we agreed that the extension of *AB* falls, there should be infinitely many other points before it on which the

[295] For example, Pines (1936/1997, 21–31); Wolfson (1976, 132–182), and, arguing with Wolfson, Frank (1967, 249–253).

extension would fall. Therefore one should, necessarily, imagine a first point of it [and thus] it must be finite. And they say that this is a decisive geometrical demonstration of the impossibility of an infinite distance, whether in plenum or in void.

Terms

תכלית וקצה The reason for the repetition "an extremity and end" is not clear. The same expression תכלית הקו וקצהו is used by Iṣḥaq Israeli in his rendering of Euclid's *Elements* I definition 3, that the point is the extremity of a line.[296]

להיות על יושר נקדה The literal translation is "to be on a straightness of a point," we translate "to be on the straight extension of." A line segment *AB* is said to be "on the straight extension" of a point *P* when *P* falls on the straight extension of *AB*.

Commentary

II.1.1 offers a good example of a "mathematical proof" of an argument that originated in natural science.[297] The paradox of the rotating ray goes back to Aristotle, who used it to illustrate that the radius of the heavens cannot be infinite: Given are two infinite lines *AB* and *E* at right angles, *C* a point on *AB*. An infinite ray *CD* rotates about the point *C* but cannot cross the line *E* and cannot "complete the circle."[298] The argument had undergone transformation in the hands of Ibn Sīnā who turned it into an argument against extracosmic void space: "there cannot be circular motion in the void because the void is of the character that neither comes to an end nor exhausted."[299] To this end Ibn Sīnā's imagines the radius of a rotating finite sphere—e.g., the Aristotelian cosmos—in an infinite void.

Alfonso's argument is taken either from Ibn Sīnā or from someone who cites Ibn Sīnā.[300] His saying that *AB* rotates "in its circle" and his conclusion that there can be "no infinite distance, whether in plenum or in void" testify that his source was a cosmological argument. The citing is not verbatim but it is not impossible that

[296] *Yessod 'Olam Sha'ar Alef*, p. 3 in the online edition, p. 3b in Goldberg-Rosenkranz edition.

[297] See note @ above (reference to McGinnis).

[298] *De caelo* I.5272b25–9. Aristotle's argument is presented in detail in Ibn Rushd's Middle Commentary on *De caelo* Arabic 98.10–100.3; Hebrew Berlin ms. fol. 24a21–b2.

[299] *The Physics of the Healing* II.8, argument 8, McGinnis translation, vol. 1, 184–186. See also McGinnis (2006).

[300] McGinnis (2006, 14) testifies that "a cursory examination" of early Arabic sources "has produced nothing even vaguely similar to Avicenna's argument." He also is of the opinion that "nothing like Avicenna's argument ever explicitly appeared in Greek commentaries on the *Physics*" (ibid, 8).

Alfonso was to some extent acquainted with the text.[301] In Alfonso's version *CD* is a ray in the direction of *C*, *AB* is a finite line segment. *AB* is rotated until parallel to *CD*, *B* on the side of *C*. *AB* is then rotated another 180°.[302] An (imagined) extension of *AB* on the side of *B* must meet *CD*, but this is impossible because there is no first point on *CD* which the extension of *AB* meets. The conclusion is that infinite distance is impossible.

Both Aristotle and Ibn Sīnā used the paradox of the rotating ray (or the imagined extension of a rotating finite line) in order to prove that the universe must be finite. Alfonso does not accept it as a legitimate argument.

[II.1.2. Explaining the Error]

We say that this is not a demonstration at all. It is altogether false because it relies on a premise [103b] whose falsity has been established in the *Physics*, namely the existence of a first part of motion *qua* motion. Since motion is from a *terminus ex quo* to a *terminus ad quem*, when these two [termini] are not present there will be neither motion nor a definitive *terminus ad quem*, but there are other points before it [the *terminus ad quem*] towards which the motion is, because between any two points there is a continuous line that can have infinitely many points. Therefore, there is no first *terminus ad quem*. Similarly, it is established that there is no last *terminus ex quo*. How can there be in a line that was assumed to be infinite a first *terminus ad quem*, or a last *terminus ex quo*? They erred in this because they assumed concomitantly with motion something that are impossible to be imagined by the intellect, namely the existence of a first *terminus ad quem* or a last *terminus ex quo*.

Terms

בטל, שקר, נמנע Alfonso uses these three terms to denote impossibility. Shmuel Ibn Tibbon in his glossary distinguishes between two of the terms: בטל is a stronger term than שקר, and is used to denote logical impossibility. Perhaps Alfonso was aware of this distinction, though it is not certain. We translate בטל as "false," נמנע as "impossible," and שקר as either "false" or "impossible," according to the context. Apparently נמנע and בטל are used to denote logical impossibility whereas שקר is used also in *per impossibile* arguments to indicate contradictory conclusion.

---

[301] According to McGinnis (2006, 15) there are different versions of this argument: "every text has at least some variant reading of the lettering of the points, such that there are multiple ways to construct the diagram, each different and sometimes significantly so."

[302] *Prima facie* in the argument quoted by Alfonso only a finite line segment *AB* rotates, but since the argument involves the straight extension of the line segment, the assumption of this argument is not truly weaker than that of Aristotle's original argument.

Commentary

Alfonso adduces against this argument, that originated in Aristotle's *De caelo*, a counter-argument from Aristotle's *Physics*: assuming the existence of "a first part of motion" is unfounded. In *Physics* VI.5 Aristotle contends that there is no such first part. Ibn Sīnā, on whom Alfonso relies in this context, also insists that there can be no first moment of motion.[303]

[II.1.3. Second Error: Moving a Ray Along Itself]

They erred in yet another way saying that if the existence of an infinite line could be accepted, then we consider a line *AB* infinite in the direction of *B*, and mark a point *D* anywhere on it. If *DB* is infinite, when we imagine the superposition of *DB* on *AB* then either [i] *DB* and *AB* would be superposed on one another, and the smaller would be equal to the greater which is

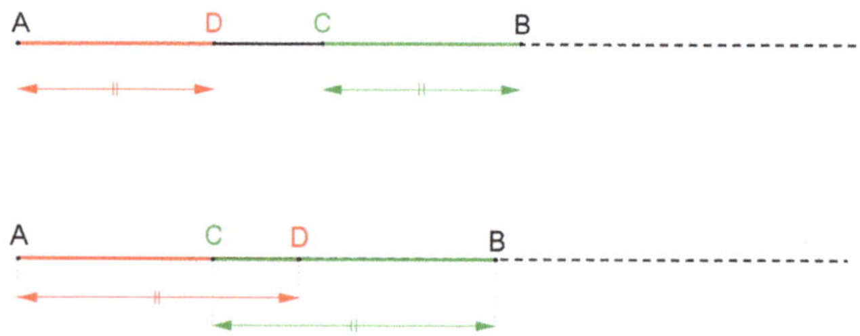

impossible; or [ii] *DB* would fall short of *AB* in the direction of *B* by *AD*. Let it be *CB*. When *CD* moves on *AB* and we add to it *CB AD* which are finite, then the whole line *AB* would be finite, whereas it was assumed to be infinite, and this is impossible. Therefore, *AB* is necessarily finite.

Commentary

AB is an infinite ray, *D* a point on it. *DB* is "pulled" to the beginning *A*. If *AB* and *DB* coincide it means that "the less is equal to the greater." If they do not coincide then there should be a finite line $CB = AD$ at the far side. $AB = AD + DC + CB$ which must be finite, contrary to the assumption.

This argument goes back at least to al-Kindi[304] and, as Wolfson has shown, can be found in several medieval texts[305]: Ibn Sīnā's *al-Najāt*,[306] Ibn Daud's *Emuna Rama*,[307]

[303] Ibn Sīnā', *Treatise on the Now*, Mcginnis (1999, 76–8); *Physics* VI.5236a11–16.

[304] Mcginnis (2006, 6–7).

[305] Wolfson (1929, 346–7). The example was used late also by Christian scholars. For example, Murdoch (1982, 571).

[306] Mcginnis (2006, 18–9).

[307] Eran, forthcoming.

Baḥya Ibn Paquda's *Ḥovot ha-Levavot*,[308] and al-Ghazali's *Maqāṣid*.[309] Of the extant texts known to us, Alfonso's statement is fairly close to al-Ghazali's. The use of the reference "they" at the beginning of this example may, perhaps, suggest that this and the previous example were taken from the same direct source that could, perhaps, have been Ibn Sīnā's *Oriental Philosophy*, a lost text that was known to Alfonso.[310]

If al-Ghazali was Alfonso's direct source, two observations should be made: first, as usual Alfonso abbreviates and omit sentences that do not seem essential to him. Second, there is no indication that he used one of the three Hebrew translations. Key terms and style are different.[311]

[II.1.4. Explaining the Error]

We say that this is false too because they associated with motion something that is impossible [104a] to imagine by the intellect, namely that the moved magnitude is infinite. This is because motion is from place to place, and the place from which the motion of[312] the line is should be a limit point, which does not exist on a line assumed to be infinite. A fortiori, according to Aristotle who says that place is that which encompasses, and that which encompasses is nothing but the extremity and the limit of that which is encompassed in it.

Commentary

In order to explain the absurdity of the situation imagined in the second example, Alfonso adduces two arguments, both relying on Aristotelian tenets: First, local motion is a change from place to place,[313] from a *terminus ex quo* to a *terminus ad quem*, but the endpoints of an infinite line are not well defined. Second, place is the inner boundary of the encompassing body, which cannot be defined for an infinite magnitude.[314] Alfonso apparently commits here a category mistake, applying a notion of place to a geometrical object (in this case infinite line). Aristotle, with one exception,[315] associates place with physical objects.[316] Mathematical magnitudes are not in place,[317] but they have position.[318]

---

[308] I.5, Yehuda Ibn Tibbon's translation, Zifroni's edition (1949, 115).

[309] Cf. for instance Yehuda ben Shlomo Natan's translation. Paris ms. fol. 42b20–43a5.

[310] Szpiech (2010).

[311] See Appendix to the introduction.

[312] Literally "to."

[313] *Phys*. V.2.

[314] *Phys*. IV, 4, 212a5, 212a28.

[315] *De gen. et corr*. I.6, 322b32–323a3.

[316] *Phys*.III.5, 205b32; IV.1, 208b28.

[317] *Meta*. XIV.5, 1092a18–20, Annas' translation.

[318] *Phys*. IV.1, 208b23–5.

[II.1.5. Third Error: Moving an Infinite Line Parallel to Itself (the Attempt of Moshe of Seville and Others to Prove the Parallels Postulate)]

Similarly, others, among them R. Moshe of Seville, erred when they endeavored to prove the premise stated in the beginning of Euclid's book: "If a straight line falls on two straight lines, and the two [interior] angles formed on one side are less than two right angles, then when the two straight lines are extended on this side they would necessarily meet." They say, let us first agree on that which is self-evident:

[Premise of Moshe of Seville's and others] that when two infinite straight lines intersect on a plane, say *AB* and CD, and we assume one of these two lines to move on the plane that contains them, then they would never separate, but would either always intersect, or the one be superposed on the other. It cannot be imagined that the one would separate from the other, unless it is terminated before the intersection; but each of them was assumed to be infinite, so it is impossible that the one will be separated from the other. Then they said: we assume line *ABC* falling on two straight lines *AE* and *BG*, and assume that the two angles *EAB* and *GBA* are less than two right angles; we say that if lines *AE BG* be extended in the direction of *E G* then they would necessarily meet on this side.

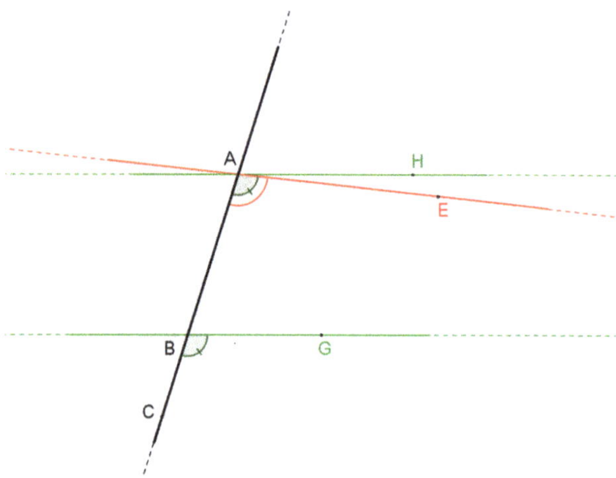

The demonstration: the two angles *GBC GBA* together equal two right angles, and the two angles *GBA EAB* are less than two right angles. Therefore angle *GBC* is greater than *EAB*. We construct angle *BAH* equal to angle *GBC*, and assume three infinite lines *AE*, *AH*, [104b] *BG*. We assume *AH* moving so that the point *A* moves along the line *AB* keeping the magnitude of angle *HAB* fixed. We assume *AE* fixed. Since each of the lines *AH*, *AE*, *BG* is infinite therefore *AH* is not separated from *AE* and since the line *AH* keeps the magnitude of the angle *HAB*, it is impossible that *AH* is superposed on *AE*. Therefore *AH* would always intersect *AE*. Therefore, when *A* reaches point *B* via the motion of *AH*, then *AH* would be superposed on *BG*, because angle *CBG* equals angle *CAH*, and line *AH* always keeps the magnitude of angle HAC fixed. Therefore when *AH* is superposed on BG and AE intersects AH it would also intersect BG. Therefore lines AE, BG necessarily meet.

Historical Commentary

From a medieval perspective, this "proof" of Euclid's postulate 5 is less sophisticated than the one offered by Ibn al-Haytham (paragraph II.2.1 below), as it allows moving infinite lines. Alfonso thus lists it among the arguments that involve an essential error. In paragraph II.1.7 below Alfonso refers to it as "the demonstration that R. Moshe brought in the name of the previous scholars and which he accepted as valid." Gad Freudenthal identifies Moshe as the thirteenth-century physician and philosopher Moshe ben Yosef, known in Arabic as Abu 'Imrān Musa' al-Lawī al-Ishbilī.[319] As to the identity of "the others," we have some evidence that the argument circulated in Andalusia from the eleventh century and was included in Ibn Hud's famous geometrical encyclopedia, the *Istikmāl*. The first book of the *Istikmāl* is now lost, but Ibn Sartāq al-Marāghī's quotes it in his *Al-Ikmāl fi al-handasah,* which follows Ibn Hud's text.[320] He could thus have been one of the "others" referred to by Alfonso. Several of the attempts to prove the parallels postulate by Arab mathematicians have not survived,[321] and it is possible that Moshe followed one of them. Moshe's argument had been studied by Tony Lévy, who found a similar argument in an anonymous Hebrew manuscript fragment.[322] It may be a copy of Moshe's original "proof" or some presentation of it. The structure is the same, but the language and the lettering of the diagrams are not identical.

Mathematical Commentary

The argument relies on an added premise that two coplanar intersecting infinite straight lines cannot be separated by motion. Assuming this premise and allowing the moving of an infinite line the argument is simple:

$$\sphericalangle GBC + \sphericalangle GBA = 180^\circ \ \text{(complementary angles)}$$
$$\sphericalangle GBA + \sphericalangle EAB < 180^\circ \ \text{(according to the assumption)}$$
$$\text{Therefore } \sphericalangle GBC > \sphericalangle EAB.$$

[319] See Introduction, Section 6.

[320] Hogendijk (2003, 22–23).

[321] Sabra (1968, 15).

[322] London Beth Din ms. 138/16, IMHM 4799, fols. 89a–b. After this passage follows a construction for finding two mean proportionals, also using motion, that may be by the same author. Alfonso offers a different construction of two mean proportional using conchoid in III.31.

Through *A* construct a line *AH* || *BG*. This can be done using *Elements* I.28 without relying of the parallels postulate. Let *AH* move in parallel to itself, so that the point *A* moves along the line *AB*. According to the additional premise *AH* and *AE* should remain intersecting. When *A* falls on *B*, *AH* coincides with *BG* because ∢*CBG* = ∢*CAE*. *AE* intersects *AH* and *AH* coincides with *BG*; this means that *AE* intersects *BG*.

Alfonso located Moshe's error in his allowing infinite lines and even moving them. Modern geometry allows infinite lines and moving them, so we have to find Moshe's error elsewhere. Unlike Ibn al-Haytham's added premise, the added premise stated here is not equivalent to the parallel's postulate[323] and holds in non-Euclidean geometries as well. The problem with Moshe's proof is that whereas it is emphasized that line *AH* is moved along *AB* keeping the angle between the two lines fixed, it is taken for granted that the angle between *AH* and *AE* also remains fixed. In order to prove this, however, one has to rely on I.32, hence on the parallels postulate.

[II.1.6. Moshe of Seville's Justification of the Use of Motion in Geometry]

The aforementioned R. Moshe affirmed that it is not inappropriate to imagine motion of lines in such contexts, but it is inappropriate to practice it. This is because the geometer abstracts the lines as well as other magnitudes from matter, whereas these magnitudes do not exist outside the imagination, either essentially or accidentally. A line cannot be moved outside the imagination unless accidentally, i.e., qua being in a moved body. Since the geometer can imagine lines in a way that do not exist outside the imagination either essentially or accidentally being abstracted and devoid of matter, *a fortiori* he can assume lines that exist accidentally outside the imagination being moved.

He said that *many* geometers commonly use in this science the moving of lines, and *all* geometers use what is similar to motion, namely superposition of lines or areas on one another, in the way that Aristotle uses the moving [105a] of lines in his demonstrations in natural science.

Commentary

Both Thabit ibn Qurra and Ibn al-Haytham provided justifications for their use of motion in geometry,[324] and Moshe of Seville followed this tradition. He offers two arguments. The first argument is logical and somewhat awkward: [i] being abstractions, planes and lines have no existence outside the mind either essentially or accidentally[325]; [ii] planes and lines can be moved only accidentally (as parts of bodies).[326] The desired conclusion is, perhaps, that the geometer, who

---

[323] II.2.1–2.

[324] See I.2, introductory note.

[325] Aristotle, *Phys.* II.2, 193b31–194a5; *Meta.* M 3.

imagines planes and lines would not further complicate his discipline by imagining them in motion. The second is historical: several previous geometers used the moving of lines in their proofs; all geometers use superposition, which is a kind of motion[327]; even Aristotle used the moving of lines in his natural science.[328]

[II.1.7. Explaining the Error]

We say that this demonstration that R. Moshe brought in the name of previous scholars, and which he himself accepted as valid is not valid at all; not because imagining motion [in geometry] is inappropriate in itself, but because they associated with it the existence of an infinite line; and an infinite line in act is impossible, whether in motion or at rest. And if you argue that the geometer should not reject the possibility of an infinite line, since Euclid himself admitted a concept of infinite line, I shall reply that the geometer does not mean the existence of infinite line in act, which is absolutely impossible. I shall prove this in this chapter,[329] because I have not seen a previous satisfactory proof [באור] of it, and because it is very useful in this study in which we are engaged.

The geometer means to say that there exists a limited infinite line, namely a line that has an end, but the end is not actually limited or actually determined in such a way that extending it cannot be imagined. Rather, no matter how long the line is, one can imagine an extension, and an extension of the extension, so that it would always, with all the extensions, remain finite in act. They mean the same when they say that the line is divisible into infinitely many parts, they do not mean that division results in infinitely many parts in act, because this is absolutely impossible. Rather they mean that the line is potentially divisible into an infinite number of parts, though actually it has a determined number of parts. Similarly, the line cannot be extended until it becomes actually infinite; but it can be extended indefinitely[330] so that remains always finite in act.

---

[326] Aristotle, *Meta.* XI.21060 b12–16; XIII.31078a10.

[327] Since common notion 4 is a basic principle of geometry.

[328] The reference to Aristotle at the end may be to *Physics* IV. A possible example may be Ibn Rushd's presentation of *Physics* IV.4212a23–8: "When a radius of a circle rotates about the center, every point on the radius describes a circle and this is true also of the center point itself. The circle generated by the center point must be either in a void or in a spherical body. The first alternative is ruled out. The circle is therefore in a spherical body [i.e., corporeal]. This spherical body must be either in motion or at rest. In the first case the argument can be repeated. We can thus assume that it is at rest. Furthermore, the center cannot be a point because a point can be in motion or at rest only accidentally." *MC Phys*. IV.1.9, Hamburg MS, 39b14–25.

[329] II.4.1 below.

[330] Two unreadable words in the text.

This is equally so with the big and to the small parts, due either [i] to that which sets the actual limit, or [ii] to the specific capacity of the material substratum [105b] that receives the actual limit, or [i + ii] to both of them together, or [iii] to the fact that first quantity is not an attribute of the first body. It is not due to the division of the capacity that the geometer is used to imagine,[331] since he does not deal with material distance or with a specific substratum. Had he done so, his science would not be applicable to all matters and substrata. That which has no qualities has no upper or lower determined boundary. This is the reason of the geometer's assuming "to extend a finite line"[332] until it becomes infinite, and "to draw a straight line between any two points."[333] Therefore the aforementioned R. Moshe erred when he extended the line *AH* to infinity *in actu* when on *BG*, *a fortiori* when he assumed it to be [both] in motion and infinite.

Terms

קצוב The adjective קצוב, translated here as "determined," appears three times in *SMA*, all three in this paragraph in the combinations מספר קצוב, תכלית קצובה. The expression תכלית קצובה was used also by Falaquera.[334]

Commentary

Alfonso recapitulates his point that it is not motion but the actual existence of an infinite line that should be excluded from geometry. This pertains also to the argument cited by Moshe. Alfonso may have followed here Ibn al-Haytham who argues in his *Commentary on the Premises of Euclid's Elements* that "every single body, whether sensed or imagined, must be finite, since if we assumed an infinite body there would be no way for us to imagine all of it,"[335] yet, "The line can be imagined separate from the surface (that can be imagined separate from the body)."[336]

The geometer may always extend or divide a given finite line in the imagination,[337] but physical factors can impose limitations on the extension or division of lines in nature. These may be [i] restrictions due to actual dimensions (perhaps

---

[331] The sentence is difficult and cannot be translated literally.

[332] *Elements* I Postulate 2.

[333] *Elements* I Postulate 1.

[334] See Klatzkin, *Thesaurus*, vol. 3, 294.

[335] Hooper-Sude 23.13–15.

[336] Hooper-Sude 26.24.

[337] *Elements* I, postulate 2, Proposition 10.

alluding to the size of the universe), or [ii] restrictions set by the specific material (perhaps alluding to the theory of *minima naturalia*), or [iii] restrictions dictated by the very nature of physical body which is not essentially quantified (perhaps alluding to the subtle body).[338] All these restrictions, however, are irrelevant to the geometer who studies magnitudes *qua* magnitudes.

[II.2. Motion in Geometry: Accidental Errors]

The second type of external accidents which prevent the imagining of motion from being a first principle of geometry, but cannot qualify as first principles of it [geometry] because they are secondary to it.[339]

[II.2.1. Ibn al-Haytham's and Others' Attempt to Prove the Parallels Postulate]

For instance, what Ibn al-Haytham and many others wrote when they commented on Euclid's postulate mentioned above about a straight line falling on two straight lines [i.e., the parallels postulate]. He said: we consider a straight line *AB*, erect on it a line *AC* at a right angle, and assume that line *AC* is moved while keeping *A* always a right angle. We assume that the point *A* is moved along the line *AB* and that line *AC* is moved until it becomes as *BD*, and point *C* describes a straight line *CD* equal to *AB*. And we join the straight line *AD*. Since the three sides of the triangle *ABD* are equal to the sides of the triangle *ACD* respectively, then the triangle would be equal to the triangle and each of the angles to the corresponding angle. Therefore the right angle *C* is equal to the right angle *B* and angle *ADB* equals the angle *DAC* and angle *BAD* equals angle *CDA*. Therefore the two angles *BDA* [106a] *CDA* are equal to the two angles *DAC BAD* together that form one right angle. Therefore the polygon *ABDC* has four right angles and its parallel sides equal one another. The remaining part of his demonstration is built on the premise that point *C* describes by its motion a straight line *CD* equal to the line *AB* opposite to it.

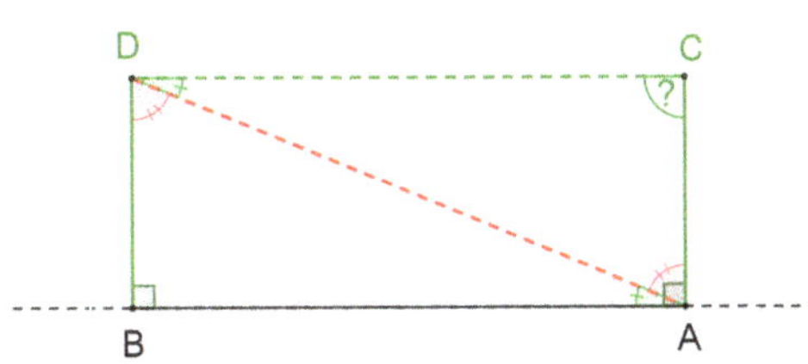

[338] See Introduction Section 3, argument iv, and Section II.4.2 below.

[339] There is a problem in the syntax: it is the imagining of motion, rather than the accidents, that is said to be a principle of geometry.

Historical Commentary

We already had reasons to suspect[340] that Alfonso was familiar with the *Commentary on the Premises of Euclid's Elements (Sharḥ muṣādarāt kitāb Uqlīdis fi al Uṣūl),*[341] but only here Ibn al-Haytham is mentioned by name. Was Alfonso familiar with this text first hand? The rendering of the argument is deficient and over-simplified. Tony Lévi contends that Alfonso missed the subtle point in Ibn al-Haytham's argument; that he either misread or distorted the argument of his great predecessor.[342] Alfonso indeed offers no overview of Ibn al-Haytham's argument, but he apologizes in advance that he will follow it until he pinpoints what he considers to be the "accidental error."

As noted in the Part I, Ibn al-Haytham was aware of the problem of the distortion that can be caused to magnitudes by motion. Commenting on the definition of parallel lines (*Elements* I definition 23), he imagines the motion of a finite line at a right angle to another line as "one and simple not composed of movements or of motion and rest."[343] He further introduces an added premise on which he will later rely in his "proof" of the parallels postulate:

> **Ibn al-Haytham's Premise**: When a finite straight line is moved uniformly along a fixed straight line maintaining a right angle with it throughout the motion, the other extremity of the moved line describes a straight line parallel to the fixed line.[344]

The justification is that the endpoint is moved with *the motion of the whole line*, which is one and simple.[345] Ibn al-Haytham takes the "rigidity" of the motion of the straight line to follow from the very nature of uniform motion that is "one and simple."[346] On the basis of this premise he proves a theorem which is equivalent to the parallels postulate that was proved earlier by Thābit Ibn Qurra[347]:

[340] I.4.2, II.1.7 above.

[341] See Jauiche (1986), commentary 57–64, translation of the text 161–175; Huper-Sude (1974), 93.12–106.19; Commentary 109–119.

[342] Lévy (1992, 47).

[343] Hooper-Sude 50.26–51.1.

[344] Hooper-Sude 53, 1–3. It should be noted that Ibn al-Haytham's premise is not valid in hyperbolic geometry.

[345] Hooper-Sude 53.6–14.

[346] He repeats again and again that points on the moved lines produce by their motion "similar lines" (e.g., Hooper-Sude 51.24–52.7, 52.11–16, 52.20–25).

[347] Thābit first proof, fifth proposition: Sabra (1968, 25); Jaouiche (1986, 148–9).

Thābit's theorem: If two perpendiculars are drawn from the two endpoints of a given straight line in the same plane, then any perpendicular drawn from a point of one of the two perpendiculars will be perpendicular also to the other and equal to the given line.[348]

Both Thābit and Ibn al-Haytham proceed from Thābit's theorem to the parallels postulate, but their proofs of Thābit's theorem are different. It is possible that Alfonso was familiar with Thābit's version,[349] but here he refers to Ibn al-Haytham's.

Mathematical Commentary

Alfonso is actually discussing just the first part of Ibn al-Haytham attempt to prove the fifth postulate, which is as follows:

He assumes a line *AB* with two lines perpendicular to it, *AC* and *BD*. From a point *C* he draws a perpendicular *CD* on line *BD* so that $\sphericalangle CDB = 90°$. *CA* is extended in a straight line and duplicated to point *E*. A perpendicular *ET* is drawn to the straight extension of *DB* so that $\sphericalangle ETB = 90°$.

Δ $CAB \cong \Delta EAB$ ($CA = AE$ according to the construction, *AB* is common and $\sphericalangle CAB = \sphericalangle EAB = 90°$). Therefore $CB = EB$, and $\sphericalangle CBA = \sphericalangle EBA$.

Δ $CDB \cong \Delta ETB$ ($CB = EB$, $\sphericalangle CDB = \sphericalangle ETB = 90°$, and $\sphericalangle CBD = \sphericalangle EBT$ as complementing $\sphericalangle CBA$ and $\sphericalangle EBA$ to right angles (*Elements* I.26, not relying on Postulate 5)). Therefore $CD = ET$.[350]

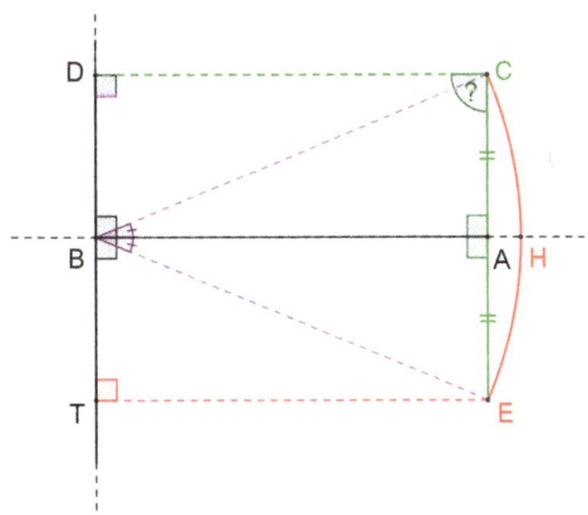

Ibn al-Haytham proceeds to show that *AB* can be neither less than nor greater than *CD* (=*ET*). He assumes that $ET > AB$ and, at this stage, in order to demonstrate the impossibility of the assumption he uses motion: "Let us imagine line *ET* moving along line *TB* while during its motion it is perpendicular to it, so that angle *ETB* is a right angle throughout the motion. When point *T* by the movement of line *ET* ends up at point *B* line *ET* must coincide with line *BA*."[351] Ibn al-Haytham assumes that point *E* is mapped on

[348] Ibn al-Haytham's proof of the theorem relying on this premise is summarized below.

[349] See Section II.3.6 below.

[350] Jauiche (1986, 163); Hooper-Sude 94–5.

[351] Jaouiche, p. 167; Huper-Sude, p. 95.

point H (above A due to the assumption that $ET > AB$). Line BH continues to move and coincides with DC. The contradictory conclusion is that there are two lines joining C and E that enclose a space between them, contrary to (the post-Euclidean) Postulate.[352] Therefore $ET = CD = AB$.

Alfonso skips Ibn al-Haytham's construction and offers a simplified abridged scheme, whose aim is to pinpoint the error. He assumes lines AB AC perpendicular to one another and "constructs" BD simply by moving AC along AB keeping the right angle constant, until A coincides with B. Triangles ABD and ACD are congruent (s.s.s.) and consequently all four angles of $ABCD$ should be right angles.

[II.2.2. Explaining the Error]

This cannot be the case unless the two lines that join the endpoints of the two parallel lines equal one another. This means that all the points of a line that is moved in a linear motion describe straight lines equal to one another. This premise [הקדמה] was proved by Euclid only in Proposition 34 of the first book, after he had already accepted the postulate [הקדמה] about the straight line falling on two straight lines [i.e., the parallels postulate].[353] Therefore his argument is circular and thus logically false.

Commentary

Alfonso charges his predecessor of *petito principii* in assuming that point E, which is moved with the motion of line ET, describes a straight line. Ibn al-Haytham by no means ignored this issue. He discusses it at great length when commenting on the definition of parallel lines (*Elements* I def. 23).[354] He contends (but does not prove) that when a finite straight line is moved along another straight line keeping the right angle between the two lines fixed throughout the motion, then the line, which the upper end of the finite moved line describes, "is a straight line because it coincides with a straight line."[355]

Unlike us (who know that the Parallels Postulate was proved to be independent of the other postulates) Alfonso had no a priori reason to suspect the validity of a

[352] The postulate is not included in Heiberg's edition but is mentioned by Heath (1908/1956 I, 232).

[353] Ibn al Haytham relies on Euclid's Proposition 34 ("In parallelogramic areas the opposite sides and angles are equal to one another") which relies on Proposition 29, in which Postulate 5 is used.

[354] Hooper-Sude 96.1–2.

[355] Hooper-Sude 53.1–3.

proof by a highly qualified well-established mathematician like Ibn al-Haytham. It could have been his attentive study of continuous, yet non-isometric superpositions[356] that led him to see that rigidity of uniform motion cannot be taken for granted.

Alfonso was not the first to criticize Ibn al-Haytham. This was done before Alfonso by Omar Khayyām (who objected to the use of motion in geometry), and following him by al-Tūsī.[357] We do not know whether Alfonso was aware of Khayyām's criticism.

[II.2.3. Al-Nayrizi's Attempt to Prove the Axiom of Parallels]

Since the geometer al-Nayrizi saw, following what Simplicius wrote, that it was not right to prove the postulate [הקדמה] about the straight line falling on two straight lines [the parallels postulate] using motion, he proved it using a different argument of his own. He started stating two lemmas [הקדמות]:

[Al-Nayrizi's First Lemma]

He said that when there are two parallel straight lines, then the shortest distance between them is a perpendicular on any one of them.

How? We consider two parallel lines *AB CD*, assume that the shortest distance between them is *EG*. I say that the line *EG* is perpendicular to each of the lines *AB CD*.

Its demonstration: If *EG* were not perpendicular to each of the lines *AB CD*, we consider the two non-right angles at *E*, *AEG* being the acute angle. We draw from *G* a perpendicular *GH* on *AB*, which falls on the side of *A*. It [follows] from Euclid's statement that EG [assumed to be] the shortest of all the lines that join *AB* and *CD* [106b] [is greater than *GH* and this is impossible]. Similarly, it can be shown that that the shortest distance between two nonparallel lines is the perpendicular on one of them.

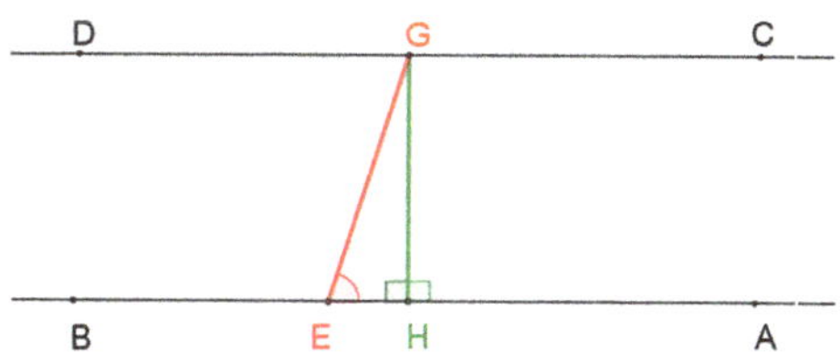

[356] I.2.1 above, II.5 below.

[357] Rashed and Vahabzadeh (1999, 273–4).

[Al-Nayrizi's Second Lemma]

He further said: we consider two lines *AB* and *CD* and line *EG* falling on each of them in a right angle. I say that *AB* and *CD* are parallel and that *EG* is the shortest line between the lines *AB* and *CD*.

Its demonstration: If *CD* were not parallel to *AB*, we pass through a point *G* a line *HI* parallel to the line *AB* so that *EG* is the shortest distance between *AB* and *HI*.[358] According to Proposition 19 mentioned above, *EGH* is a right angle.[359] But it was already assumed that *EGC* is a right angle,[360] and this is impossible. Therefore the lines *AB* and *CD* are parallel and *EG* is the shortest distance among them.

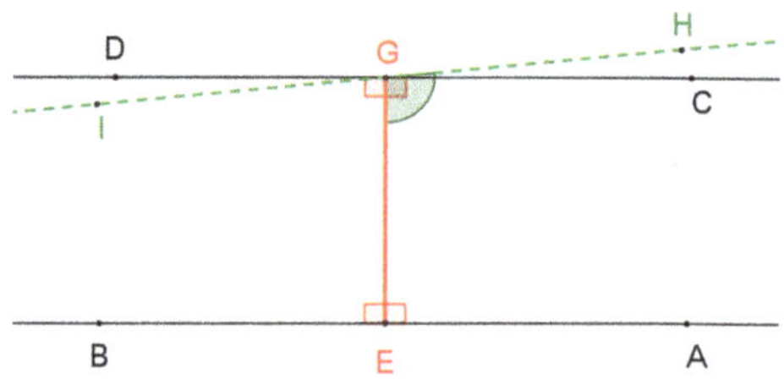

Terms

הקדמה The word *haqdama* (Arabic مقدمة) is the standard term for a premise (or "postulate" or "principle").[361] Here it is used to translate al-Nayrizi's شكل,[362] namely proposition. In Part II Section 3 below Alfonso uses the Hebrew תבנית for the Arabic شكل. Perhaps he uses *haqdama* rather than *tavnit* here because he considers al-Nayrizi's two propositions as preparatory to the main argument. We thus translate it here a "lemma."

כיצד-מופתו The pair of words כיצד (how) and מופתו (its demonstration) are used in Part II and Part III of *SMA* as titles for the exposition and the demonstration parts of the Euclidean proposition. The common Arabic titles are *mas'alah* and *burhān-hu*. The Provençal translators Moshe Ibn Tibbon and ya'aqov ben Makhir sometimes skip these titles and sometimes use the terms המשל (or המשל בו) and המופת. כיצד is not used in the Hebrew translations of Euclid, but In the *Hebrew Geometrical Compendium* (*HGC*) the word מופתו is frequently used and the word כיצד a few times.

Historical Commentary

In the first lemma the enunciation, setting-out, and proof are copied word for word from al-Nayrizi, omitting two sentences.[363] The second lemma, also copied from al-Nayrizi, is somewhat abbreviated. Alfonso merges the enunciation and

[358] Two sentences of Nayrizi's text are missing here.

[359] The reference should be to the first lemma that, in turn, relies on *Elements* I.19.

[360] Al-Nayrizi mentiones the angle *DGE*, adjacent to *EGC*.

[361] II.2.1; II.2.2 above.

[362] Al-Nayrizi *Commentary*, Besthorn 120.13.

[363] Al-Nayrizi *Commentary*, Lo Bello 159–60 (English). Besthorn 120–123 (Arabic and Latin).

exposition of al-Nayrizi into one sentence. The lettering of the diagram is identical to al-Nayrizi's. The reference to Simplicius also follows al-Nayrizi, who ascribes to Simplicius the view that the parallels postulate "was not manifest, but needed demonstration with lines."[364] The two lemmas that Alfonso quotes are attributed by al-Nairizi to Aghanis, who has been identified in the old Arabic-Latin edition as Geminus and is now believed to be Agappius.[365]

Mathematical Commentary

The First Lemma

$AB \parallel CD$. EG assumed to be the shortest distance between them. It should be shown that $EG \perp AB$ and $EG \perp CD$. Assume that $\sphericalangle AEG < 90°$. Construct $GH \perp AB$ in the direction of A. $GE > GH$ as the side opposite to the right angle (*Elements* I.19), contrary to the assumption that GE is the shortest distance between AB and CD.

The Second Lemma

AB, CD lines. $EG \perp AB$ and $EG \perp CD$. It should be shown that AB and CD are parallel. Assume that they are not. At point G Construct $HI \parallel AB$. According to the first lemma EG is the shortest distance between AB and HI and $EG \perp HI$, but it was assumed that $EG \perp CD$.

[II.2.4. Explaining the Error]

On these two lemmas al-Nayrizi built the remaining part of his argument proving Euclid's postulate mentioned above. When you read attentively you'll find that he did not prove that when a line is perpendicular to one of two parallel lines it is perpendicular also to the other, nor that it is the shortest distance between them. The sceptic would concede neither that a perpendicular on both lines [can] be drawn from any point between them (or on one of them) nor that the distance between the two straight lines is always the same. [He would concede] only that they do not meet in either of the two sides. Only this is the definition of parallel lines, namely that they do not meet in either of the two sides, not that the distance between them is always equal (at least until this be proved).

Since *EG* was assumed in the second lemma to be perpendicular to each of the lines *AB CD*, and a line *HI* was drawn between them [through G parallel to *AB*], then *EG* would not be perpendicular to *HI* but would remain

[364] Al-Nayrizi *Commentary,* Lo Bello 158.

[365] Al Nayrizi *Commentary*, Lo Bello 158–166. See Lo Bello's introduction, p. 17 and note L5, pp. 224–9.

perpendicular to *AB*. Therefore it is the shortest distance from *G* to *AE*, not the shortest distance from *E* to line *HI*, [107a] nor from any point [on *AB*] other than *E* to line *HI*.[366] One may argue that there may be a line from *E* to *HI*, shorter than *EG*, that is perpendicular to *HI* in the direction of *I*, and that lines AB and *HI* do not meet because they were assumed to be parallel, and lines *AB* and *CD* do not meet because *EG* encloses with them right angles. [Thus] if *EG* is not the shortest distance from *E* to line *HI*, then the angle *EGH* would not be a right angle, although *CD* is parallel to *AB*. Since the remaining part of al-Nayrizi's deduction [הקש] of this lemma relies on what I mentioned, that I have shown to be doubtful, there is no need to write it here, especially as it is long.

Commentary

As in the case of Ibn al-Haytham's proof, Alfonso abridges the text and presents only the part necessary for him to spot the error. He points out two errors in the second lemma: [i] Al-Nayrizi assumes without justification that *GE* is perpendicular to both lines *AB* and *HI*. [ii] Consequently Al-Nayrizi erroneously deduces that *GE* is the shortest distance from *E* to *HI* and, moreover, that it is the shortest distance between the two lines *AB* and *HI*. He does not justify his assumption that the distance between two parallel lines is always the same. Both errors result from al-Nayrizi's taking for granted a concept of distance between two lines. Such a concept was not introduced by Euclid, and introducing it obviously requires the use of the parallels postulate. Al-Nayrizi does not use Euclid's definition of parallel lines (as lines that do not meet when produced),[367] but an alternative definition in terms of equal distance. According to Proclus' testimony, Poseidonios defined parallel lines as those "that have all the perpendiculars equal that are drawn to one of them from the points on the other."[368]

[II.3. Alfonso's "Proof" of the Parallels Postulate by Means of Absolute Geometry with No Added Premises]

Introductory remark

After having dismissed three "proofs" of the parallels postulate, Alfonso ventures to offer one of his own, which he claims to be based on absolute geometry and "free of any doubt." In Section I.5.2 above, referring to the Parallels Postulate, Alfonso contends that "since Euclid did not want to disclose that motion is one of the first principles of geometry… he had to postulate a premise as self-evident, but which is not

---

[366] In the text *CD*. It should probably be *HI*, because the parallelism of *CD* to *AB* follows from *Elements* I.28, but *HI* is a parallel to *AB* through *G* but it does not necessarily follow that it is perpendicular to *EG*.

[367] *Elements* I, def. 23.

[368] Proclus, *Commentary* 176 Friedline, 138 Morrow's translation.

[self-evident] and should be established by demonstration." When he comes to prove the Parallels Postulate himself, Alfonso uses superposition three times (II.3.2, II.3.4, II.3.5). He undoubtedly wants to emphasize that this method is perfectly legitimate, and one can say that his use of superposition illustrates the visual intuitive meaning of equality, especially the congruence of angles. In none of the three contexts, however, superposition is essential. In all three cases the proof could be completed using simple absolute geometry. Alfonso's proof, thus, does not support the claim that he made in Part I.

The scheme of Alfonso's attempted proof was rather simple. He starts from a construction of "completing" a triangle into a quadrilateral whose opposite sides and angles are equal (Fig. 1).[369] In Fig. 2 he applies the construction to a right-angled triangle and tries to prove that the resulting quadrilateral is a rectangle by examining all the possible configurations that can result from his construction. He fails, however, to accomplish this task. Did he delude himself that it is just a matter of some more hard work to complete the "missing details"? Or did he recognize the inherent difficulty and realized that he cannot complete his task? These questions cannot be answered with any certainty. In Figs. 3 and 4, relying on Fig. 2, he proves Thabit's theorem, and in Figs. 5 and 6 the parallels postulate (which is equivalent to both Figs. 2 and 4).

[II.3.0. Foreword]

I present the demonstrative deduction [הקש מופתי] of this [i.e., the parallels] postulate that I worked out which, in my opinion, is clear of any doubt. I say that it is well known to anyone who read Euclid that he had used this postulate only after 28 propositions (according to al-Hajjāj's version). Since we prove this postulate relying on these previous [28] propositions, our proof [באור] is valid, not circular. Here I prove it [מבארה] using six alternative figures.

Terms

תבנית, See Introduction Section 7.

Commentary

The geometry of *Elements* I.1–28 is independent of the parallels postulate (Postulate 5), and is often referred to as "absolute geometry." Euclid uses Postulate 5 for the first time in Proposition I.29,[370] paving the way to the proof that the sum of angels

[369] Namely a parallelogram. See commentary on II.3.1.

[370] The fact that Euclid does not use postulate 5 until Proposition I.29 was noted by Proclus (*Commentary* 364 Freidline, 284 Morrow). Al-Nayrizi ascribed this comment to Simplicius (Lo Bello 158).

in a triangle is 180° (I.32) and to the theory of parallelograms (I.33–46). In absolute geometry weaker statements about the angles in a triangle can be proved (I.16 and I.17).

[II.3.1. The First Figure]

When a side of a rectilinear triangle is bisected at a point, and a line is drawn from this point to the opposite angle and extended outside the triangle until the part of it that falls outside the triangle equals the part of it that falls inside the triangle, and two lines are drawn from its vertex to the two remaining angles of the triangle, then these two lines together with the triangle form a quadrilateral, so that every two opposite angles and every two opposite sides equal one another.

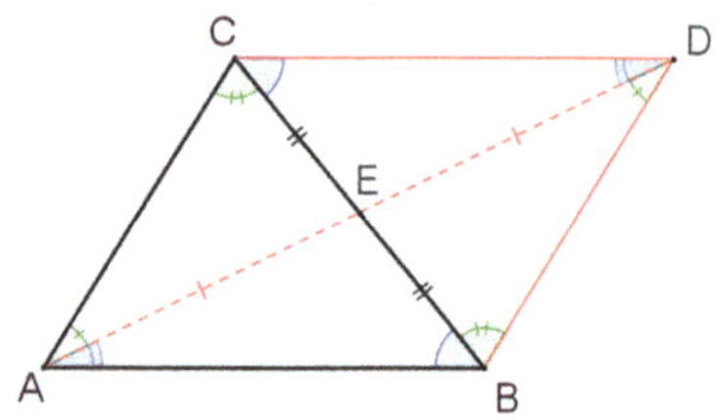

How? We consider a rectilinear triangle *ABC*, bisect *BC* at a point *E*, join line *AE* [107b], and extend it outside the triangle until line *ED* equals line *AE*. We join the lines *DB DC*. I say that line *BD* equals line *AC* opposite to it, and that line *CD* equals line *AB* opposite to it, and that angle *BAC* equals angle *BDC* opposite to it.

Its demonstration: As the two lines *AD BC* intersect, the opposite angles are equal to one another. Therefore the two lines *EB ED* in the triangle *EBD* are equal to the lines *EC EA* in the triangle AEC, respectively: *EB* to *EC*, *ED* to *EA*, and angle *BED* equal to its opposite angle *AEC*. Therefore *BD* equals *AC* and angle EDB equals angle *EAC* and angle *EBD* equals angle *ECA*. It is also proved that line *CD* equals line *AB* opposite to it and angle *EDC* equals angle *EAB*, and that angle *ECD* equals angle *EBA*. Since it was proved that angle *EBD* equals angle *ECA* and that angle *EBA* equals angle *ECD*, then the whole angle *ABD* equal to the whole angle *ACD* opposite to it. Similarly, it is proved that angle *BAC* equals angle *BDC*. Q.E.D.

Commentary

Alfonso starts with the basic construction of **finding the fourth point**: a given triangle is completed into a quadrilateral whose opposite sides are equal. This is easily achieved in absolute geometry: Given triangle *ABC*, *BC* is bisected. $BE = EC$, *AE* extended so that $AE = ED$.

$$\sphericalangle BED = \sphericalangle AEC \ (Elements \text{ I.15})$$
$$\Delta DEB \cong \Delta AEC \ (Elements \text{ I.4}) \rightarrow \sphericalangle EDB = \sphericalangle EAC, \sphericalangle EBD = \sphericalangle ECA$$
$$\Delta DEC \cong \Delta AEB \ (Elements \text{ I.4}) \rightarrow \sphericalangle EDC = \sphericalangle EAB, \sphericalangle EBA = \sphericalangle ECD$$
$$\rightarrow \text{By addition } \sphericalangle BDC = \sphericalangle BAC, \sphericalangle ABD = \sphericalangle ACD$$

The two pairs of opposite angles are equal and so, according to *Elements* I.27 the quadrilateral *ABCD* is a parallelogram. Alfonso, however, does not say this. In the next figure he will apply his construction to a right-angled triangle and try to prove that the resulting quadrilateral is a rectangle.

[II.3.2. The Second Figure]

Introductory note: textual problems

From this point the text becomes brief and difficult. In his dissertation Avinoam Baraness attempts to explain this figure without assuming discontinuity in the text.[371] In the interpretation suggested here it is assumed that the text was copied from a draft with corrections and insertions. Alfonso distinguishes three cases I, II, III, but deals with them in a confused order: II, III (divided into four subcases a–d), I, and III again (briefly).[372]

When the hypotenuse of the right angle in a right-angled triangle is bisected at a point, and a line is drawn from this point to the right angle and extended outside the triangle until the part of it that falls outside the triangle equals the part of it that falls inside the triangle, and two lines are drawn from its vertex to the remaining two angles of the triangle, then these two lines together with the triangle form a quadrilateral with four right angles.

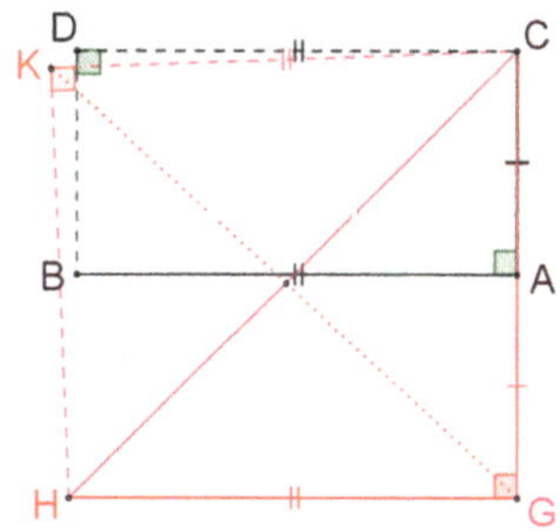

How? We consider a rectilinear triangle *ABC* [108a] with a right angle *A*, and bisect line *BC* at point E and join the straight line *AE* and extend it outside [the triangle] until *ED* equals *EA*, and join the straight lines *DB* *DC*. I say that each of the angles *ABD ACD* is a right angle.

Its demonstration [construction]: We extend line *AC* in the direction of *A* until *AG* equals *AC*, and erect from point *G* a perpendicular *GH* on line

[371] Baraness (2018).

[372] Part of the passage that claims to belong to case II seems actually to deal with case III.

CAG equal to each of the equal lines *AB CD*, as in the construction[373] of the 11th figure.[374] We join *HC* and construct on the triangle *GHC* with right angle *G* a quadrilateral *GHKC* as we did with triangle *ABC*. Then the point *K* either falls on *D* or not. We extend *CD* indefinitely as Euclid postulated,[375] and I explain his words. We also extend *BD* indefinitely in the direction of *D*. If point *K* does not fall on point *D* it should fall either **[I]** on another point on the line *CD*, or **[II]** on another point on the line *BD*, or **[III]** not on one of the two lines *BD DC*. There is no other alternative.

Commentary

Alfonso's applies the "finding of the fourth point" construction (II.3.1 above) twice. His initial construction is not identical to Ibn al-Haytham's, but it is possible that in using the basic construction twice and working with a diagram of two adjacent quadrilaterals, he adopts the idea of Ibn al-Haytham.[376]

Construction I: completing the right-angled triangle *ABC* into a quadrilateral *ABCD*. $\sphericalangle BDC = \sphericalangle BAC = 90°$.

Construction II: extending *CA* so that $CA = AG$, erecting on *G* a perpendicular (to *CG*) $GH = AB = CD$, and then completing the triangle *CGH* into a quadrilateral *CGHK*. According to the construction $CK = GH = AB = CD$, and therefore *CDK* is an isosceles triangle. Also, according to the construction $\sphericalangle CDB = \sphericalangle CKH = 90°$.

Alfonso proceeds in two steps. The first step is to show that the fourth point *K* of the second construction and the fourth point *D* of the first construction must coincide. The second step is to prove that all the four angles of *ABCD* are right angles. The difficult step is the first. Alfonso endeavors to show that $D = K$ by indirect proof. He examines three different cases: [I] $K \neq D$ on *CD*, [II] $K \neq D$ on *BD*, [III] *K* is neither on *CD* nor on *BD*.

[The Examination of the Three Cases]

[II] If it [point K] falls on another point on the line BD, as in the first diagram, then the line HK falls on the point K and on point BD.[377] In this case, the triangle CKD is isosceles, CD CK are equal, and each of the angles CDK CKD is a right angle, because CKH is a right angle being equal to the angle

[373] See comment on the term מעשה in Section I.2.5 above.

[374] Euclid, *Elements* I.11.

[375] Book I Postulate 2.

[376] Ibn al-Haytham applied twice his technique of drawing two perpendiculars on a line and then dropping a perpendicular from a point on one of them to the other.

[377] Presumably a point on line *BD*.

G opposite to it [in CKHG], and BDC is right angle being equal to BAC opposite to it [in ABCD] as I have shown. And it is impossible that there be two right angles in one triangle [CKD] according to [Euclid's] Proposition I.13.[378]

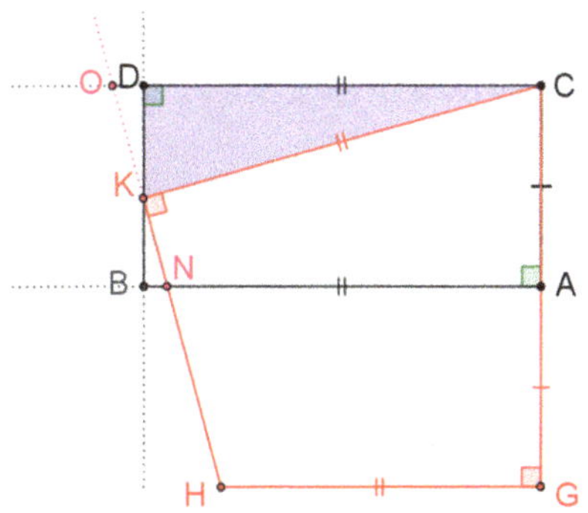

Case II.a

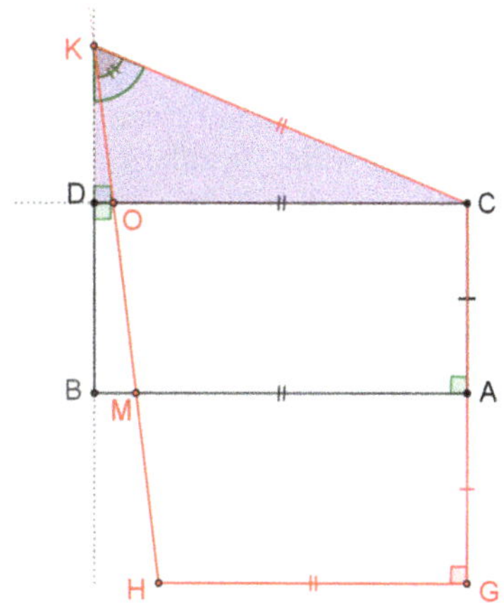

Case II.b

Commentary

$\sphericalangle CDB = \sphericalangle CKH = 90^\circ$ (according to the construction). ΔCDK is an isosceles triangle with two right angles which is impossible even in absolute geometry (*Elements* I.17). The proof rules out all points on the line *BD* (extended indefinitely in both directions) except point *D*.

The following passage seems to be corrupt. We translated it literally without trying to correct it.[379]

[III] Or line *HK* does not fall on a point [108b] *B*, but outside. Let it be HMKO. In this case either **[III.a]** is equal and then the right angle *CDK* equals the angle *CKD*, which is greater than the right angle *CKO*, and this is impossible; or **[III.b]** it is beyond *D* and right angle *CDK* would be equal to angle *CKD* which is less than the right angle *CKH*[380] and this is also impossible. Or line *HK* falls below point *B*. Let it be line *HNKO*. This can be either **[III.c]** to the side of *B* and then the right angle CDK equal to the angle *CKD* which is smaller than the right angle *CKO* and this is impossible. Or **[III.d]** it be beyond *D* and then the right angle *CDK* be equal to the angle *CKD*, which is greater than the right angle *CDK* and this is impossible.

[378] Euclid, *Elements* I.17. See introduction about proposition numbering.

[379] For another interpretation see Baraness (2018).

[380] In the text *CHK*. Accepting Gluskina's correction.

Commentary

In case III two possibilities should be examined in order to complete the proof: either line *HK* can pass to the right of *B* or to the left of *B*. *O* is probably the intersection of *HK* with *BD* and *M* or *N* the intersection with *DC*. *K* can in principle fall in each of the four quarters of the plane defined by the axes *BD* and *DC* (indefinitely extended).

The information in the manuscript is not sufficient to identify the subcases: the definition of III.a is not given, III.b and III.d are defined as "beyond *D*." and III.c as "to the side of *B*." In subcases III.a–d *CDK* is referred to as a right angle, but the case that *CDK* is a right angle is covered in case II.[381]

[I] And if point *K* falls on another point on line *CD* as in the second diagram,[382] then *CD* would equal *CK*, the smaller to the greater [assuming that *K* does not coincide with *D*] and this is impossible.

Terms

צורה The word appears in *SMA* four times, in three of which (twice here and in III.15) it seems to be used to refer to a diagram.[383]

Commentary

This case is simple. *CD* = *KD* according to the original construction. This is impossible if *K* falls on *CD* unless *K* coincides with *D*.

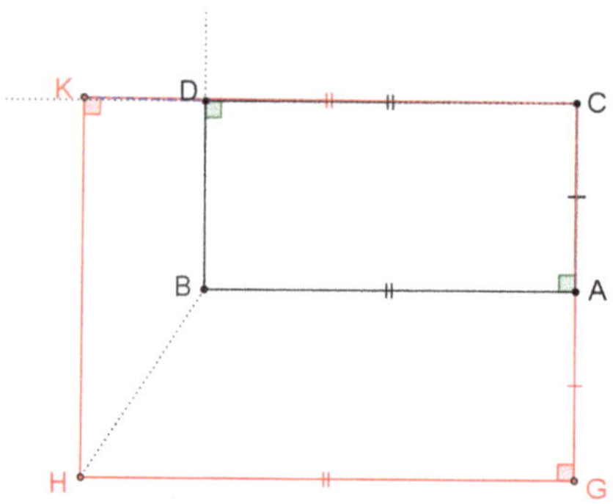

Case I.a

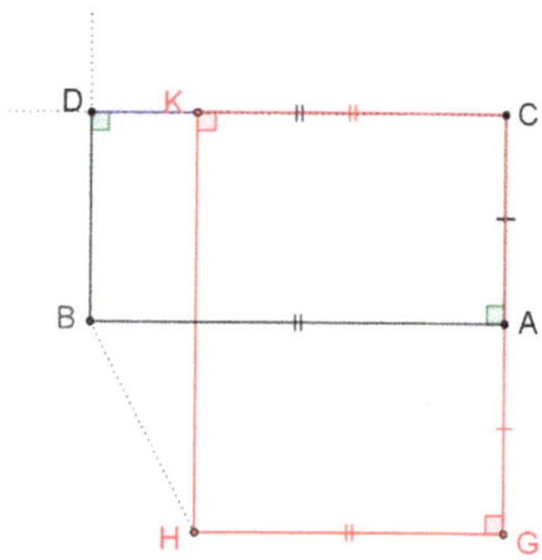

Case I.b

[381] A possible understanding is that a–d are subcases of II. This interpretation is discussed in Avinoam Baraness (2018).

[382] Probably referring to a diagram in the original manuscript.

[383] Cf. Klatzkin, III, pp. 238–9. In *Sefer Yessod 'Olam* (by Alfonso's Castilian contemporary Iṣḥaq Israeli) the word is used to refer to a proposition.

[III′] And if point *K* elsewhere, not on lines *BD CD*, and we join line *DK*, then the angle *CDK* which is greater than a right angle equals angle *CKD*, which is less than right angle. […][384] equal to *CKD* which is greater than right angle, and all this is impossible.

Commentary

This case may correspond to four diagrams, according to the four quarters of the plane defined by axes *DB* and *DC*. The condition that *CDK* is an obtuse angle fits the cases that *K* falls to the left of *DB* (III.a and III.d):

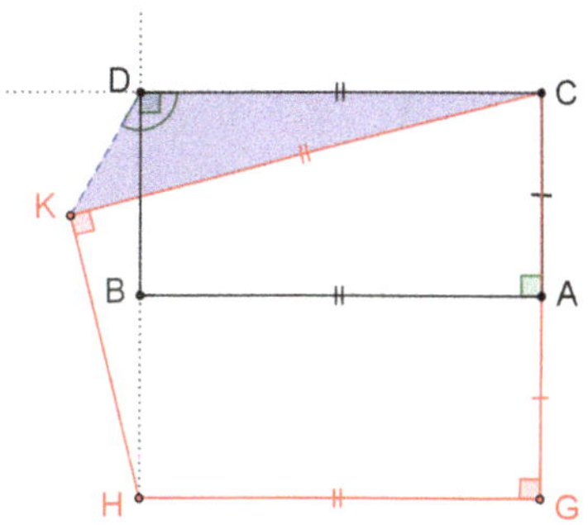

Case III.a

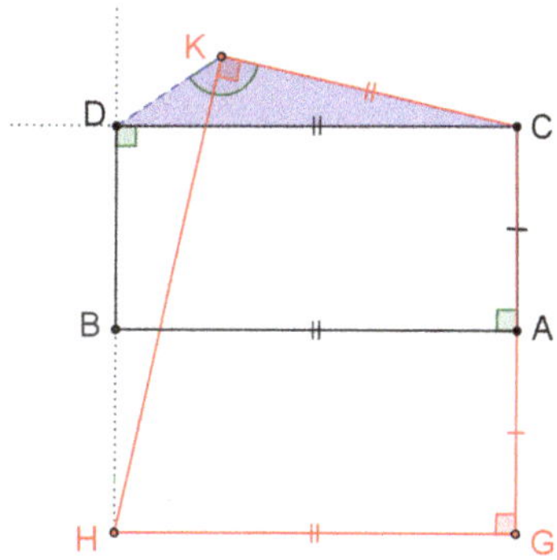

Case III.b

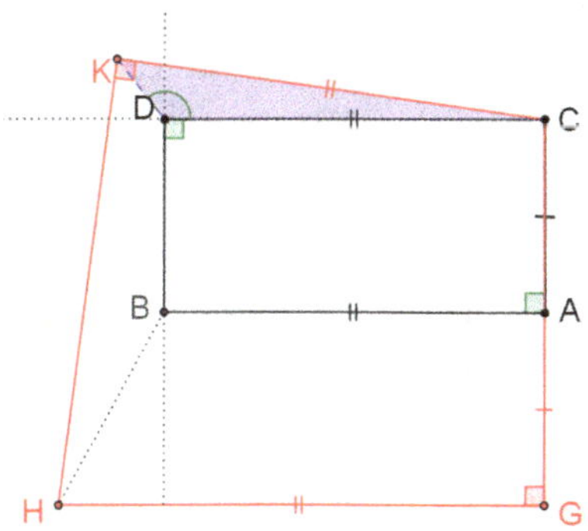

Case III.d

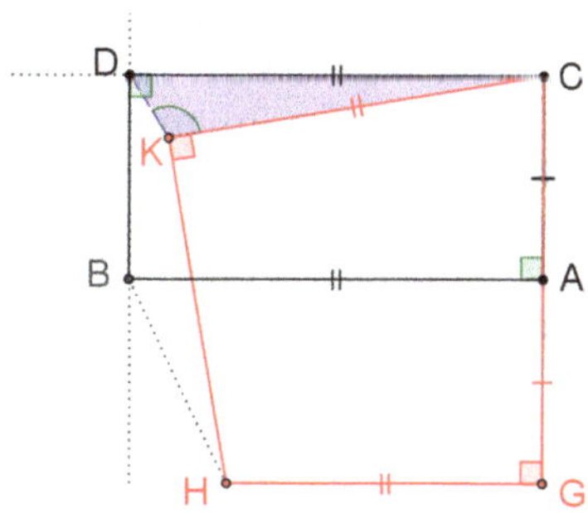

Case III.c

[384] Seems that a sentence is missing here.

The Error in Alfonso's Proof

Cases I and II are simple and do not require subdivision. The division of case III into subcases is confused. The error in the proof is in ignoring the case that ∢*ABH* is acute, *K* falls inside quadrilateral *ABDC* and *HK* intersects *BD* between *B* and *D* (i.e., intersects *BD* before intersecting *CD*). In this case ∢*CKD* and ∢*CDK* can both be acute and no contradiction follows.

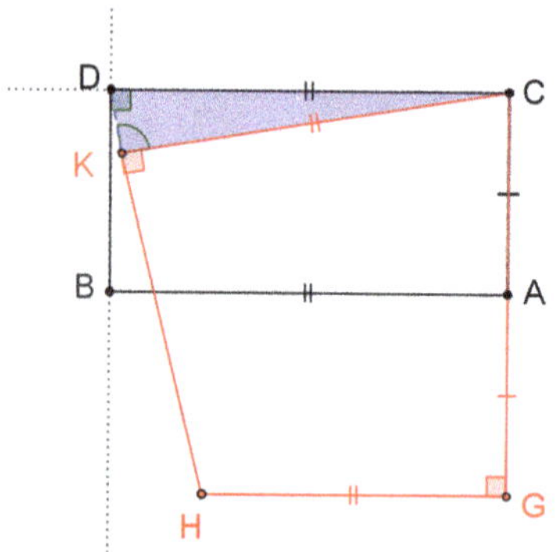

[Completing the Argument Using Simple Superposition]

Therefore the point *K* necessarily falls on *D* and *CK* is superposed on *CD* equal to it, and the right angle *CKH* on the right angle *CDB*, and line *HK* is superposed on line *BD* and they are on the same straight line. And the two angles *ABH ABD* are equal to two right angles, according to the 12th proposition of the first book,[385] and since angle ABD equals angle *ACD* opposite to it, and angle *GHD* equals *ACD* opposite to it, then angle *GHD* equals angle *ABD*.

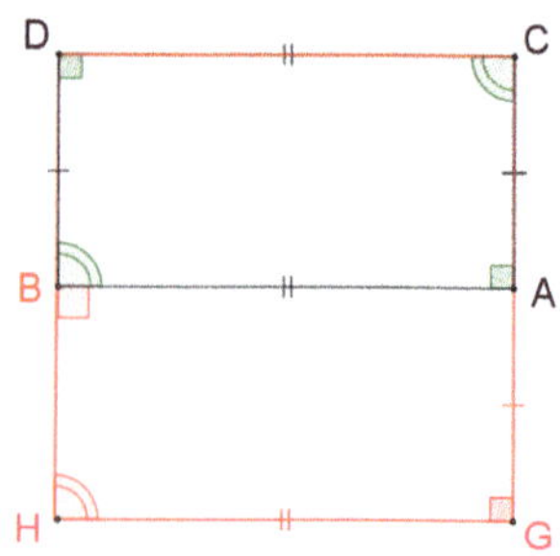

Therefore, we superpose area *GHBA* on area *ABCD*,[386] so that point *A* of *GHBA* falls on point *C* of *ABCD* [109a],[387] [1] and point *G* of area *GHBA* falls on point *A* of area *ABCD*, because lines *AG AC* are equal, [2] and the right angle *HGA* is superposed on right angle *BAC* and line *GH* is superposed on line *AB* which is equal to it, [3] and angle *GHB* is superposed on angle *ABD* which is equal to it, and line *HB* falls on

[385] Presumably the 13th proposition according to Heiberg. See introduction about Alfonso's numbering of the proposition in Euclid's *Elements*.

line *BD* because the angles *GHB ABD* are equal,[388] [4] and point *B* falls on point *D* because line *HB* is equal to *BD*, and line *HB* is superposed on *BD* because angle *GHB* equals angle *ABD*,[389] Then angle *ABH* is superposed on the right angle *BDC* and is a right angle, so the remaining angle *ABD* is a right angle, and the angle *ACD* opposite to it is also a right angle, and the two angles *BAC BDC* were assumed to be right. Therefore, the area *ABCD* has four right angles. Q.E.D.

Commentary

In the first step of II.3.2, Alfonso did not manage to establish that *K* coincides with *D*, but he pursues the argument as if he did. The straightforward way to show that *ABCD* and *ABGH* are congruent is by establishing twice the congruence of triangles:

$$AC = AG = BD \ (\text{construction and Fig. } 1)$$
$$HK = GC \ (\text{construction and Fig. } 1), \text{ but } K \text{ falls on } D \rightarrow HD = GC.$$
$$GC \text{ is twice } AC. \text{ Using subtraction } HB = BD = AC$$
$$\Delta ABD \cong \Delta GHB \ (\textit{Elements } \text{I.4}) \rightarrow DA = BG.$$
$$\Delta ADC \cong \Delta GBA \ (\textit{Elements } \text{I.8}) \rightarrow \sphericalangle ACD = \sphericalangle GAB = 90^\circ.$$

Alfonso, however, prefers (here and in II.3.4 and II.3.5) to use Euclidean superposition, which he accepts as a perfectly standard method in geometry.[390] The text is rather confused and requires correction. Gluskina introduced a few corrections. The following interpretation, offered here as a suggestion by Avinoam Baraness, introduces a few other corrections (notably substituting ה for א and ב for ג in steps 3–4).

Three sides of the two quadrilaterlas are equal respectively: *CA* = *AG*, *AB* = *GH*, *BD* = *BH*. The two enclosed angles at *A*, *G* and at *B*, *H*,[391] are also equal respectively.

[386] The correct order, here and in the sequel is *ADBC*.

[387] Redundant text in the manuscript: then point *B* of area *ABCD*.

[388] In the text "right."

[389] In sentences 3–4 the letter א was corrected to ה and the letter ג was corrected to ב. The corrections are listed in the notes to the Hebrew text.

[390] On the distinction between Euclidean and mental superposition see introductory note to Section I.2.

[391] *A*, *B* in quadrilaterals *ABDC*; *G*, *H* in quadrilateral *AGBH* respectively.

Alfonso "constructs" the superposition by following the vertices and sides of the two quadrilaterals in a clockwise direction:

[0] *A* is placed on *C* (points)
[1] *GA AC* are equal (sides), so *G* falls on *A* (points)
[2] ∢*CAB* = ∢*AGH* (angles). The intuitive meaning of equality of angles is that when placed on one another the respective sides coincide, so *GH AB* coincide (sides)
[3] ∢*GHB* = ∢*ABD* (angles), so the sides of the angles *HB BD* coincide (the verb used is נפל) and consequently lines *HB BD* of the respective quadrilaterals coincide (the verb used is רכב) (sides)
[4] *HB BD* coincide (sides), so *B* falls on *D* (points)

Consequently,
[5] *BA DC* coincide, and ∢*HBA* = ∢*BDC* = 90°.

Therefore the adjacent *ABD* is a right angle, and so is *ACD* which is equal to it (as was shown in Fig. 1).

The advantage of Baraness' interpretation is that it offers a more elegant and systematic reconstruction that illustrates how Alfonso *visualized* Euclidean superposition. Also in II.3.4 and below the text is better and the superposition is also carried out in the same manner.

[II.3.3. The Third Figure]

The two acute angles in a right-angled triangle together are equal to one right angle.

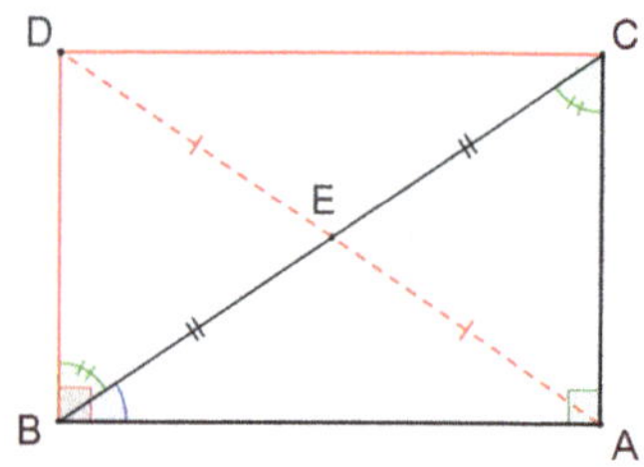

How? We consider a rectilinear triangle *ABC* with a right angle *A*. I say that the two remaining angles *B C* together are equal to one right angle.

Its demonstration: We bisect the hypotenuse *BC* on point *E*, join *AE*, extend it in a straight line until *ED* [109b] equals *AE*, and join *DB DC*. It follows from the second figure that *ABD* is a right angle, and from the first figure that angle *DBC* equals angle *ACB* alternate to it. Let us join to them angle *ABC*. Then the two angles *ABC ACB* are equal together to the right angle *ABD*. Q.E.D.

Commentary

The theorem that the sum of angles in a triangle is 180° (*Elements* I.32) is not a part of absolute geometry. Alfonso tried to prove it for the special case of a right-angled triangle. Given a right-angled triangle *ABC* (∢*BAC* = 90°). The two remaining angles are acute (*Elements* I.16). Alfonso uses once again the technique of finding the fourth point (as in Figs. 1 and 2 above) and completes the triangle into a quadrilateral *ABDC*.

$$\sphericalangle ABD = 90^\circ \ (\text{Fig. } 2).\ \sphericalangle DBC = \sphericalangle ACB \ (\text{Fig. } 1).$$
$$\sphericalangle ACB + \sphericalangle BAC = \sphericalangle DBC + \sphericalangle ABC = 90^\circ.$$

[II.3.4. The Fourth Figure [Thabit's Theorem]]

When two perpendiculars on a straight line equal to one another are erected and their edges joined by a straight line, then the joining line would be equal to the straight line opposite to it and forms with the two perpendiculars two right angles.

How? We erect two perpendiculars *AB CD* equal to one another on a given line *AC* and we join line *BD*. I say that *BD* is equal to *AC* and the angles *B* and *D* are right angles.

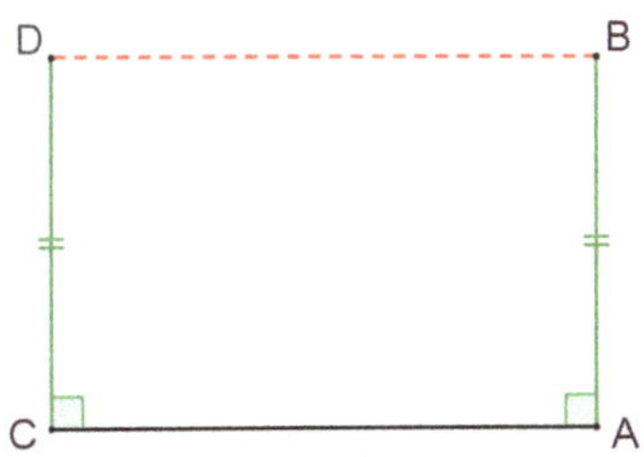

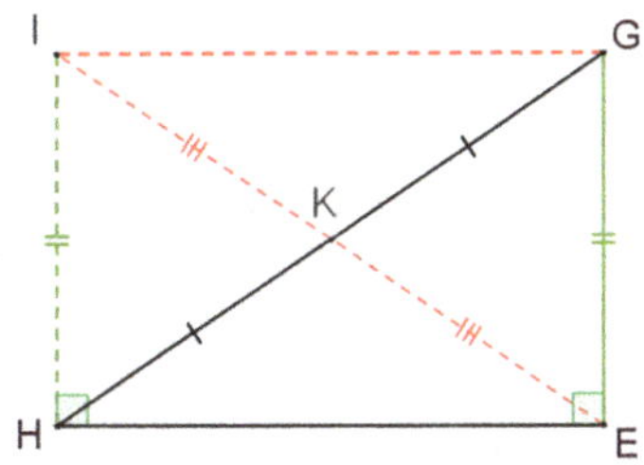

Its demonstration: We consider *EH* equal to *AC* and construct on it a perpendicular *EG*, equal to each of the perpendiculars *AB CD*. We join *GH* and bisect it at point *K*. We join *EK* and extend it outside so that *KI* equals *KE*, and join *IG IH*. It follows from the second figure that *IG* equals *EH* opposite to it, which is equal to line *AC*, and that each of the angles *EGI HIG* is a right angle.[392] Therefore when we superpose area *EHIG* on area *ACDB*, so that point *H* falls on point *C* and point *E* on point *A*, as lines *AC EH* are equal, then the right angle *GEH* is superposed on the right angle *BAC*, and point *G* is superposed on point *B*, and [similarly angle EHI is superposed on and *ACD* and] point *I* on point *D*, and line *GI* (which is equal to *AC*[393]) on line *BD*. Therefore, *BD* is equal to *AC*, and the right angle *EGI* is superposed on angle ABD and be a right angle [too]. Also right angle *HIG* is superposed on angle *CDB*,[394] which should [thus] be a right angle. Q.E.D.

392 The first part of the sentence follows from the first figure, the second part from the second figure.

393 In the text *AD*. Gluskina's correction.

394 In the text *CBD*. Gluskina's correction.

Mathematical Commentary

Alfonso's fourth figure is parallel to the fifth in Thabit Ibn Qurra's "Proofs of Euclid's Postulate 5,"[395] but the construction and proof are somewhat different.[396] He constructs a right-angled triangle on a basis equal to *AC* and using the method of finding the fourth point completes it into a quadrilateral *EHIG*. Following the first and second figures *EHIG* is a rectangle, and superposing it on *ACDB* leads to the conclusion that $\sphericalangle ABD = \sphericalangle CDB = 90°$.

The description of the superposition is similar to that described in II.3.2, and as in II.3.2, this figure can also be proved using simple geometry without the use of superposition:

$$\begin{aligned}
&\Delta ABC \cong \Delta EGH \ (\textit{Elements}\ \text{I.4}) \\
&\rightarrow \sphericalangle EHG = \sphericalangle ACB \\
&\rightarrow \sphericalangle GHI = \sphericalangle BCD,\ CB = HG, HI = CD \\
&\rightarrow \Delta GHI \cong \Delta BCD \ (\textit{Elements}\ \text{I.4}) \\
&\rightarrow BD = GI, \sphericalangle CDB = \sphericalangle HIG = 90^{\circ}. \\
&\text{Similarly, } \sphericalangle EGI = \sphericalangle ABD.
\end{aligned}$$

Alfonso could also complete triangle *ABC* to a parallelogram (as in Fig. 1) and show that the fourth point must coincide with *D*, but he preferred to use superposition.

The quadrilateral constructed here—by erecting two equal perpendicular on the endpoints of the base—is known as the Khayyam-Saccheri quadrilateral.[397] In this rectangle the "summit angles" are necessarily equal, but the assertion that they are right angles is equivalent to the parallels postulate.

[II.3.5. The Fifth Figure]

When the hypotenuse of the right angle in a rectilinear [right-angled] triangle is bisected at a point, and a perpendicular is drawn from this point to one of the sides enclosing the right angle, then this perpendicular bisects the side on which it falls.

How? We consider a triangle *ABC* with right angle *BAC*, and bisect hypotenuse *BC* at point *D*, and draw from *D* a perpendicular *DE* on *AC*. I say that line *AE* equals line *EC*.

---

[395] Sabra (1968, 26–8); Jaouiche 159–160.

[396] In Thābit's construction the use of letter is different: the given line is *AB*, the perpendiculars *AC* and *BD* from a point *E* on *AC* a perpendicular is drawn that meets *BD*. See Jaouice (1986, 157–8), Sabra (1968, 25).

[397] After the Persian mathematician Omar Khayyam (1048–1131) and the Italian mathematician Giovanni Girolamo Saccheri (1667–1733).

Its demonstration: Since *BCA* is an acute angle and BAC is a right angle, the perpendicular *ED* would not fall outside the triangle neither in the direction of *C* nor of *A*, so that there would not be two right angles or more than two right angles in one triangle. We cut from line *AB* a line *AG* equal to *ED* and join *DG*. From the fourth figure it follows that GD equals *AE* and each of the angles *EDG AGD* is a right angle. *EDC* is an acute angle and together with angle *EDB* they amount to two right angles. Therefore, angle *EDC* together with *GDB* amount to one right angle. Therefore, line *DE* falls inside the triangle *ABC*. And since the two angles *EDC GDB* together amount to one right angle, and the two angles *GDB GBD* in the triangle *GBD* together equal one right angle (according to the third figure), then angle *GBD* equals angle *EDC*. It can also be established that angle *GDB* equals angle *ECB*.

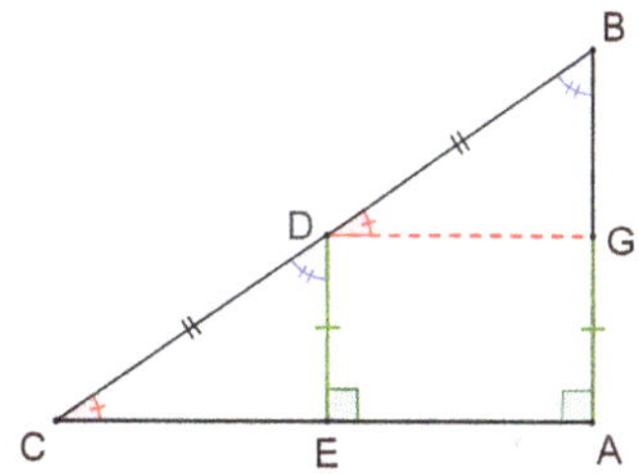

Therefore, if we superpose triangle *GBD* on triangle *EDC* so that point *D* of line *BD* falls on *C* from line *DC*, and point B on point *D*, and since lines *BD DC* are equal and angle *GBD* is superposed on angle *EDC* equal to it, and angle *GDB* is superposed on angle *ECD* equal to it, therefore point *G* is superposed on point *E* and line *GD* on line *EC*. And it has already been proved that line *GD* equals line *AE*. Therefore, *AE* equals *EC*. Q.E.D.

Mathematical Commentary

The fifth figure is a lemma that will be used in the sixth figure. Point *E* must fall inside the triangle (otherwise either in ΔDCE or in ΔDEA the sum of angles would be greater than 180°).

$$AG = ED \ (\text{construction}).$$
$$GD = AE, \sphericalangle AGD = \sphericalangle EDG = 90^{\circ} \ (\text{fourth figure}).$$
$$\sphericalangle EDC + \sphericalangle EDB = 180^{\circ} \ (\textit{Elements} \ \text{I.13}).$$
$$\text{Subtracting angle } EDG \text{ we get } \sphericalangle EDC + \sphericalangle GDB = 90^{\circ}.$$
$$\sphericalangle GDB + \sphericalangle CGBD = 90^{\circ} \ (\text{third figure}).$$
$$\text{Therefore } \sphericalangle EDC = \sphericalangle GBD. \text{ Similarly } \sphericalangle ECD = \sphericalangle GDB = 90^{\circ}.$$

At this stage, Alfonso uses superposition. The argument by superposition seems to assume *Elements* I.26: the hypotenuses are equal and the two acute angles are equal, therefore point *G* must be superposed on point *E*. As in the two previous cases he could have proceeded by using simple geometry:

$$\Delta GBD \cong \Delta EDC \ (\textit{Elements} \ \text{I.26}) \ GD = EC. \text{ Therefore } AE = EC.$$

[II.3.6. The Sixth Figure]

When a straight line falls on two straight lines and forms with them on one side angles less than two right angles, if the two lines be extended in this direction indefinitely – they would necessarily meet.

How? We consider a straight line *AC* falling on the two straight lines *AB CD*, forming with them two angles *ACD CAB* that together amount to less than two right angles. I say that if the two lines *AB CD* be extended in the direction of *B* and D they would meet.

Its demonstration: One of the two angles *ACD CAB* must be acute. We draw it – angle *ACD*. We draw from *A* a perpendicular *AE* to line *CD*. The perpendicular *AE* should fall in the direction of *D*, since otherwise it would have formed with *AC* a triangle whose two angles together amount to more than two right angles, which is impossible. And according to what was said and proved in the third figure the two angles *EAC ECA* equal together to one right angle, and angle *GAE*[398] is acute.

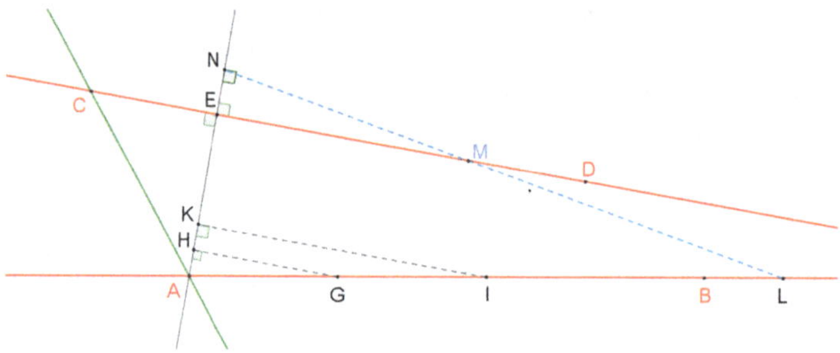

We mark a point *G* somewhere on the line AB, and we draw a perpendicular GH to AE.[399] It would fall in the direction of *E* because the angle GAE is acute. We cut *GI* from *AB*, equal to AG and draw from *I* a perpendicular *IK* to *AE*. Since the triangle *AIK* has a right angle *K*, and since from G)which bisects the hypotenuse *AI*) a perpendicular *GH* was drawn to *AK* that according to the fifth figure bisects it, then *AH* equals *HK*. We further cut *IL* from *AB* equal to *AI* and draw from *L* a perpendicular *LM* to *AE*. For the same considerations that we used above, line *KM* equals line *KA*. Therefore, *AK* is twice *AH* and *AM* is twice *AK*.

---

[398] In the text *GAH*.

[399] In the text *AB*. Gluskina's correction.

When we repeat this procedure many times we will eventually obtain a line [111a] greater than *AE*.[400] Let it be *AM*. This is according to one of the first principles in this science, which Euclid mentioned in a premise to the fifth book, namely that magnitudes between which there is a ratio are of the same genus, and when one of them is multiplied many times it will [eventually] become greater than the other.[401]

If *AB* and *CD* never meet, then line *LM* should have intersected line *CD*, because [according to the construction] the two points *L M* are on the two sides of *CD*. Let it intersect it at *N* if possible. Therefore, *E* and *M* in the triangle *ENM* are right angles, and this is impossible. Therefore, the lines *AB CD* necessarily meet. Q.E.D.

Mathematical Commentary

The sixth figure draws on the seventh proposition of Thabit Ibn Qurra's "Proofs of Euclid's Postulate 5."[402] The lettering in the diagram is a little different. The argument is similar:

$\sphericalangle ACD + \sphericalangle CAB < 180°$. Assume $\sphericalangle ACD < 90°$.
$AE \perp CD$ (*E* falls between *C* and *D*).[403]
$\sphericalangle EAC + \sphericalangle ECA = 90°$ (third figure). Also, $\sphericalangle GAC + \sphericalangle ECA < 180°$.
Subtracting a right angle we get that $\sphericalangle GAE < 90°$.
G is an arbitrary point on *AB*.
$GH \perp AE$ (since $\sphericalangle GAE < 90°$, *H* falls between *A* and *E*).
On line *AB* draw $GI = AG$ and a perpendicular $IK \perp AE$.
ΔAIK is right angled ($K = 90°$) and $HK = HA$ (fifth figure).

The process of duplicating the segment on *AB* and drawing a perpendicular to *AE* is continued until $AN > AE$ (*Elements* V definitions 3–4). *LM* must intersect *CD* because according to the construction *L* and *M* are on the two sides of *CD*. If *LM* intersects *CD* at point *N*, then *MEN* would be a triangle with two right angles (*MEN* and *CEA* are vertical angles). So necessarily *AB* intersects *CD*.

[II.3.7. Concluding Remark]

And the six figures that I mentioned can be included anywhere from the end of Proposition 16 to the end of Proposition 26 of the first book of Euclid's *Principles of Geometry*, because they are proved in their way, using no other concept of motion that may raise doubt.

[400] In the text *AK*. Gluskina's correction.

[401] *Elements* V definition 4.

[402] See note to II.3.4 above.

[403] Assuming that *D* is far enough.

Terms

עקרי חכמת השעור Alfonso's refers to Euclid's *Elements* as עקרי חכמת השעור, namely *The Principles of Geometry*. This title is not common. Moshe Ibn Tibbon and Ya'aqov ben Makhir refer to it as ספר השרשים or ספר היסודות.

Commentary

Alfonso suggests to place them after Proposition I.16 and before I.27. The lower limit raises again the question whether the numbering of the propositions in the version of Euclid's *Elements* that Alfonso used was identical to that of Heiberg's version that we use. We cannot give a decisive answer. In the proof of the second figure Alfonso uses Proposition I.17 (according to Heiberg) to which he refers as 13. The parallels postulate is actually used for the first time only in I.29. Alfonso himself mentions in the introduction to Sectoion I.3 that Euclid used this postulate "only after 28 propositions (according to al-Hajjāj's version)." Defining the upper limit before I.27 is, however, reasonable. The division of the first book of the *Elements* into two parts—Propositions 1–26 (the theory of triangles) and 27–47 (the theory of parallelograms)—goes back to Proclus. The notion of parallelism is used only in the second part. It is thus reasonable to open the second part with the "proof" of the parallels postulate.

[II.4. Motion in Geometry: Essential Problems (Continued)]

Introductory remark

In II.1.7 Alfonso "promised" to prove the impossibility of actual infinity. This task is carried out in II.4.1 after the supposed proof of the parallels postulate in II.3. Alfonso relies on this postulate. The examples examined in II.1 showed that motion and actual infinity cannot go together. Alfonso's choice was to exclude infinity.

[II.4.1. Dismissing the Possibility of Actually Infinite Magnitudes]

Now I will prove what I intended: that an infinite line in act is impossible.

Its demonstration. If this were possible we consider a straight line *AB* infinite in act. We erect on it a perpendicular *AC* and extend from a point *C* a line *CD* parallel to *AB*, as in the construction of Proposition 31.[404] Therefore angle *ACD* is a right angle like *BAC*. Since lines *AB CD* are parallel, the distance between them is always equal *AC* [=] *BD*, as Euclid proved in Proposition 33. We rotate line *CA* about the center *C* until it is superposed

[404] *Elements* I.31. As to the term מעשה see I.2.5 above.

on line *CD*. We assume a point moving on line *AB* [111b] which is always on the straight extension of the moving line *CA*. When the moving line *CA* is extended in a straight line, it would always meet *AB* before it [*CA*] is superposed on *CD*, as I explained [when I discussed] the straight line falling on two straight lines.[405]

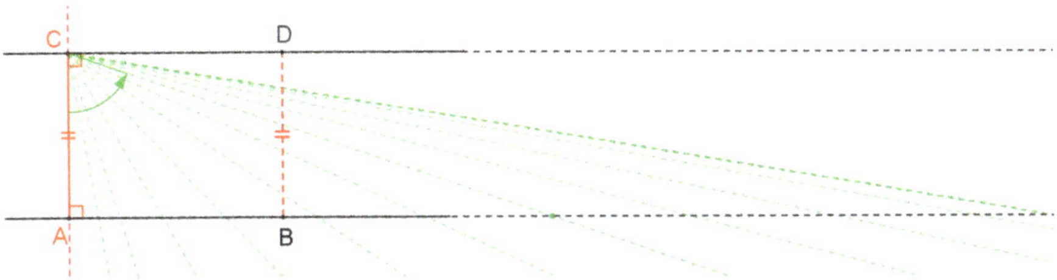

And since the moved point was [first] on the line *AB* and then on *CD*, and since two parallel lines never meet but, being parallel, the distance between them remains constant, it follows that it [the moving point] cannot cross from line *AB* to line *CD* unless instantaneously, because every motion [over distance] must be temporal, as the moved body is at half the distance before it is at the end, viz. it [the motion] measures the "before" and "after."[406] If this were so, then at the beginning of this time[407] the moving point would have been on the line *AB* but would not be there afterwards during all this time, and at the end of this time it would be on *CD*. If line *AB* were infinite in act, the moving point would never have separated from it until the end of this time. But it has been shown that it separated from it at the beginning of this time, and this is impossible. Therefore, *AB* cannot be infinite in act. Q.E.D.

Commentary

Given infinite line *AB*. Construction: erect $AC \perp AB$, draw from *C* a parallel $CD \parallel AB$ (*Elements* I.31, which relies on Postulate 5). $\sphericalangle ACD = \sphericalangle BAC = 90°$. Line *AC* rotates on *C*, its extension falls on *AB* until the minute that *CA* coincides with *CD*. If *AB* were infinite the extension of *AC* could not have separated from it.

[II.4.2. Theological Remark]

From here one can (if one wishes) come near to the understanding of the existence of God, as he sets actual specific limits on distances after having been potentially unlimited. He said to Job: "Where were you when I laid the

[405] II.1.1 above.

[406] Cf. Aristotle *Phys.* IV.11219a30–b2.

[407] That is, the time interval in which the point crosses over the distance between the two parallel lines.

foundations of the earth? Tell me if you understand. Who marked off its dimensions? Surely you know! Who stretched a measuring line across it,"[408] and he said, "This far you may come and no farther; here is where your proud waves halt,"[409] and other similar verses.

Commentary

Using metaphorical language and biblical verses, Alfonso lets us feel that this elliptic passage is significant, but provides no explanation. The meeting points of Alfonso's geometry and his theology are the concepts of body, dimensionality and measurement. The words "potentially unlimited" may refer to the Aristotelian first matter,[410] but more likely refer to the subtle body (*cuerpo glorificado*) which in itself has no quantity or dimensions,[411] and is endowed with dimensions by God.[412] Identifying the subtle body with the body of Christ, Alfonso finds means to interpret the sacrament of the Eucharist. His notion of *hitlabbeshut* stands for the Christian incarnation,[413] as well as for a more general concept of divine immanence.[414] In this vein the notions of שעור ומדידה (evaluation and measurement), which play a major role in Alfonso's geometry, play a role also in his theology. The Jewish concept of knowing God through knowing his dimensions (developed in Sefer *Shi'ur Qoma*) adds the geometrical numerologiical flavor to the Christian concept of incarnation.[415]

[II.5. Motion in Geometry: Accidental Problems (Continued from II.2)]

In the first part (I.2.1) Alfonso mentioned two examples of non-isometric superposition. Now he comes back to study in depth the second example, based on the paradox of Aristotle's wheel (*Rota Aristotelis*). The paradox was first presented in the Pseudo-Arisotelian *Mechanica*: "There is a question why a large circle traces out a

[408] Job 38, 4–5, NIV translation.

[409] Job 38, 11, NIV translation.

[410] In medieval Aristotelian philosophy the origin of dimensionality is discussed in terms of "prime matter" and "corporeal form." See Wolfson (1929, 101, 579–590); Elior (2012).

[411] Gershenzon (1984, 140–155). See above, Introduction, Section 3 argument iv.

[412] Sadik (2007, 136–7; 2011, Chap. 1.6.2.2).

[413] Gershenzon (1984, 137).

[414] Gershenzon (1984, 137, and Chap. V); Sadik (2011, Chap. 1.6.2).

[415] The style of the following passage from *Minhat Qena'ot* illustrates this point:

> וכתב ר׳ אברהם בן עזרא (שמות ל״ג, כ״א) שעל כן היה השם הנכבד בעל שלש אותיות להורות על גוף בעל שלשה מרחקים... וזה מן הסודות הנפלאים אצלו, כלומר היות הגוף מונח אל האלוה, וכי מזה הצד נקרא שכינה ,שוכן עד ומקום ומעון ויושב הכרובים ואלים וצבאות. ואמר שזאת היתה הכוונה בשיעור קומה, כמו שאמר ר׳ ישמעאל: כל היודע שיעורו של יוצר בראשית מובטח לו שהוא בן העולם הבא. אמר ר׳ עקיבא, ואני ערב בדבר. ונראה מזה כי ר׳ עקיבא שנכנס בשלום ויצא בשלום ידע את הענין כמו ר׳ ישמעאל, רצוני לומר שיתלבש האלוה בגופות העולם וישוער בשעורים בדרך מקרה.

Quoted from Baer (1959, 201); See also Sadik (2011, 76).

path equal to that of a smaller circle, when they are placed about the same center, but when they are rolled separately, their paths are to one another in the proportion of their dimensions."[416] This text was partly translated into Arabic, and it is not known whether this question was included.[417] The paradox was presented in yet another ancient text, Hero of Alexandria's *Mechanica*,[418] which was translated into Arabic and is extant in Arabic.[419] In the Latin west the paradox came up again and attracted much attention in the sixteenth and seventeenth centuries.[420] Drabkin in a classical article on this paradox does not mention any evidence from the fifteen centuries that separate Cardano from Hero. He remarks that "with few exceptions the work of the commentators prior to Galileo adds little to what the author of the *Mechanics* has given us."[421]

These "exceptions" deserve more attention. Cardano testifies that Ibn Rushd mentioned the wheel paradox,[422] but we were unable to locate the passage. Rashed has shown that the paradox was known and used in the school of al-Ṭūsī: an anonymous mathematician criticizes the suggestion to measure the circle by rolling it on a tangent and used Aristotle's wheel paradox as an argument against this suggestion.[423] Presumably Alfonso heard about this critical argument this way or another.

Alfonso's interest in this paradox could have been explained by his interest in atomism. The renewed interest in the paradox in the seventeenth century was associated with the rise of atomism.[424] Atomism was at the center of interest also in Alfonso's time and this subject occupied Alfonso's thought intensively. The controversy between the indivisibilists and the divisibilists that took place at the universities of Oxford and Paris in the first half of the fourteenth century was, undoubtedly, a major intellectual issue.[425] There is tacit, but suggestive evidence in *SMA* that Alfonso knew about this controversy and was interested in it.[426] We are not aware of

[416] Pseudo Aristotle, *Mechanica*, Chap. 24, Forster's translation. On the history of this problem see Drabkin (1950); Palermino (2001).

[417] Abattouy (2001, 103).

[418] Chapter I. 7.

[419] See editors' introduction to Hero's *Opera*, vol. II, p. xviii.

[420] Drabkin (1950); Palermino (2001).

[421] Drabkin (1950, 172).

[422] Abattouy (2001, 103, notes 18, 19). See Cardano 221–2 (240–1 in online version).

[423] Rashed (2015, 470) has no doubt that Hero's text was available to the anonymous mathematician who criticized al-Shīrāzī. On Alfonso's acquaintance with al-Ṭūsī's schools see Introduction Section 6.

[424] Palermino (2001, 383).

[425] Since Murdoch's (1957) dissertation and subsequent articles much has been written on this controversy. For a recent brief presentation of the subject see Grellard and Aurélien (2009, 1–14); Murdoch (2009, 16). Although the controversy had strong underlying theological motives (Sylla 1997), a major part of the arguments adduced against indivisibilism by Aristotle, al-Ghazali, Duns Scotus, Bradwardine and by other divisiblists were mathematical to different degrees. We are grateful to Liran Gordon for sharing with us his digitalized version of Duns Scotus' *Commentaria Oxeniensia*, vol. II.

[426] See Introduction, Section 6.

a scholastic discussion of Aristotle's wheel paradox in the fourteenth century, but Alfonso's discussion was "in context" in the cultural environment of his time, where similar arguments were used against the assumption that the continuum can be measured by indivisible parts.

[II.5.1. Measuring the Circumference of a Circle by Rolling]

Among the secondary accidents of geometry that prevent qua secondary causes [112a] the imagining of motion from being a first principle of geometry is the claim of some scholars that they found a straight line equal to the circumference of the circle in the following way: if one rolls a circle, for instance, the base of a cylinder, on a straight line until the point on the circumference of the circle that touches the straight line at the beginning touches it a second time, then the straight line between the two points of contact equals the circumference, because each point on the circumference of the circle touches a corresponding point of the straight line on which it rolls. This is like a superposition of the whole arc on the whole straight line which implies their equality. We say that if the line were composed of points, this would have been a true knowledge, but since this is not the case, it is a false knowledge.

Commentary

The scholars to whom Alfonso alludes could, perhaps, be al-Ṭūsī and his student al-Shīrāzī who offered such practical method to measure curved lines.[427] The suggested method is to roll a cylinder on the floor and to measure the distance between two touching points of one point on the floor.

[II.5.2. The Continuity of the Rolling Motion]

In order to prove that it is false we consider two circles *AB* and *DE* on a center *C* in one plane and one [rigid] body, and we extend from points *A D* two straight lines *AHL DM* touching the circumferences of the circles. While the circumference of circle *AB* rolls on the line *AL* until point *A* on the circumference of the circle falls on point L of line AL, point D on the circumference of circle *DE* would fall on a point *M* of line *DM*, the center *C* would fall on the center *K*, and line *CDA* on line KML. We pick a point *B* somewhere on the circumference of the circle AB, and join the straight line *CEB*. I say that when point *B* meets line *AL* on point *H*, then point *E* would meet line *DM* on point *I*.

[427] Rashed (2015, 469).

Its demonstration. We assume that *G* is the center of the circles when point *B* is on point *H*. We join line *HG*. Then line *CB* falls on line *GH* and line *CE* [112b] on line *GI*. Since *AH* touches the two circles *AB NH*, the two angles *CAH GHA* are right angles and lines *AC GH* parallel and equal. Also the two lines *CD GI* are parallel and equal to each other, and the straight line that joins the two points D I parallel to each of the lines *CG AH*. Therefore angle *IDC* is a right angle. But angle MDC is a right angle. Therefore point *I* is on line *DM*. But point E falls on point I. Therefore point *E* meets line *DM*. It can similarly be established that when point *E* on line *DM* meets point *I*, then point *B* on line *AL* meets point *H*. Therefore when an arc[428] of one of the two circles rolls on a straight line, a similar arc of the other circle would roll on a straight line, equal to the other straight line.

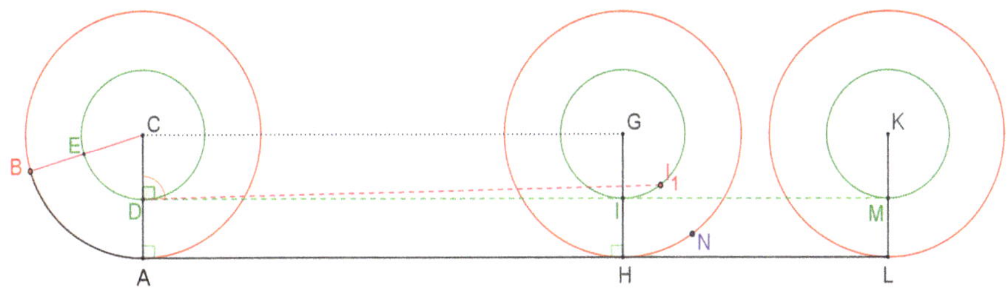

Historical Commentary

Alfonso may have followed here the anonymous Arab mathematician who criticizes al-Shīrāzī and defended the Aristotelian position. The anonymous Aristotelian adduced several arguments against measuring the circle by a line, including the argument of Aristotle's wheel. The argument is that the movement of rolling (حركة دحرجة) does not necessarily imply the equality of the circular line and the rectilinear line defined by the rolling.[429] Alfonso's argument is not a verbatim quotation of that of the anonymous mathematician and the lettering of the diagrams are not identical, but the two arguments are basically the same.

Terms

נשליך Alfonso uses this term for picking at random a point on a line or on a circumference of a circle. We translate it "pick" although literally the meaning is "throw."

[428] We ignored the word שוה (equal) that appears in the text here.

[429] Rashed (2015, 578–582).

Mathematical Commentary

When a rigid cylinder rolls on the floor, the rigidity implies that a radius of the basis (or of any section parallel to the basis) moves as a straight line.[430] *CDA* is the radius perpendicular to the two tangent lines *AL* and *DM* at the beginning. After completing one revolution *CDA* is superposed on *KML*. The center *C* and the points *A* and *D* (the points on the large and small circles that touch the tangents in the initial position) are moved the same distance in the horizontal direction, i.e., $CK = DM = AL$.

It could be similarly argued that any radius *CEB* at some stage of rolling becomes perpendicular to the **two** tangents.[431] Alfonso shows that when any radius *CB* of the external wheel coincides with radius *GH*, the collinear point *E* on the internal wheel coincides with a point *I*, which is the tangent point of the internal wheel to line *DM* (after one revolution), collinear with points *G* and *H*. He assumes point *I* to be somewhere on the internal wheel (for instance, I_1; see diagram above), and shows that it is necessarily on line *DM* [and therefore on radius *GH*]. Since *AC*, *GH* and *CD*, *GI* are equal and parallel to one another, *CAHG* and *CDIG* are parallelograms, hence $DI \parallel CG$, $DI \parallel AH$, so $ID \perp DC$, and because $MD \perp DC$ as well, point *I* must be on line *DM*—and therefore $DI = AH$.

The opposite direction of the assertion can be shown similarly.

[II.5.3. It Is Impossible to Measure the Circumference of the Circle by Rolling (the Wheel Paradox)]

Apparently, if two circles rotate on a center *C* that is fixed to one place, then the whole circumference of the circle *DE* would pass on the point *D* in the line *DM*. But if the center *C* moves to point *K* while the two circles complete one rotation, then from the two motions [the rotation and the translation] another motion is generated, superposed of them, which is called rolling. And the circumference of the circle AB rolls on the line *AL* and the circumference of the circle *DE* on the line *DM*. Now, if the line *AL* were equal to the circumference of circle *AB* measuring their touching each other while rolling, and line *DM* were equal to the circumference of circle *DE* measuring[432] their touching each other while rolling, then the circumferences of the two circles should be equal and this is impossible. Similarly, if center *C* moves to point *G* while the two circles complete one rotation, then the whole circumference of circle AB would roll on line AH [113a] and the whole circumference of circle DE on the line DI.

---

[430] When speaking of the rolling circle Alfonso adds "for instance the base of a cylinder," thus guaranteeing the rigidity of the motion. According to Aristotle one or two-dimensional entities have no independent existence and can be said to be moved only qua sections of three-dimensional bodies (e.g., *Meta.* III.5, 1002a15–b10). The Arabic anonymous mathematician adduces another argument against al-Shīrāzī by comparing the rolling of a cylinder on the floor to that of rolling a cone with the same base on the floor (Rashed 2015, 588–592).

[431] For a dynamic illustration see: http://mathworld.wolfram.com/AristotlesWheelParadox.html.

[432] Word not certain. Corrected after the first appearance in this sentence.

If rolling could evaluate the straight line by equating it to the circumference of the circle rolling on it, as those people said, then the smaller line *AH* would equal the greater line *AL* which is impossible.

Commentary

The paradox of the rotating wheel shows that the circumference of the circle cannot be measured by means of rolling the circle along a tangent. Alfonso elegantly explains that the motion of rolling is continuous, being composed of two continuous components: rotation and translation. If a straight line could be measured by the rolling of a circle on it, then the equality of lines *AL* and *DM* would imply the equality of the circumferences the circles *AB* and *DE*, and actually the equality of all circles and all straight lines.

[II.5.4. Rolling Defined a One-to-One Correspondence Between a Circle and a Line Segment, but Such a Correspondence Does Not Imply Equality]

In order to complete the imagining of rolling as it is, without misunderstanding:

[First direction] we pick a point *H* on line *AL*.

[Existence] I say that a point of the circumference of circle *AB* would meet it [*H*] when rolling as described above.

Its demonstration: We draw a perpendicular *GH*.[433] Since area *AG* is a parallelogram, *GH* equals *CA*. And since the center *C* moves on the straight line *CG*, it would meet point *G*, and the line that joins *G* with the meeting point of line *AL* and the circumference of the circle would be perpendicular to line *AL* and equal to line *CA*. It would thus be line *GH*. Therefore, point *H* would be on the circumference of the circle.

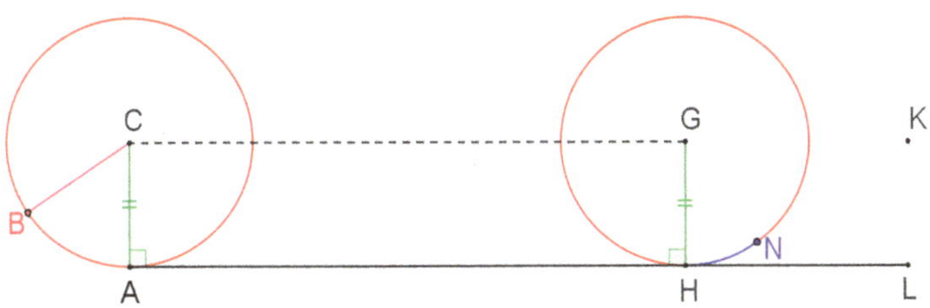

[Uniqueness] I say that no other point of the circumference of the circle would be translated on the point *H* on line *AL* when rolling as described above. Had another point *N* been translated on it [then this motion would not be rolling]. For if the arc *HN* rotated on point *H* which is on line *AL*, the center would not move away from point *G*. Therefore this [motion] would not be rolling.

[433] *G* is on the line *CK* as in the previous section.

[Second direction] We pick a point *B* on the circumference of the circle *AB*.

[Existence] I say that a point of the line *AL* would meet it in the course of this rolling. And this is because if center *C* were fixed and the circle rotates on it, then point *B* would be translated on point *A*. [But if center *C* is not fixed but moves along CK] we consider a line *HA* touching the circle on point *A* and moving with it on line AL while the center C moves on line *CK*. When

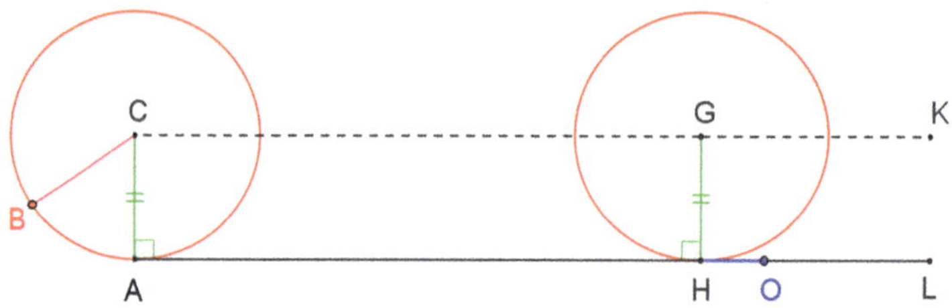

point *B* arrives at point A on line AL, then point *A* on the line *HA* would arrive at the point *H*. Thus, point *B* would arrive at the point *H* on the line *AL*.

[Uniqueness] I say that no other point of the line *AL* touches point *B*. For if it were possible, another point [113b] *O* would touch it. Then the motion of point *B* on line *OH* would consist of translation without rotation, and this would not be rolling.

Therefore every point of the straight line on which the [circle] rolls touches a corresponding point of the circumference of the circle and every point on the circumference of the circle touches in the [course of] rolling a corresponding point of the straight line. This is what misled those people to think that the straight line equals the circumference of the circle. They thought of the superposition of an arc on a straight line as a superposition of all the arc's points on all the line's points, each on its corresponding point. They have not realized that this is the case, no matter whether the straight line is long or short. The circumference of the circle can never be equal to the straight line on which it rolls, unless the linear motion of the center [i.e., the linear component] and the concomitant circular motion of the point on the circumference [i.e., the circular component] were equal, where each of these two motions in itself is continuous and uniform. We know that if a straight line is not equal to the circumference of a circle, then the uniform motion on the straight line will not be equal to the concomitant uniform motion on the circumference of the circle, because distance, time, and motion are proportional to one another and are measured by one another as is established in the *Physics*.[434] These scholars [mentioned in the beginning of paragraph 5] take the consequence of their deduction as a premise to it, and this is logically false.

[434] *Phys.* VI.1231b18, VI.4.235a15.

Historical Commentary

Not long before Alfonso, Henry of Harclay (died 1317) uses the rolling example:

> [Between] a straight [line] and a circle, or a spherical body and a circular [body], there is nothing in common but a point, and contact is always at something common. But the sphere is continuously moved along the plane. Therefore, it touches continuously, point after point, and through its motion describes a line. Therefore, point after point will be continuously in that line, and consequently, such a line is composed of points.[435]

At any moment the circle touches the tangent at a point. This thought experiment supports Harcklay's indivisibilist belief. The circle and the tangent are made of points and the rolling motion defines what we call today a one-to-one correspondence between them. Harclay's arguments raised many objections among scholastic scholars.[436] Alfonso's analysis of rolling is much more sophisticated than Harclay's.

Mathematical Commentary

Rolling is a continuous motion that is composed of two continuous motions: straight and circular. Such a motion defines a one-to-one correspondence between a circle and line segment, but not necessarily an equality between them. To any point *H* on the tangent corresponds a unique point *B* on the circle and vice versa. Two tacit assumptions about continuity underlie the argument. Assumption [i]. A designated point (e.g., the center *C*) would necessarily meet another designated point on a line (e.g., *G*) when moved along line (in the direction of *G*). Assumption [ii]. The rolling motion does not involve slipping or sliding. In other words, the ratio between the two components of the rolling motion—translation and rotation—is always the same. At no point can rotation cease when translation continues or vice versa.

The first part of the argument leads from an arbitrary point *H* on the tangent to the corresponding point *B* on the circumference of the rolling circle that meets it; the second part leads from an arbitrary point *B* on the circumference of the circle to the corresponding point *H* on the tangent. In both parts Alfonso proves that the corresponding point exists and is unique.

First part: a point *H* on the tangent *AL* is given. *G* is constructed by drawing a perpendicular from *H*. Assumption [i] implies that when the center *C* of the circle moves along the line *CK* at a certain time it would coincide with *G*. The rest of the argument follows from simple geometrical considerations. Uniqueness follows directly from assumption [ii].

Second part: a point *B* on the circumference of the circle is selected. In the case of rotation without translation (the center *C* is fixed) the point *B* will touch the tangent *AL* at the same point that point *A* did. In the case of rolling (i.e., combined

---

[435] Zupko (1993, 160).

[436] Harclay was criticized by several of his scholastic contemporaries, most notably by Thomas Bradwardine in his *De continuo* written between 1328 to 1335. See Murdoch (1957, 200; 1987, 114–5).

translation and rotation) the points of tangency moves. At the beginning *A* is the tangency point of the circle and the line *AL*, and point *B* is somewhere else on the circumference. While the circle rolls *B* approaches the tangent *AL*, but the center *C* and the tangency point move a distance *AE* to the right (in the diagram) along line *CK* and tangent *AL*, respectively. The resulting point of tangency would be *H*. As in the first part, uniqueness follows directly from the assumption [ii].

The conclusion of the proof reflects Alfonso's deep understanding of the example of rolling and its implications, perhaps following the anonymous Arabic commentator on al-Shīrāzī.[437] There is a one-to-one correspondence between the circle and the line, but this does not imply isometry between the two. Since the motion of rolling is a superposition of two motions, the circumference of the circle can be mapped on infinitely many straight lines, and thus one-to-one correspondence cannot be used for measuring.

[II.5.5. It is Impossible to Measure the Circumference of the Circle by Thread]

One should know, and this had not escaped the scholars, that people bind straight thin threads of silk or linen or the like around a solid circle and so they measure its circumference by a straight line. But they [the scholars] wished to know whether the thing itself is so, [not] in the sense perception [114a] of the person who applies approximation, because sense perception is insufficient, as it involves approximation. Every two unequal lines can be cut into very small parts, so that they are perceived as having a rational ratio to each other, although it is possible that there is no rational ratio between them, since there are infinitely many kinds of irrational ratios, as was proved in Euclid's tenth book. One can consider the roots of all numbers either precisely or approximately.

Furthermore, a serious doubt can be raised about [the method of] the thread that is bound around the circle. Let the thread when straight be line AB. It is a body [so] let its width be the straight lines CD EG HF erected on its length in right angles. When the body AB is bent in the direction CEH, then the points C E H and the other points on the same side come nearer to

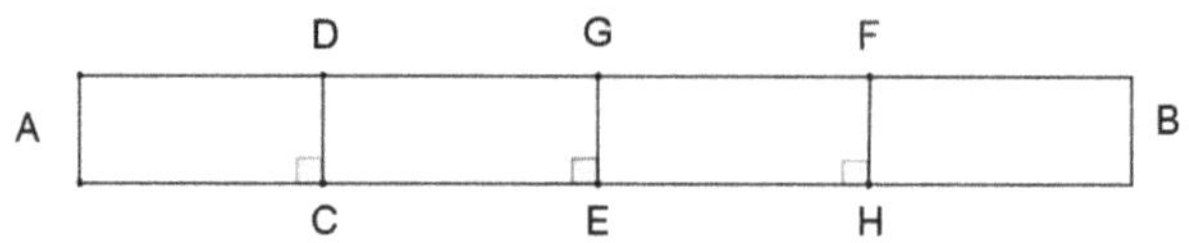

the center and to one another, and points D G F and the other points on the side of the circumference move away from each other. And the lines bent in the direction of the center where points CEH are, are shorter than what they were when they were straight, and the lines bent in the direction of D G F are

[437] Cf. Rashed (2015, 592–3). See also Section II.5.1 above.

longer than what they were when they were straight, since the encompassing is greater than the encompassed. And since these lines do not maintain their specific measure whether straight or bent, assuming equality between them is impossible. The philosophers who contended that the curved line or curved surface can never become straight (as I mentioned above[438]) perhaps said so because of the getting nearer of points CEH to [114b] and the getting away from each other of point D G F. We will talk about it.[439]

Terms

מדובר, אלם Alfonso uses these two terms to denote rational and irrational ratios, respectively.[440] The Greek *rhētos* that is used in the definitions of *Elements* X is translated into Arabic *muntaqah* by Iṣḥaq-Thābit, and into Hebrew מדבר in Moshe Ibn Tibbon.[441] The two words المنطقة, الصم, namely מדובר, אלם appear together in the Arabic commentary on *Elements* X attributed to Pappus,[442] and in later texts.

גדר This term is used in II.2.4 above in the usual sense of definition. Here it is used in the sense of root. This is one of the senses of גדר coined by Bar-Ḥiyya.[443]

Commentary

The idea of measuring the circle by a flexible ruler was ancient. Aristophanes in his satiric manner attributes such an idea to the astronomer Meton.[444] Measuring by thread is mentioned by Ibn al-Haytham,[445] and was most probably used in everyday life.[446] Alfonso offers two arguments against this method of measuring. First, relying on perception, it is insufficient for the geometer. For example, it is impossible to distinguish using perception a rational from an irrational ratio. Second, when bent the length of the inner circumference of the thread is shorter than that of its straight length, and the length of the outer boundary is longer than that the straight length. This illustrates that the method cannot be precise.

The reference to the tenth book of Euclid's *Elements* provides a strong argument for supporting the identification of Alfonso as Abner.[447]

---

438 Referring to the views of Ibn Rushd and Ibn Sīnā, I.1.1 above.

439 Alfonso does not return to this point in the extant part of *SMA*.

440 See Szpiech (2006, 525); Smith (1925/1953, vol. II, 252).

441 Translating the definitions of *Elements* X, Moshe Ibn Tibbon rendered the passive expression *yuntaq bi-hi* as ידובר בו, ידובר בהם. See Paris ms. fol. 83a9–10. Cf. Judeo-Arabic manuscript fol. 94a.10–11.

442 Pappus, Commentary on book X, p. 191, #1, line 2.
See http://www.wilbourhall.org/pdfs/pappus/PappusBookX.pdf.

443 Sarfatti (1968, 82, 128).

444 Heath (1921 I, 220–221).

445 Hooper-Sude 85.

446 Knorr (1986, 44 n. 46; 284 n. 88).

447 See more on this example in the Introduction Section 6, (vii), example [i].

[II.5.6. Concluding Remark]

Also in the first part[448] I mentioned that the general knowledge that was achieved by the intellect on determining the ratio between the circle and the rectilinear area does not provide the geometer with sufficient means to construct this polygon. Moreover, one may argue that the intellect cannot recognize it [this ratio] in actual reality, whereas it can recognize it in the intermediate existence between potentiality and actuality, in the same way that it cannot recognize a line infinitely divided in actual reality, but it can recognize it in the intermediate existence between potentiality and actuality. And it is my main wish and intention in this book to construct this polygon [that is equal to a circle] in actual reality, in the way that geometers deal with areas and magnitudes. But that which a Christian named Campanus, who boasted to have squared the circle, wrote does not meet our criterion of precision,[449] because he relied on Archimedes' approximations of the circumference of the circle as three and a seventh and less than a tenth times its diameter. This is appropriate for *Ḥokhmat ha-Tishboret*, which is practical and considered to be approximate, not for *Ḥokhmat ha-shi'ur*, which is considered to be truly precise.

Terms

חכמת השעור, חכמת התשבורת Alfonso identifies here חכמת השעור as theoretical geometry and חכמת התשבורת as of practical geometry. In I.5.3, however, he associates חכמת התשבורת with שברים, a concept which plays a major role in his theoretical geometry.[450] Bar-Ḥiyya uses חכמת השעור for the science of geometry,[451] and תשבורת to denote area or product.[452]

Commentary

Knowledge that was acquired using the method of exhaustion, should not be sufficient for the geometer. The geometer should seek a constructive method to square the circle. The method should be rigorous and not approximate.

Somewhat contemptuously Alfonso dismisses Campanus' attempt to achieve this goal. Clagett, who published the text *Quadranta circuli*, also found faults with this text and questions its ascription to Campanus.[453] Albert of Saxony distinguished several senses of squaring the circle, and attributed to Campanus a quadrature of the circumference, not of the area.[454]

---

[448] I.2.5 above.

[449] Gluskina identifies him as Johannes Campanus. See Introduction, Section 6.

[450] See comment on terms after I.5.3.

[451] Sarfatti (1968, 69, 91).

[452] Sarfatti (1968, 68, 88, 171).

[453] The text *Quadranta circuli* is published in Clagett (1964, 591–605).

[454] Grant (1974, 172. Col. 1).

III Part III

Introductory note

I. The structure of part III

Part III of *SMA* differs in genre from the first two parts. It is a collection of geometrical theorems, referred to as "properties" (*segulot*).[455] These "properties" are post-Euclidean theorems, and—with the exception of the last property—the presentation is strictly Euclidean. Part III does not follow closely any geometrical compilation known to us. Only one theorem (III.12) explicitly refers to its author (Ptolemy). Several of the properties of Part III are well-known theorems, several are similar to known proposition, some others we could not identify. We will be grateful to anyone, who can throw additional light on the sources of Part III.

Part III originally included 33 properties, but due to a large lacuna Property 1 is truncated, 2–9 and the enunciation of 10 are missing. The extant part includes:

1. Trigonometry. Property 1 is a geometrical presentation of a theorem that can be understood as an introductory lemma to trigonometry. It is possible that at least some of the missing Properties 2–9 were trigonometrically oriented geometrical propositions.
2. Plane geometry. Properties 10–23 are theorems in plane geometry, some of which deal with relations between circles and polygons (often with inscribed or circumscribing triangles and quadrilaterals).
3. Theory of proportions. Properties 24–27 offer an extension of Euclid's theory of proportions (*Elements* V). Properties 24–5 are simple, but deal with the notion of compound ratio that was not defined by Euclid; Properties 26–7 are much more complex and provide an elaboration on Euclid's groundbreaking definition of proportion (Definition V.5).
4. Spherical trigonometry. Property 28 is a geometrical presentation of a theorem imported from spherical trigonometry.
5. Advanced geometry. Properties 29–32 deal with special geometric constructions not restricted to the use of ruler and compass: Properties 29–33 deals with the construction and uses of the conchoid, Property 33 deals with the mechanism of the Ṭūsī couple, and shows how a linear motion can be produced by a superposition of circular ones. The style is not strictly geometrical as in the rest of Part III. It includes a philosophical argument with Aristotle on the issue of *quies media* introduced in *Physics* VIII.

[455] Alfonso's several terms for geometrical propositions are discussed in the Introduction, Section 7.

It is possible that Part III prepares for Parts IV–V, and that Alfonso had in mind a specific use of each of the properties of Part III in the following parts of *SMA*. It is equally possible that Part III is a selection from as a scratchpad in which Alfonso copied, during the years, theorems that interested him from various post-Euclidean sources. Some properties of Part III bear directly on the main theme of *SMA*, namely the relation between straight and curved lines,[456] but we cannot know how Alfonso intended to integrate them into the arguments of Parts IV–V. Whatever the sources and the criteria of selection of the propositions of Part III were, Alfonso adapted everything to his specific style and perhaps added some lemmas of his own.

II. The geometrical "ideology" of Part III: geometrical operations

Most indicative of the "ideology" underlying Part III is the notion of רבוי of two lines, basically meaning half their product (definition [c] below).[457] The term רבוי in this sense does not occur in Parts I–II, and occurs 172 times in Part III. Fairly consistently Alfonso translates the geometrical theorems that he includes in Part III into his distinctive language of *ribui*. By replacing the Greek language of homogeneous ratios and proportions by a language of "expansions" (=half-products)[458] Alfonso gains more flexiblility in geometry without adopting full-fledged algebra. His idea was, most probably, to define a flexible notion of product without abandoning the strict formalism of geometry. *Ribui*, however, is not a product, and we translate it using an English word that is somewhat close to its literal meaning: "expansion."

The language of expansions was presumably meant to replace the Greek language of ratios and proportions.[459] The geometrical basis for this "revolution" could be Euclid's Proposition VI.16: "If four straight lines are proportional, then the rectangle contained by the extremes equals the rectangle contained by the means; and, if the rectangle contained by the extremes equals the rectangle contained by the means, then the four straight lines are proportional."[460] This theorem is the geometrical equivalent of the arithmetical rule of four (i.e., cross multiplication) that was commonly used in numerical computations.[461] Alfonso went beyond *Elements* VI.16, and led the way to a free use of multiplication of lines *in geometry* without providing a construction in which the products are represented by concrete areas (whether rectangles or triangles).[462] It is likely that in adopting this approach he followed the lead of Ptolemy.[463] Alfonso keeps to strict formal geometry, but extends it by introducing "geometrical operations." He uses **two sets of terms**: one for geometrical objects and one for the associated geometrical operations. The terms are summarized in the table below.

---

[456] Property18 (following Archimedes), 23 (following Hippocrates), 33 (following al-Ṭūsī), and 29–32 which deal with asymptotic lines and mechanical constructions.

[457] Definitions [d] and [e] add two variants to this basic concept.

[458] See introductory comment to Part III, Section II.

[459] For example, in Properties III.11–13.

[460] Heath' translation.

[461] Smith II, 488.

[462] See Saito (1985) on the use of invisible diagrams.

[463] See commentary on Property 13 below.

Geometrical objects	משלש נצב הזויות		משלש	שטח (בעל ד' צלעות) נצב זויות	מרובע	מוצק	מושגם
Meaning	right-angled triangle	right-angled-isosceles triangle	triangle	rectangle	square	pyramid	polyhedron
Geometrical operations	רבוי קו ישר בקו ישר[464]	רבוי קו ישר בעצמו	רבוי קו ישר בקו ישר בזוית מונחת[465]	רבוע קו ישר בקו ישר[466]	מרובע קו[467]	רבוי קו ישר בקו ישר נקבץ בקו ישר[468]	רבוי קו ישר בשטח[469]
Meaning	$\frac{1}{2}a \cdot b$	$\frac{1}{2}a^2$	$\frac{1}{2}a \cdot b \cdot \sin(\theta)$	$a \cdot b$	$a \cdot a$	$\frac{1}{3}a \cdot b \cdot c$ (in the case of right pyramid whose base is a right angled triangle)	$a \cdot b \cdot c$ (in the case of boxes)
Symbol	$a \otimes b$	$a^{\textcircled{2}}$	$a \otimes b\ (\theta)$	$a \boxtimes b$	a^2		

[464] This terms appears many times in III.10–23, 31.

[465] III.10.

[466] In I.1.4; I.2.4.

[467] III.17, III.23 and III.29.

[468] Three-dimensional expansion according to Definition [e].

[469] Three-dimensional expansion as used in Property 32.

The great advantage of the geometrical operations is that they can be flexibly applied to lines and areas without the need to actually construct the resulting geometrical objects. This flexibility could have been achieved by Lemmas 1 and 2 listed below. The two lemmas could have either been taken for granted or proved among the missing Properties 2–9:

Lemma 1 The expansion $a \otimes b$ is equal to the area of any triangle whose base is a and whose height is b. This lemma follows immediately from *Elements* I.38 (used in III.14, III.15).

Lemma 2 Distributive Laws for Expansion

[i] $(a + b) \otimes c = a \otimes c + b \otimes c$ (used in III.11, 12, 14, 15)
[ii] $(a - b) \otimes c = a \otimes c - b \otimes c$

These lemmas can be also extended to expansion in a given angle (definition [d] below):

[i′] $(a + b) \otimes c\ (\theta) = a \otimes c\ (\theta) + b \otimes c\ (\theta)$
[ii′] $(a - b) \otimes c\ (\theta) = a \otimes c\ (\theta) - b \otimes c\ (\theta)$ (used in III.10)

Why did Alfonso prefer to define "expansion" of two lines, his basic geometrical operation, as the right-angled triangle enclosed by the two lines, rather than as the rectangle enclosed by the two lines, as in definition [c*] that he quotes? Why did he prefer to define the three-dimensional expansion as the pyramid erected on the plane expansion, rather than as the right prism, as in definition [e*] that he quotes? Perhaps thinking of area in terms of triangles could have reflected Plato's conception of the triangle as the basic component of reality or Archimedes' describing the area of a circle to a triangle.[470] In both cases the triangle is perceived as a basic unit of area. The choice of triangles reflects a geometrical-trigonometrical, but not an explicitly algebraic approach. For the most part, expansion is compared to another expansion and the reader can think of expansion as a product. Still five times expansion is compared to a purely geometrical magnitude, a triangle or a square (Properties 10, 14, 15) or mixed entities (sum of expansion and area in Property 22).

The Third Part, on the Properties of Rectilinear Magnitudes and Areas which are Useful in this Science

[III.0. Preface]

Alfonso said: at the beginning of this chapter I list some commonly accepted terms that I find appropriate for use [but] not in the way that they are used in other books. This is in order [115a] to facilitate understanding and also to provide the needed materials so that no other geometrical book will be needed for the understanding of this book except Euclid's *Elements*. And I say:

[470] *The Measurement of the Circle* Proposition 1, quoted by Alfonso in I.2.4 above.

[a] I call **an arc of a rectilinear angle** [קשת הזוית] the arc of a circle that subtends this angle when it is on the circumference.

[b] I call **a chord of a rectilinear angle** [יתר הזוית] the chord of the arc which subtends the angle. It [the chord] is twice its sine [משען], which is called *jayb alqaus* in Arabic, while its double is called *wathar alqaus*.

[c] And I call **an expansion** [רבוי] **of a straight line by a straight line** the rectilinear right-angled triangle enclosed by the two lines.

[d] And I call **an expansion of a straight line by a straight line at a given angle**″ the rectilinear triangle enclosed by the two straight lines at the given angle.

[e] And I call **an expansion of a straight line by a straight line** at a given angle[471] expanded [נקבץ][472] by a straight line the pyramid[473] whose base is the expansion of the two lines by one another and its height is the third line.

In analogy to naming

[c*] The **product** [רבוע] **of a straight line [by a straight line]** is the figure with four right angles [i.e., rectangle] whose distances are the two lines. And to naming

[e*] The **product of a straight line by a rectilinear area** is the right prism whose base is the rectilinear area and whose height is the straight line.

[f] And we call a **rectified area** [שטח מיושר] the rectilinear area [שטח ישר הקוים], namely that straight lines enclose.

[g] Also we call **rectified solid** [מוגשם מיושר] the solid [מוגשם] that rectified areas enclose.

Commentary

The aim of the preface, as Alfonso announces it, is to define terms. Its placement at the beginning of Part III points out that this part was a separate unit. Some of the concepts were used before being defined, and some were not used in the extant part of *SMA* after being defined. The text of the preface is somewhat defective, and this may indicate that patches have been added to the original manuscript. The preface includes three groups of definitions:

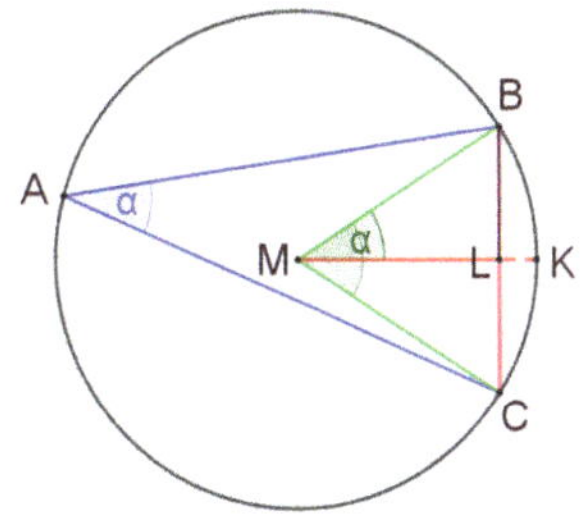

Definitions [a]–[b] introduce the notions of arc and chord *of an angle*. Definition [a] indicates that Alfonso associates the arc and the chord with the inscribed, rather than with the central angle.[474] The term "chord of an angle" was used in Part II before being defined and in Part III after having been

[471] Literally: "the expansion of a straight line by the hypotenuse of a given angle." The word "hypotenuse" seems to be out of place.

[472] See Terms below.

[473] On the meaning of מוצק see I.3.2 comment on terms.

[474] So it is used in Properties 11–13. In Property 1 *DH* and *CG* are half chords of central angles, which equal chords of inscribes angles.

defined.[475] The term "arc of an angle" does not occur in the extant part of *SMA* at all. The simple terms "arc" and "chord," however, are used several times before and after the definitions.

In the diagram the inscribed angle $\sphericalangle BAC = \alpha$ subtends chord [יתר] BC and arc $\widehat{BC}$, and it is twice the sine [משען] BL.

Definitions [c]–[e] are the most relevant, as they introduce the key terms of Part III expansion, expansion at a given angle and three-dimensional expansion. The importance of this term is explained in the introductory note to Part III above. The notion of expansion is very frequent in Part III; that of expansion at a given angle is used only in Property 10[476]; three-dimensional extension is used in Property 32, where it is not clear whether Alfonso indeed thought of expansion in the sense of definition [e]: it is more likely that he thought of it as a box, than as a pyramid.

Definitions [f]–[g] assign new names to the basic geometrical concepts of polygon and polyhedron: מיושר הקוים instead of שטח ישר הקוים (a term that was used 17 times in parts I–II), and מוגשם מיושר[477] instead of מושגם ישר הקוים (that will be used in III.32 below). The two new terms are no where used in the extant part of *SMA* (except for definition [f] in definition [g]).

Terms

רבוע, רבוי, מרובע See Introductory Note to Part III above.

שטח אשר יקיפו בו The first definition of *Elements* II states: "Any rectangular parallelogram is said to be contained by the two straight lines containing the right angle." Moshe Ibn Tibbon translates השטח אשר יקיפו בו. For the product of two line segments Banu Musa use the term تضعيف (i.e., multiplication), whereas al-Ṭūsī uses سطح (i.e., area).[478]

משען This is Alfonso term for *jayb alqaus*, namely sine. Abraham Bar-Ḥiyya used the term מיתר מחוצה (half chord) following the Arabic نصف وتر[479]; Iṣḥaq Israeli, Alfonso's contemporary, used the term בקע. The Provençal translators used other terms. The word משען means something to lean on, something that supports.[480] It is used in III.28 below in the sense of half a chord, not associated directly with a specific angle. The term משען is used also in the astronomical tables by Juan Gil of Burgos who lived about the same time.[481]

נקבץ This term was used by Bar-Ḥiyya for the result of both addition and multiplication.[482] The word נקבץ appears in *SMA* only once in definition [e], where the sentence

[475] II.3.5; III.1, III.20. קשת and יתר without reference to an angle are used many times.

[476] Perhaps Alfonso intended to use it in the following parts that are lost.

[477] See Appendix B.

[478] See Rashed (2012, 25–9).

[479] Sarfatti (1968, 107–8).

[480] We are grateful to Akiva Kadary, who gave a lecture on the Hebrew terms for sine.

[481] Goldstein Chabas (2000, 152); Goldstein (2019, 134).

[482] Sarfatti (1968, 82).

is corrupt, so its meaning is not certain. Presumably it is used for the result of the geometrical operation of the expansion of an area by a line. It is used in the same sense in the Hebrew translation of Nahmias' *The Light of the Lord*, as the equivalent of מג׳תמע.[483] See also comment on the word קבוץ in III.26 below.

מושגם ישר הקוים, מושגם, מוגשם The term מוגשם is the loan translation of the Arabic مجسم and was commonly used by the Provençal translators.[484] The word, however, was not used by Bar-Ḥiyya or by Iṣḥaq Israeli, and it is possible that Alfonso was not familiar with it. In *SMA* מוגשם occurs only once in definition [g], probably a mistake (see Introduction, Section 7, (ii) example [ii]). The word מושגם occurs 13 times, once here in definition [g], once in II.5.5, and 11 times in III.32. In II.5.5 מושגם means solid, namely three-dimensional body in a general sense. In III.32 it is used in the expression מושגם ישר הקוים, namely polyhedron. It is not clear what led Alfonso to translate مجسم as מושגם.

The Properties

III.1 Properties 1, 10–13: Expansions

[1. The First Property]

If two rectilinear triangles have two sides equal [115b] to two sides, respectively, the ratio of one triangle to the other is as the ratio of the chord of the angle enclosed by the two sides in the first triangle to the chord of the angle enclosed by the two sides in the second triangle.

How? We consider two rectilinear triangles *ABC ABD* [such that] the two sides *AB AC* of the first triangle are equal to sides *AB AD* of the second triangle, respectively. We describe a circle *CDE* with center *A* and distance [i.e., radius] *AC AD* which are greater [than *AB*] and equal to one another. We extend *AB* to point *E* on the circumference of the circle, and from *D* we draw a perpendicular *DH* to *AE*, which is the sine of angle *BAD* and its double is its chord. We also draw from *C* a perpendicular *CG* to *AE*, which is the sine of angle *BAC* and its double is ….

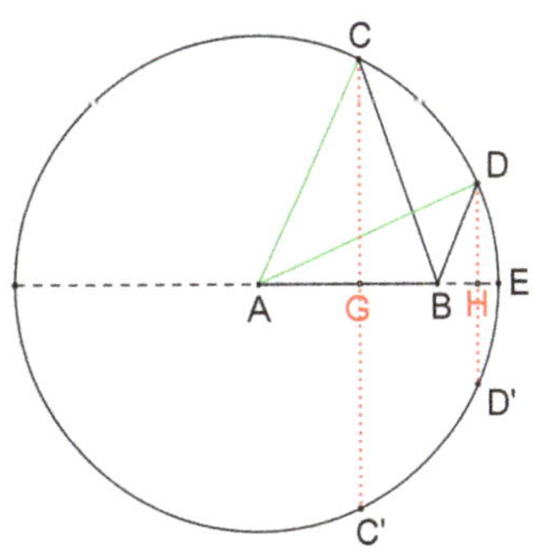

[At this point the text is interrupted and resumed in the annunciation of Property 10. Folio 115b is interrupted in the middle and folios 116 and 117 are missing.]

[483] Naḥmias, A.II.8, Hebrew, p. 196, Arabic, p. 60.

[484] See Sarfatti (1968, 195).

Terms

מרחק See commentary on terms after I.4.1.

כערך Typical of the style of Part III is the comparison of ratios or of expansions using the comparative prefix כ (meaning "as") rather than using the explicit expression "equal to." This style is typically Euclidean. כערך occurs in Part III 54 times, שוה לערך only six times. When expansions are compared כרבוי and שוה לרבוי are used more or less equally frequently.

Mathematical Commentary

The missing proof can be reconstructed. The ratio between the triangles with the common basis *AB* equals the ratio of their heights (*Elements* VI.1) *CG*:*DH* and therefore of twice their heights *CC*′:*DD*′. By definition [b] above *CC*′ and *DD*′ are the chords of the inscribed angles that are half the central angles *CAB* and *DAB* (*Elements* III.20).

III.[10] **Property 10**[485]

[118a] If two opposite sides of a quadrilateral are not parallel and the diagonals intersect at a point, then the expansion of these two sides with one another at the angle on which they meet is as the excess of the triangle formed by the two diagonals with the greater of the two remaining sides over the triangle formed by [the two diagonals] with the smaller of the two remaining sides.

How? We consider a quadrilateral *ABCD* of which *AB*, *CD* are opposite and nonparallel, and when extended will meet at point *G*. We draw the diagonals *AC BD* which intersect at point *E*. I say that the expansion of *AB* and *CD* at angle *G* is as the excess of triangle *AED* over triangle BEC.

The demonstration. The expansion of *AG* by *GD* at the angle *G* is triangle *AGD*, and the expansion of *BG* by *GD* at angle *G* is triangle *BGD*. The remainder – the expansion of *AB* by *GD* at angle *G* – is triangle *ABD*. Similarly, the expansion of *AB* by *GC* at angle *G* is triangle ABC. So the expansion of *AB* by *CD* which is the excess of the expansion of AB by GD over the expansion of *AB* by *GC* is as the excess of triangle *ABD* over triangle *ABC*, which is as the excess of triangle *AED* over triangle *BEG*. Q.E.D.

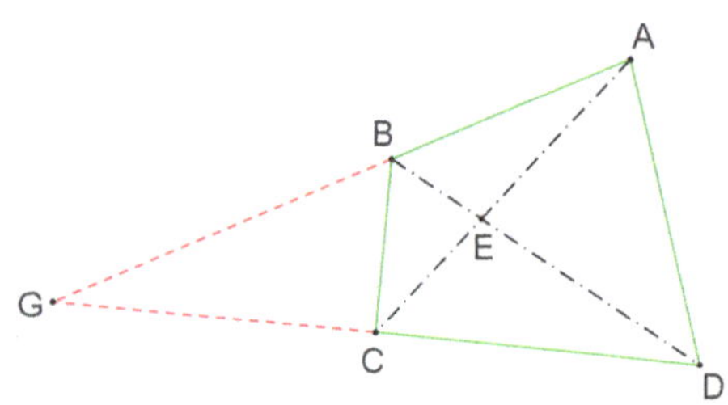

[485] The text is resumed here after a long lacuna. The title "Proposition 10" is missing but it seems that the text of Proposition 10 is complete.

Mathematical Commentary

This proposition is rather sophisticated. As noted above expansions are usually compared to expansion in *SMA* III, but in some cases they are compared to triangles or to other geometrical magnitudes. This property is a good example. The expansion of two lines at a given angle is compared to a geometrically defined area obtained by the subtraction of two triangles. Both magnitudes are not visible in the initial diagram.[486] The theorem could be proved by providing additional constructions specific to this diagram. Alfonso does not provide any such proof but, rather, uses freely the operation of subtraction. This means that he takes for granted the distributive laws (in this case Lemma 2 [ii′],[487] which could have been either taken for granted or proved in the lost part of Book III). Anyway, using them, the proof of Property III.10 becomes simple and, more importantly, the flexibility achieved by using "expansion" as a geometrical operation is illustrated.

ABCD is a quadrilateral; nonparallel opposite sides *AB CD* meet at *G*. Let us denote angle *AGD* by γ.

The expansion $AG \otimes GD\ (\gamma)$ is ΔAGD (III.0, definition [d])

$$BG \otimes GD(\gamma) \text{ is } \Delta BGD$$

By subtracting $(AG - BG) \otimes GD\ (\gamma) = AB \otimes GD\ (\gamma) = \Delta ABD$ (III.0, Lemma 2 [ii′]).

For similar considerations, $AB \otimes GC\ (\gamma) = \Delta ABC$.

$$\begin{aligned}\text{Consequently, } AB \otimes CD(\gamma) &= AB \otimes GD(\gamma) - AB \otimes GC(\gamma) \text{ (Lemma 2 [ii}'\text{])}\\ &= \Delta ABD - \Delta ABC\\ &= \Delta ABD - \Delta ABC.\end{aligned}$$

Q.E.D.

III.[11] Property 11

If two straight lines are drawn from the two extremities of a diameter of a circle, intersecting [each other] and cutting the circumference of the semicircle, then the expansion of the part of the first line from the extremity of

[486] On visible vs. invisible diagrams see Saito (1985).

[487] See introductory note to Part III above.

the diameter to the meeting point of the lines and the part of it which is a chord in the circle, together with the expansion of the second line's part from the [other] extremity of the diameter to the meeting point of the lines, and the part which is a chord in the circle, together they equal to the expansion of the diameter of the circle by itself.

How? We consider *AB* the diameter of circle *ABCD*, and draw lines *AE BE* intersecting at *E* and cutting the circle's circumference at two points *C*, *D*. I say that the expansion of *AC* by AE together with the expansion of *BD* by BE together equal the expansion of *AB* by itself.

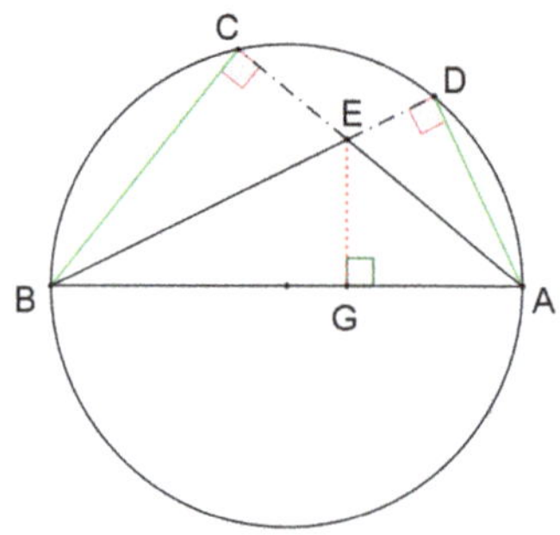

The demonstration. We draw from [point] *E* a perpendicular *EG* on the diameter and join *AD BC*. Then triangles *BEG ABD*[488] are similar and angle *BEG* is equal to angle *BAD* whose chord is *BD*. Therefore, the expansion of *BE* by the chord of angle *BEG* which is *BD* is as the expansion of *BG* and the chord of right angle *G*, which is the diameter *AB*. Similarly, the expansion of *AE* by the chord of angle *AEG* which is *AC* is as the expansion of *AG* by the diameter *AB*. Therefore, the expansion of *AC* by *AE* together with the expansion of *BD* by *BE* equals to the expansion of the diameter *AB* by itself. Q.E.D.

This property is more general than Proposition 47 of book I of Euclid's Elements, asserting that in every right-angled triangle the expansion of hypotenuse by itself is equal to the expansion of the two remaining sides each by itself, together.

Mathematical Commentary

The proofs of Properties 10 and 11 illustrate the flexibility of Alfonso's language of geometrical operations that allows multiplication of any two lines, no matter how they are situated in the diagram.

Since $\sphericalangle BDA = 90°$, $\Delta BEG \sim \Delta BAD$, then $BE{:}BG = BA{:}BD$,

and thus $BE \otimes BD = BG \otimes BA$ (Elements VI.16).

Similarly, $\sphericalangle BCA = 90°$, $\Delta AEG \sim \Delta ABC$, $AE{:}AG = AB{:}AC$,

and thus $AE \otimes AC = AG \otimes AB$.

Using Lemma 2 [i] we get:

$$BE \otimes BD + AE \otimes AC = BG \otimes BA + AG \otimes AB = \left(BG + AG\right) \otimes AB = AB \otimes AB.$$

Alfonso fails to distinguish between the two cases, that in which *AEB* is obtuse (*E* falls inside the circle) and that in which *AEB* is acute (*E* falls outside the circle).

[488] The correct order of the letters should be *BAD*.

A few hints suggest that he had in mind the former case.[489] In the case that *E* falls on the circle the two right-angled triangles coincide and Proposition 11 is reduced to the Pythagorean Theorem (*Elements* I.47). Alfonso himself notes that this proposition can be considered as a generalization of the Pythagorean Theorem. In the *Elements* we find two approaches to such generalizations: In II.12, II.13 Euclid generalizes the theorem for obtuse-angled and for acute-angled triangles. In VI.31 he offers a general theorem for similar figures (not necessarily squares) constructed on the sides of a right-angled triangle.

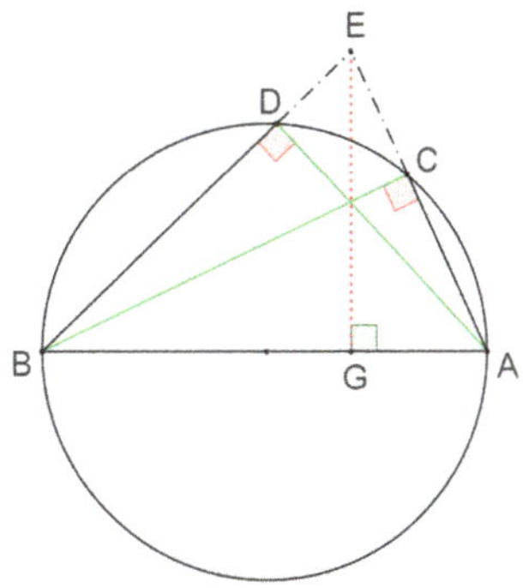

III.[12] Property 12 [Ptolemy's Theorem]

In a quadrilateral[490] inscribed in a circle, the sum of the expansions of opposite sides with one another is as the expansion of its two diagonals with one another.

How? We consider a quadrilateral *ABCD* inscribed in a circle *ABCD*, and draw [lines] *AC BD*. I say that the expansion of *AB* and *CD* and that of *AD* by *BC* together is as the expansion of *AC* by *BD*.

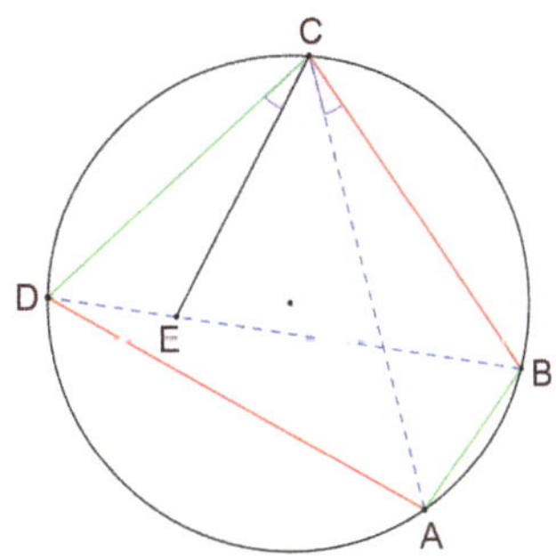

Ptolemy had already stated this demonstration, i.e., we consider an angle *ECD* equal [to angle *ACB*, and angle *BAC*] equal to angle *BDC*. [119a] The remaining angle *CED* is equal to angle *ABC*. Therefore, the expansion of *CD* by the chord of angle *DCE* which is *AB* is as the expansion of *DE* by the chord of angle *E* which is *AC*. [Similarly] since angle *ACD* equals angle *ECB*, and angle *CAD* equals angle *EBC*, and the remaining angle *CDA* is equal to angle *BEC*, then the expansion of *BC* by the chord of angle *BCE* which is *AD* is as the expansion of *BE* by the chord of angle *E* which is *AC*. But the expansion of *AB* by CD is as the expansion of *AC* by *ED*, therefore the expansion of *AD* by CB and [the expansion] of *AB* by *CD* together is as the expansion of *AC* by *BD*. Q.E.D.

---

[489] The wording "intersecting one another and cutting the circumference" and the presentation of the circle as "*ABCD*" accord with the situation that *E* falls inside the circle.

[490] Literally "a figure with four straight sides."

Mathematical Commentary

This nice theorem, known as "Ptolemy Theorem" characterizes cyclic quadrilaterals (i.e., quadrilaterals inscribed in circles) in terms of the relations between their sides and diagonals.

$\sphericalangle DCE = \sphericalangle ACB$ (construction); $\sphericalangle CAB = \sphericalangle CDE$ (inscribed angles on the same arc BC), $\Delta CDE \sim \Delta CAB$. Thus $CD \otimes AB = DE \otimes AC$.

Similarly, $\sphericalangle ACD = \sphericalangle ECB$ (according to the construction $\sphericalangle DCE = \sphericalangle ACB$, adding ACE on both sides) see also note below. $\sphericalangle CAD = \sphericalangle CBE$ (inscribed angles on the same arc DC), $\Delta CEB \sim \Delta CDA$. Thus $BC \otimes AD = BE \otimes AC$.

Using Lemma 2 [i] (III.0 above) we get:

$$CD \otimes AB + BC \otimes AD = DE \otimes AC + BE \otimes AC = (DE + BE) \otimes AC = BD \otimes AC$$

Note. The equality $\sphericalangle ACD = \sphericalangle ECB$ follows from adding $\sphericalangle ACE$ to the two sides of the equality $\sphericalangle DCE = \sphericalangle ACB$, or subtracting it from the two sides, depending on the location of point E on line BD. In the specific case that point E is the meeting point

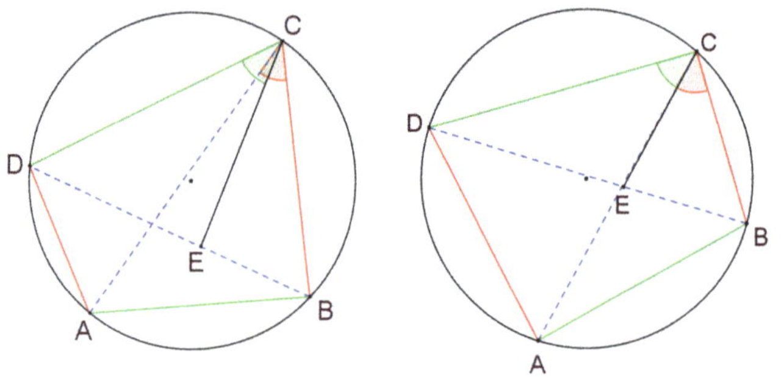

of BD and AC, then these two equalities are identical.
See also note after Property 13 below.

Historical Commentary

We already had reasons to assume that Alfonso was acquainted with the first book of Ptolemy's Almagest,[491] and he could have learned this theorem directly from Ptolemy. In the spirit of his geometrical agenda Alfonso substitutes ⊗ (רבוי) for ⊠ (רבוע).[492] Ptolemy uses for "product" the term ορθογώνιο,[493] namely rectangle, translated into

[491] I.2.1. See introductory note to I.2. There is a Hebrew translation (from the Arabic) by Ya'aqov Anatoly (cf. Wien ms. 194/40, IMHM 1317, fol. 20b20–12a11).

[492] See introductory note to Part III.

[493] Ptolemy (1816, 29).

Hebrew by Ya'aqov Anatoly as מרובע.[494] Ptolemy's theorem is also quoted in *HGC*, the *Hebrew Geometrical Compendium*, where the term שטח, i.e., area is used.[495]

III.[13] Property 13

The expansion of two sides of a rectilinear triangle by one another is as the expansion of the perpendicular drawn from [the vertex] between them to the base by the diameter of the circle encompassing the triangle.

How? We consider a rectilinear triangle ABC [inscribed] in a circle, and from point *B* we draw a perpendicular *BD* on the base, *AC*. I say that the expansion of *AB* by *BC* equals the expansion of the perpendicular *BD* by the diameter of the circle.

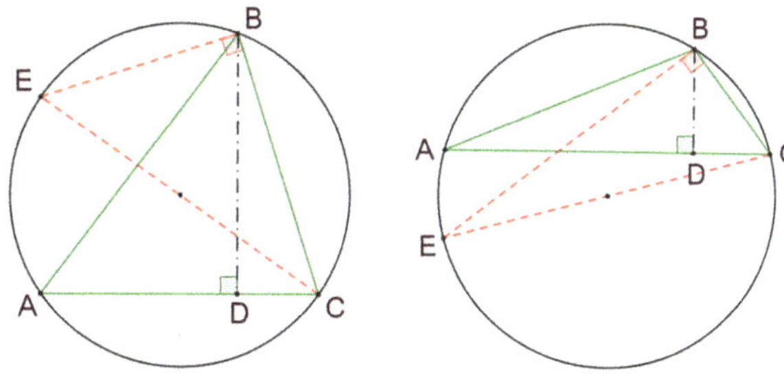

Because rectilinear triangle *ABD* is the expansion of *AB* by the chord of angle *A* on the base which is *BC* equals the expansion of *BD* by the chord of angle *D* on the base which is the diameter. Q.E.D.

Mathematical Commentary

From point *E* draw a diameter *CE*.

Since $\sphericalangle ADB = \sphericalangle EBC = 90°$ (inscribed angle on diameter) and $\sphericalangle BAC = \sphericalangle BEC$ (inscribed angles on the same arc *BC*), $\Delta ABD \sim \Delta ECB$ then $AB{:}BD = EC{:}BC$.

Thus $AB \otimes BC = EC \otimes BD$ (*Elements* VI.16).

Property 13 holds for all three types of triangles. For a right-angled triangle, the proposition is obvious. In the case (ignored by Alfonso) of a triangle with an obtuse angle the diagram and the argument are somewhat different. If Properties 2–9 included some early version of the law of sines, one proof could have sufficed for all cases. This remark pertains to Property 11 too.

From a perspective of modern trigonometry, Property 13 can be perceived as a variant on the laws of sines. The similarity between the triangles implies $\frac{AB}{BD} = \frac{EC}{BC}$ and

[494] Ptolemy, Hebrew text, Paris ms. fol. 7a. On this term see Bar-Ḥiyya *Hibbur ha-Meshiḥa*, 10. For a summary of Alfonso's terms see table in the commentary on III.0.

[495] Mantova MS fol. 66a.

consequently $2R = EC = \frac{BC}{(BD/AB)} = \frac{BC}{\sin\sphericalangle BAC} = \frac{a}{\sin\alpha}$. If an early version of the law of sines was included among Properties 2–9, Property 13 could have been a special case of this early version. Property 13 is used in the proof of Property 18.

Properties 14–22: Right-Angled Triangles and Circumscribed Polygons

III.[14] Property 14

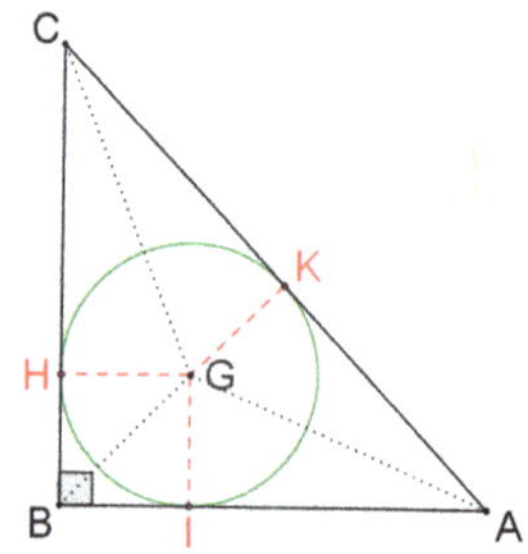

In every rectilinear right-angled triangle, the expansion of half the excess of the two sides enclosing the right angle over the remaining side by itself [together with] the remaining side is equal to half of the expansion of the two sides by one another, which is half of the triangle.

How? Let *ABC* be a triangle with right angle *B*. I say that the expansion of half of the excess of *AB BC* together over the hypotenuse *AC* by itself and by the hypotenuse *AC* is equal to half of triangle *ABC*.

The demonstration. Inside the triangle we describe a circle with a center *G*, touching the sides of the triangle at points *H, I, K*, and we join lines *GA*, *GB*, [*GC*], *GH*, *GI*, *GK*. [119b] Lines *GA*, *GB*, *GC* bisect the angles of the triangle. Since *GKA* is a right angle and *B* is a right angle *HB* is equal to *HG*, *BI* is equal to *GI*. Also, *CH* is equal to *CK* and *AI* is equal to *AK*. Hence the excess of *AB*, *BC* together over *AC* equals twice *GH*. But the expansion of *GK* by all the sides of the triangles is equal to the triangle which is the expansion of *AB* by *BC*. So the expansion of *GH* by twice itself which is *BH* [+] *BI* and by twice *AC*, which is *AI* [+] *CH* together, is equal to the triangle. So the expansion of *GH* by itself and by *AC* is equal to half of triangle *ABC*. Q.E.D.

Mathematical Commentary

Property 14 expresses the area of a right-angled triangle in terms of half the excess of the two legs over the hypotenuse and of the hypotenuse:

$$(*) \quad ½(AB \otimes AC) = ½\left[(AB + BC - AC) \otimes \left(½(AB + BC - AC) + AC\right)\right]$$

The construction follows *Elements* IV.4, the center of the circle being the meeting point of the three angle bisectors.

$$GH = GK = GI\,(\text{radiuses}) = HB = BI\,(HBIG \text{ is a square})$$

$$AK = AI, BI = BH, CH = CK \left(\text{tangents from the same point}\right)$$

Hence it is easy to see that ½(*AB* + *BC* – *AC*) is the side of the square *BHGI* in the diagram.

$$\begin{aligned}\text{But}\quad AB + BC - AC &= \left(AI + BI\right) + \left(BH + HC\right) - \left(CK + KA\right)\\ &= BI + BH\\ &= GI + GH\\ &= 2GH\end{aligned}$$

Therefore ½(*AB* + *BC* – *AC*) = *GH*.

Substituting in (*), it remains to show that ½(*AB* ⊗ *BC*) = *GH* ⊗ (*GH* + *AC*).

According to the construction Δ*ABC* = Δ*ABG* + Δ*ACG* + Δ*BCG*

$$\begin{aligned}AB \otimes BC &= AB \otimes GI + AC \otimes KG + BC \otimes GH \text{ (III.0, Lemma 1)}\end{aligned}$$[496]

$$\begin{aligned}&= (AB + AC + BC) \otimes GH \text{ (III.0, Lemma 2 [i])}\\ &= (BH + BI + AC + HC + AI) \otimes GH\\ &= (2GH + 2AC) \otimes GH\\ &= 2(GH + AC) \otimes GH\end{aligned}$$

Consequently ½(*AB* ⊗ *BC*) = (*GH* + *AC*) ⊗ *GH*.

Q.E.D.

The diagram may remind the reader of Hero's formula for an area of a triangle:

$$A = \sqrt{s\left(s-a\right)\left(s-b\right)\left(s-c\right)}$$

where *a*, *b*, *c* are the sides of the triangle, and *s* its semi-perimeter. Hero's formula was known in the Arab world,[497] it is quoted by Abraham Bar-Ḥiyya, and was also known in the Latin west.[498] Alfonso's formula applies only to right-angled triangles. The parameter used is not half the perimeter but rather half the excess of the two legs over the hypotenuse, $s = \frac{1}{2}(a + c - b)$. The area, which is the expansion of the two legs, is given by $a \otimes c = s \otimes (s + b)$. This formula is less symmetrical and less elegant than Hero's, but is simpler to use because it does not involve the extraction of square root.

III.**[15] Property 15**

In every rectilinear right-angled triangle circumscribing a circle, the expansion of the two parts of the hypotenuse, as divided by the tangent of the circle, by one another is equal to half of the triangle.

---

496 See introductory comment to Part III.

497 See Rashed (2012, 46 and 767).

498 For example, in Fibbonaci. See Barnabas (2008, 80–81).

How? Let *ABC* be a right-angled rectilinear triangle circumscribing a circle *KHI* such that point *K* is tangent point of the circle and line *AC* the hypotenuse. I say that the expansion of *AK* by *KC* is equal to half the triangle.

Its demonstration. We join the lines as in the former diagram. So the expansion of *AB* by *BC* is as the expansion of *AI* by *BI* and by *CH* [together] with the expansion of *BI* by *BH* and by *HC*. But the expansion of *AI* by *BH* is as the expansion of *AI* by *BI*. Therefore, the expansion of *AI* by *BI* and by *CH* [together] with the expansion of *BI* by *BH* and by *HC* is equal to triangle *ABC*. And we discard[499] the expansion of *BI* by *BH* and by *HC* and by *AI*, which equals half of triangle *ABC*, then the remainder is [the expansion of] *AI* by *CH*, so [the expansion of] *AK* by *CK* is equal to half of triangle *ABC*. Q.E.D.

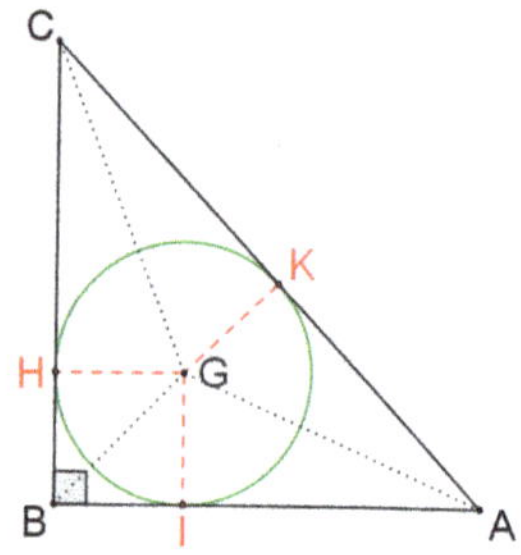

Terms

משׁשׁת Alfonso's word for tangent is משׁשׁת; Bar-Ḥiyya uses מוּשׁשׁ,[500] and the Provençal translators ממשׁשׁ.[501] Although Properties 14 and 15 refer to the same diagram, the term appears only in Property 15. In Property 14 Alfonso used the verb נגע to refer to the tangent points *H*, *I*, *K*.

מיתר / מייתר (v.) A verb formed of the word יתר, namely to be the side opposite the right angle (i.e., to be the hypotenuse). The word appears twice, here and in III.17.[502]

Mathematical Commentary

The statement to be proven is: $AK \otimes CK = ½\ AB \otimes BC$.

As before $GH = GK = GI = HB = BI$ and $AK = AI$, $BI = BH$, $CH = CK$.

$$
\begin{aligned}
(1)\quad AB \otimes BC &= (AI + BI) \otimes (BH + CH) \\
&= AI \otimes (BH + CH) + BI \otimes (BH + CH) \quad (\text{III.0, Lemma 2 [i]}) \\
&= AI \otimes (BI + CH) + BI \otimes (BH + CH) \\
&= BI \otimes (AI + BH + CH) + AI \otimes CH \\
&= GI \otimes (AI + BH + CH) + AK \otimes CK
\end{aligned}
$$

[499] Accepting Gluskina correction of נשלים (complete) to נשליך (discard).

[500] Sarfatti (1968, 85); e.g., *Sefer ha-Meshiha we-ha-Tishboret* 3, 4, 18.

[501] Sarfatti (1968, 178).

[502] The expression שתי צלעות מיתרות זויות שוות appears in the version of Euclid's *Elements* in Firenze 137 ms. in the proof of Proposition I.26. We are grateful to Ofer Elior for this comment.

But, as in the previous proposition

$$(2)\quad AB \otimes BC = GI \otimes (AB + BC + AC) = 2GI \otimes (AI + BH + CH)$$

From (1) and (2) we get ½($AB \otimes BC$) = $AK \otimes CK$.

From the perspective of modern trigonometry: if r is the radius of the circle, then

$$\frac{r}{AK} = \tan\left(\tfrac{1}{2}\measuredangle A\right); \quad \frac{r}{CK} = \tan\left(\tfrac{1}{2}\measuredangle C\right); \quad \frac{r}{BK} = \tan\left(\tfrac{1}{2}\measuredangle B\right)$$

$$\rightarrow AK \cdot CK = \frac{r \cdot r}{\tan\left(\tfrac{1}{2}\measuredangle A\right) \cdot \tan\left(\tfrac{1}{2}\measuredangle C\right)}$$

III.**[16] Property 16**

We consider a triangle *BED* with a right angle *D* inscribed in a circle *ABDE* whose center is G and whose two diameters are intersecting at right angles *AC*[503] *BE*. [120a] We describe around point A a circular sector *AEKB* whose distance is [i.e, radius] *AB*. I say that the center of the circle inscribed in triangle *BED* falls on arc *EKB*.

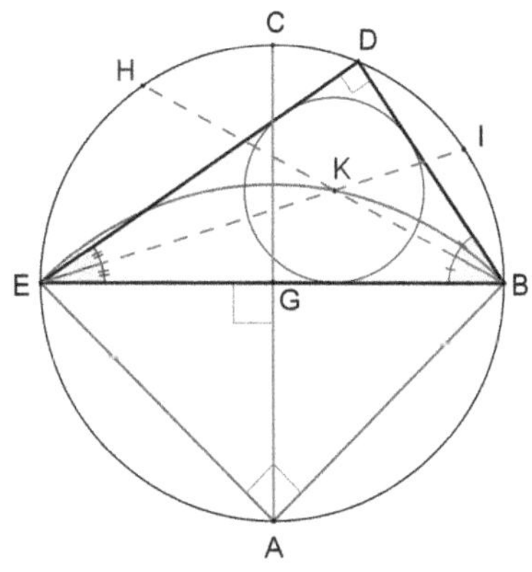

Its demonstration. Arc *EKB* is a quarter of the circumference of its circle, and we bisect angle *EBD* by a line *BH*, intersecting arc *BKE* on *K*. Therefore, arc KE is similar to arc *EH*, which is a half of arc *ED*, and the remaining arc *KB* is similar to semi-arc *BD*. So when we join [line] *EK* and extend it to point I on arc BD, it bisects arc *BD*, and hence it bisects also angle *BED*. Therefore, point *K* is the center of the circle inscribed in the triangle *BED* and touching its sides. Q.E.D.

Mathematical Commentary

$\measuredangle EAB = 90°$ is an inscribed angle on diameter *BE* in the circle whose center is *G*, and is also a central angle in the circle whose center is *A*; therefore $\overset{\frown}{BKE} = 90^{\circ}$.

[503] Not accepting Gluskina's correction AD. Her correction, as well as her diagram, indicate that she considered only the case that points *C* and *D* coincide.

$\sphericalangle HBE$ is an inscribed angle on arc *HE* in circle *AEB*, and is also an inscribed angle on arc *KE* in circle *BKE*; therefore $\overset{\frown}{HE} \sim \overset{\frown}{KE}$.

Since *BH* bisects angle *DBE*, arcs *DH*, *HE* are equal, and $\overset{\frown}{HE} = \frac{1}{2}\overset{\frown}{ED}$.

In circle *ABE* $\overset{\frown}{BDE} = 180^\circ$ and in circle *BKE* $\overset{\frown}{BKE} = 90^\circ$, therefore $\overset{\frown}{BKE} \sim \frac{1}{2}\overset{\frown}{BDE}$, so $\overset{\frown}{KE} + \overset{\frown}{BK} \sim \frac{1}{2}(\overset{\frown}{ED} + \overset{\frown}{DB}) = \frac{1}{2}\overset{\frown}{ED} + \frac{1}{2}\overset{\frown}{DB}$.

But $\overset{\frown}{HE} = \frac{1}{2}\overset{\frown}{ED}$, hence $\overset{\frown}{KE} + \overset{\frown}{BK} \sim \overset{\frown}{HE} + \frac{1}{2}\overset{\frown}{DB}$, and because $\overset{\frown}{HE} \sim \overset{\frown}{KE}$ it means $\overset{\frown}{BK} \sim \frac{1}{2}\overset{\frown}{DB}$.
Now construct a line *EK* intersecting circle *BDE* on *I*, then $\sphericalangle KEB$ is an inscribed angle on arc *BK* in circle *BKE* and on arc *BI* in circle *BDE*. $\overset{\frown}{BK} \sim \overset{\frown}{IB}$ and since $\overset{\frown}{BK} \sim \frac{1}{2}\overset{\frown}{DB}$, $\overset{\frown}{IB} \sim \frac{1}{2}\overset{\frown}{DB}$, and consequently $\overset{\frown}{IB} = \overset{\frown}{DI}$ (because points *I*, *B*, *D* are on circle *BDE*).

But in the circle *BDE*, $\sphericalangle BEI = \sphericalangle DEI$ (inscribed angles on equal arcs). *EK* bisects $\sphericalangle BED$, and *K* is accordingly the intersecting point of two angle bisectors in ΔBED. So *K* is the center of the circle inscribed in triangle *BED*.

This elegant theorem characterizes the locus of the centers of circles inscribed in right-angled triangles, which in turn is inscribed in a given circle *ABDE* on the diameter. When the vertex *D* of the triangle *EDB* traces the circumscribing semicircle *BCE* with radius *r*, the center of the circle inscribed in the triangle traces an arc whose center is *A* (i.e., on the circumference of *ABDE*) and whose radius is $\sqrt{2}r$. Although motion plays a central role in *SMA* with a special attention to the motion of rolling, Alfonso does not offer a dynamic presentation of this theorem.

III.[17] **Property 17**

If two right-angled rectilinear triangles have their sides proportional, the expansion of their (two) hypotenuses by one another is equal to the expansion of the remaining two sides of the first triangle by the remaining two sides of the second triangle—each by the one corresponding to it—together.

How? Let *ABC*, *DEG* be two rectilinear triangles (*C G* right angles) and similar: *AC* corresponds to *DG* and *CB* to *GE*. I say that the expansion of *AB* by *DE* is equal to the expansion of *AC* by *DG* and [the expansion of] *BC* and *EG* together.

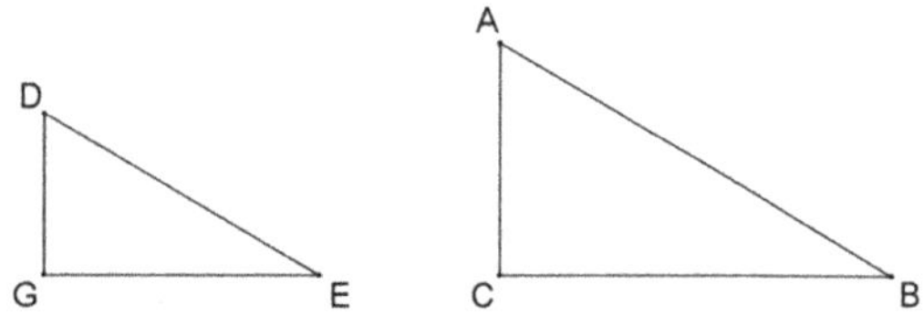

Its demonstration. Since the ratio of *AB* to *DE* is as the ratio of *AC* to *DG*, so the ratio of the square on *AB* to the square on *DE* is as the ratio of the square on *AC* to the square on *DG*. Similarly, the ratio of the expansion of *AB* by *DE* to the expansion of *AC* by *DG* is as the ratio of the square on *AB* to the square on *AC*. Also the ratio of the expansion of *AB* by *DE* to the expansion of *BC* by *EG* is as is the ratio of the square on *AB* to the square on *BC*. But the square on *AB* is equal to the squares on lines *AC BC* together. Therefore the expansion of *AB* by *DE* equals the expansion of *AC* by *DG* and [the expansion of] *BC* and *EG* together. Q.E.D.

Mathematical Commentary

In Property 17 Alfonso proves a theorem about expansion of sides of triangles by examining the squares on these sides. This theorem is stated in terms of expansion, but in the proof Alfonso switches from ratio between expansions to ratio between squares and back. To simplify the presentation, we use the modern notation of fractions rather than the Greek notation for ratios:

It is given that $\triangle ABC \sim \triangle DEG \Rightarrow \dfrac{AB}{DE}=\dfrac{AC}{DG}=\dfrac{BC}{GE}$.

$$\frac{AB}{AC}=\frac{DE}{DG} \text{ (\textit{Elements} V.16).}$$

$$\frac{AB^2}{AC^2}=\frac{DE\otimes DE}{DG\otimes DG} \Rightarrow (1)\ \frac{AB^2}{AC^2}=\frac{AB\otimes DE}{AC\otimes DG}$$ (perhaps relying on VI.22)[504]

$$\frac{AB^2}{BC^2}=\frac{DE\otimes DE}{GE\otimes GE} \Rightarrow (2)\ \frac{AB^2}{BC^2}=\frac{AB\otimes DE}{BC\otimes GE}$$

By *Elements* I.47 $AC^2+BC^2=AB^2$, hence $AC\otimes DG+BC\otimes GE=AB\otimes DE$.

Alfonso does not explain the last step. The proof, however, can be completed:

It can be achieved by inversing Eqs. (1) and (2) and by joining them correspondingly we get: $\dfrac{AC^2+BC^2}{AB^2}=\dfrac{AC\otimes DG+BC\otimes GE}{AB\otimes DE}$ (*Elements* V.24),

and by V.16: $\dfrac{AB^2}{AC\otimes DG+BC\otimes GE}=\dfrac{AB^2}{AB\otimes DE}$.

Using V.9 we get the required result: $AC\otimes DG+BC\otimes GE=AB\otimes DE$.

Like Proposition 11 above, this is a generalization of the Pythagorean Theorem but, in this case, Alfonso does not comment on it. It can be viewed as an analysis of a special case of *Elements* VI.31: "In right-angled triangles the figure on the side opposite the right angle equals the sum of the similar and similarly described figures on the sides containing the right angle." The special case examined here is that the similar figures are triangles.

[504]An alternative proof using VI.22: from the proportion of the sides of the triangles ($AB : AC = DE : DG$) it can be directly concluded that the ratio of the square on *AB* to the square on *DE* is as the ratio of the ratio of the expansion of *AB* by *DE* to the expansion of *AC* by *DG*. It isn't clear why Alfonso preferred to use the mediation of the squares on the sides.

III.[18] Property 18

If the circumference of a circle is cut [120b] into equal parts, and chords joining each two of the [cutting] points are drawn parallel to one of these chords, then the expansion of this chord by itself and by all the other chords together is equal to the expansion of the diameter of the circle by itself.

How? Let *AB, CD, EG,* etc. be parallel chords on a circle *AB DG,* such that all the arcs *AB, BD, DG, GN,* [*NM*] *ME,* are equal to one another. I say that the expansion of chord *AB* by itself and by all the chords, which are *AB,*[505] *CD, EG,* etc., are equal to the expansion of the diameter of the circle by itself.

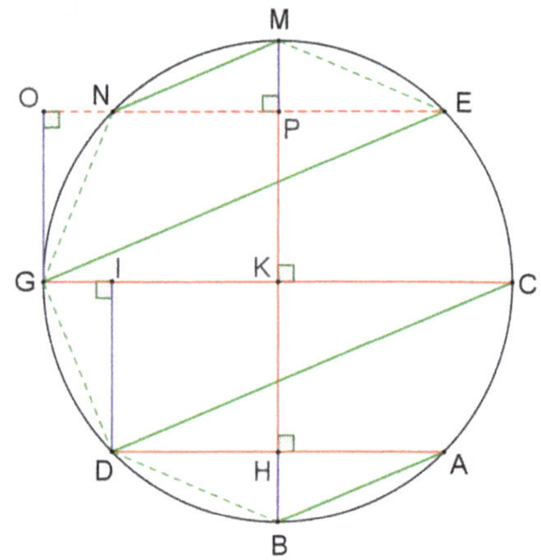

Its demonstration. We join chords *CD, DG, AD, CG,* and we draw from point *B* a diameter *BM,* cutting chords *AD, CG, EN* in right angles on two points *H, K,* [*P*]. And we draw from point *D* a perpendicular *DI* on *CG.* Because the expansion of two sides of this triangle is as the expansion of the perpendicular drawn across from [the vertex] between of them to the base by the diameter of the circle encompassing the triangle—as we established in Proposition 13 of this part—the expansion of *AB* by *BD* is equal to the expansion of the diameter of the circle by perpendicular [*BH.* Similarly the expansion of *CD* by *DG* is equal to the expansion of the diameter of the circle by perpendicular] *DI,* which is equal to *HK.* Therefore, the expansion of *AB CD* together by *BD* equals the expansion of the diameter of the circle by line *BK.* And similarly it can be shown that the expansion of *AB* by itself and by all the other chords parallel to it, i.e., *CD, EG,* etc., is equal to the expansion of the circle's diameter by itself, which is the circle's diameter *BM.* Q.E.D.

Mathematical Commentary

The theorem deals with sides and chords of a regular polygon of even order n inscribed in a circle. Alfonso proves the theorem for $n = 8$.

[505] *AB* should not be listed because the expansion of *AB* by itself was mentioned before.

Applying Property 13 (which states that the expansion of two sides of a triangles equals the expansion of the height to the third side by the diameter of the circumcircle) to the inscribed triangles ΔABD, ΔCDG, ΔEGN, ΔEMN we receive:

$$AB \otimes BD = BM \otimes BH$$
$$CD \otimes DG = BM \otimes DI$$
$$EG \otimes GN = BM \otimes GO$$
$$MN \otimes ME = BM \otimes MP$$

It is given that $AB = BD = DG = GN = MN$, so by adding the four equations and factorizing we get:

$$(AB + CD + EG + MN) \otimes AB = BM \otimes (BH + DI + GO + MP)$$

But $BH + DI + GO + MP = BM$, hence

$$(AB + CD + EG + MN) \otimes AB = BM^{②} \text{ (i.e.,the expansion of } BM \text{ by itself).}$$

Historical Commentary

This proposition is based on *The Sphere and Cylinder* where Archimedes states two similar theorems, one for an even n and one for an odd n (*SC* I.21, 22, respectively).[506] Archimedes' proposition appears also in Ibn Hud's *Istikmāl*,[507] stated for n divisible by 4. Both Archimedes and Ibn Hud refer to a diagram of 12 points of division. The two theorems are included also in *HGC*, the Hebrew Geometrical Compendium,[508] where the compiler draws directly on Archimedes and the lettering of his diagram is the same.

Alfonso offers his own version, which is close to, but different from Archimedes' original theorem. He states one theorem for an even n and proves it for a diagram of eight points. His version of the theorem is unique. He translates the theorem to his language of expansions, and thus changes the presentation significantly. In Archimedes' version the left-hand side of the equation is a sum of n ratios between lines, and in the right-hand side a ratio of two lines; in Alfonso's version the left-hand side is a sum of n expansions of pairs of chords (i.e., two-dimensional magnitudes), the right-hand side is a two-dimensional magnitude.

---

[506] Netz (2004, 113–4). The proof is analogous for any even number of points of division, but is different in the case of an odd number of points of division.

[507] Proposition 313.12 in Hogendijk analytical table (1991, 239). Also Rashed (2012, 756–7).

[508] Mantova ms. 73b–74a. See Glasner (2019).

Note

Since Alfonso's theorem is not identical to Archimedes' and since he proves it only for the case of an even number of sides, we complete his proof by examining the case of seven points of division, using modern notation:

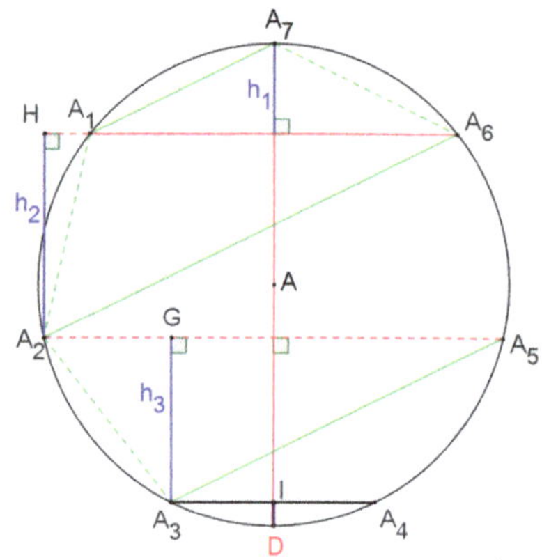

$$A_1A_7 \cdot A_7A_6 = 2R \cdot h_1$$
$$A_1A_2 \cdot A_2A_6 = 2R \cdot h_2$$
$$A_2A_3 \cdot A_3A_5 = 2R \cdot h_3$$

Following the construction $2R = \overline{A_7D}$, therefore:

$$A_1A_2 \cdot (A_7A_6 + A_2A_6 + A_3A_5) = 2R \cdot (h_1 + h_2 + h_3).$$

The heights do not exhaust the diameter, but the difference $\overline{DI} = 2R - (h_1 + h_2 + h_3)$ decreases as the number of points of division increases.

In the general case:

Let $A_1, A_2, \ldots, A_{2n}$ be the partition points of a given circle O. Then A_1A_2 is parallel to $A_{n+1}A_{n+2}$, and A_1A_{n+1} is a diameter which is perpendicular to chords A_2A_{2n}, $A_3A_{2n-1}, \ldots, A_iA_{(2n+2)-i}, \ldots, A_nA_{n+2}$.

The procedure described above can be applied to the triangles $\Delta A_1A_2A_{2n}$, $\Delta A_2A_3A_{2n-1}$, $\Delta A_3A_4A_{2n-2}$, $\Delta A_4A_5A_{2n-3}, \ldots$, $A_iA_{i+1}A_{2n-i+1}, \ldots, \Delta A_nA_{n+1}A_{n+2}$ to get the equation:

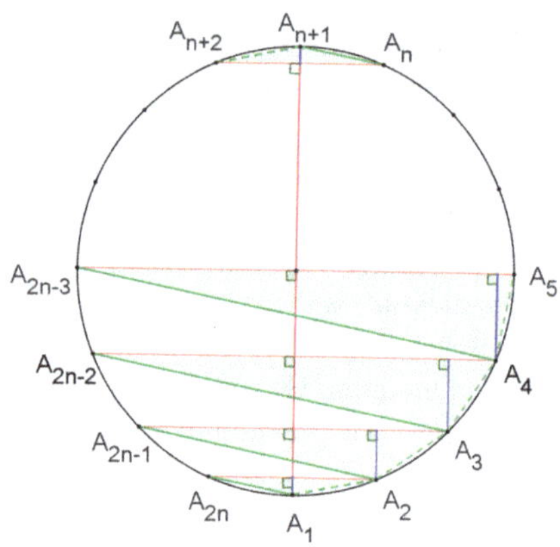

$$(*) \quad A_1A_2 \cdot (A_1A_{2n} + A_2A_{2n-1} + A_3A_{2n-2} + \cdots + A_nA_{n+1}) = (2R)^2$$

If the number of the partition points is odd, say $2n - 1$, then A_1A_2 is not parallel to $A_{n+1}A_{n+2}$, and there is no chord through A_{n+1} which is parallel to chord A_1A_2. One may use the procedure suggested above on triangles $\Delta A_2A_3A_{2n-1}$, $\Delta A_2A_3A_{2n-1}, \ldots$, $\Delta A_nA_{n+1}A_{n+2}$ to get the expansion of chord A_1A_2 by itself and by its parallels, but in this case the diameter $A_{n+1}B$ is greater than the sum of the heights.

Let us denote the excess of the diameter on the sum of heights by "d_n," and the side of the regular polygon of order $2(2n-1)$ by c_n, then: $({**})\quad c_n^2 = 2R \cdot d_n$.

For, if the second extremity of the diameter through A_{n+1} is B, join chords A_1B, A_2B, then on triangle ΔA_1A_2B the height from B completes the heights of triangles $\Delta A_2A_3A_{2n-1}$, $\Delta A_3A_4A_{2n-2}$, …, $\Delta A_nA_{n+1}A_{n+2}$ to the whole diameter $A_{n+1}B$, i.e., $BM = d_n$. Then Eq. (**) is accepted by applying Property 13 to ΔA_1A_2B.

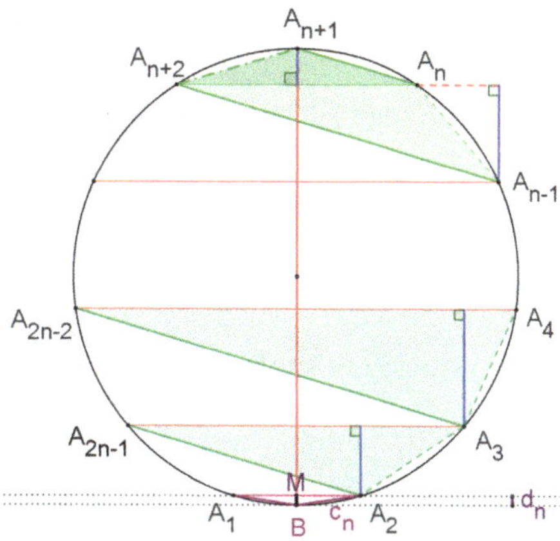

The analogous equation of expansions of the basic chord by itself and by its parallels is

$$({}^{*}_{1})\quad A_1A_2 \cdot (A_2A_{2n-1} + A_3A_{2n-2} + \cdots + A_nA_{n+1}) = 2R \cdot A_{n+1}M.$$

Jointing it with (**) one may have:

$$({***})\quad A_1A_2 \cdot (A_2A_{2n-1} + A_3A_{2n-2} + \cdots + A_nA_{n+1}) + c_n^2 = 2R \cdot A_{n+1}M + 2R \cdot d_n = (2R)^2$$

where d_n is the side of the regular polygon of order $2(2n-1)$.

This equation can be the alternative version for a polygon of odd number of sides:

$$({}^{***}_{1})\quad A_1A_2 \cdot (A_2A_{2n-1} + A_3A_{2n-2} + \cdots + A_nA_{n+1}) + c_n^2 = (2R)^2.$$

From a trigonometric point of view:

Given a regular polygon $A_1A_2…A_n$ circumscribed in a circle whose radius is R, its side would be $2R \cdot \sin(\frac{\alpha}{2})$, where α is central angle of the side.

For, by the law of sines:

$$\frac{c_1}{\sin\alpha} = \frac{R}{\sin\left(\frac{180-\alpha}{2}\right)} \Rightarrow \frac{c_1}{\sin\alpha} = \frac{R}{\cos\left(\frac{\alpha}{2}\right)} \Rightarrow c_1 = \frac{R \cdot \sin\alpha}{\cos\left(\frac{\alpha}{2}\right)} = \frac{R \cdot 2 \cdot \sin\left(\frac{\alpha}{2}\right) \cdot \cancel{\cos\left(\frac{\alpha}{2}\right)}}{\cancel{\cos\left(\frac{\alpha}{2}\right)}}$$

$$\Rightarrow c_1 = 2R \cdot \sin\left(\frac{\alpha}{2}\right).$$

Similarly, $c_i = 2R \cdot \sin\left(\frac{(2i-1)\alpha}{2}\right)$, where c_i is the ith chord paralleling to the side of the polygon.

The required equation is: $c_1 \cdot (c_1 + c_2 + \cdots + c_k) = (2R)^2$

or: $$2R\sin(\tfrac{\alpha}{2})\cdot\left(2R\sin(\tfrac{\alpha}{2})+2R\sin(\tfrac{3\alpha}{2})+\cdots+2R\sin(\tfrac{(2k-1)\alpha}{2})\right)=(2R)^2.$$ [509]

This is exactly so in the even case, approximately in the odd case.

The last expression could be simplified using the identity $\sin(x)+\sin(3x)+\sin(5x)+\cdots+\sin((2k-1)x)=\dfrac{\sin^2(kx)}{\sin(x)}$:

$$2R\sin(\tfrac{\alpha}{2})\cdot\left(2R\sin(\tfrac{\alpha}{2})+\cdots+2R\sin(\tfrac{(2k-1)\alpha}{2})\right)=4R^2\cancel{\sin(\tfrac{\alpha}{2})}\cdot\frac{\sin^2(\frac{k\alpha}{2})}{\cancel{\sin(\frac{\alpha}{2})}}=4R^2\cdot\sin^2(\tfrac{k\alpha}{2})$$

Now, if $n=2k$, then $\alpha=\frac{2\pi}{n}$, hence $\alpha=\frac{\pi}{k}\Rightarrow\frac{k\alpha}{2}=\frac{\pi}{2}$. But $\sin^2(\frac{\pi}{2})=1$, and we get the required equation.

If $n = 2k + 1$, then $\alpha=\frac{2\pi}{2k+1}\Rightarrow k=\frac{\pi}{\alpha}-\frac{1}{2}$, so $\sin^2(\frac{k\alpha}{2})=\sin^2(\frac{\pi}{2}-\frac{\alpha}{4})$. When n approaches infinity, α approaches zero; $\lim\limits_{\alpha\to 0}\sin^2(\frac{\pi}{2}-\frac{\alpha}{4})=1$, so in the odd case the expression $2R\sin(\frac{\alpha}{2})\cdot\left(2R\sin(\frac{\alpha}{2})+2R\sin(\frac{3\alpha}{2})+\cdots+2R\sin(\frac{(2k-1)\alpha}{2})\right)$ is just an approximation of the diameter's square, $(2R)^2$, which improves as n increases.

III.**[19] Property 19**

When the excess of magnitude *AB*, the first, over magnitude *CD*, the second, is equal to the excess of *DE*, the third, over *BG*, the fourth, [then] *AB*, the first, together with *BG*, the fourth, is equal to *CD*, the second, together with *DE*, the third.

Its demonstration. Consider *HB* of AB be the excess of *AB* over *CD* and *DI* of *DE* be the excess of *DE* over BG. Hence *HB DI* are equal to one another, and AH is equal to *CD*. Therefore AB is equal to *CI* and *IE* is equal to *BG*. So *AB*, *BG* together are equal to *CD*, *DE* together. Q.E.D.

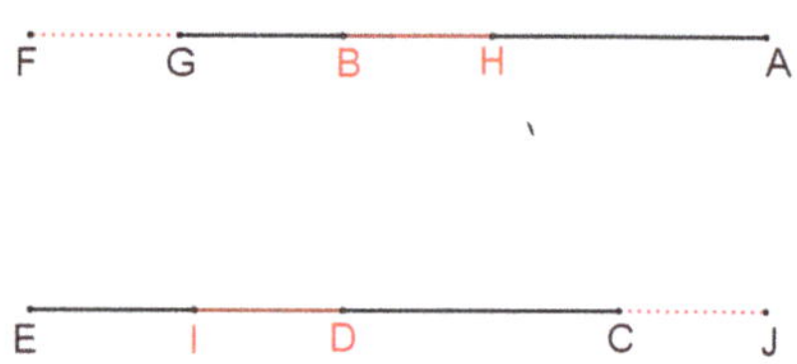

[509] Being angles of a polygon the angles are less than 360° and their halves less than 180°. Consequently the sines are positive. The equality of the corresponding chords follows from the equation $\sin\alpha = \sin(\pi - \alpha)$.

Mathematical Commentary

According to the construction: $HB = AB - CD$, $DI = DE - BG$.

$AB = AH + HB = AH + AB - CD$; $AH = CD$ (*).

Also, it is given that $HB = DI$ (**).
Adding the Eqs. (*), (**) we get $AH + HB = CD + DI$, hence $AB = CI$.

Similarly, $ED = EI + ID = EI + DE - BG$, hence $BG = EI$.

By adding the last two equations we get $AB + BG = CI + IE$.

In algebraic terms, this proposition states that if $a - b = c - d$, then $a + d = b + c$.

In this property the letters of the diagram are mentioned in the annunciation and thus the setting-out is omitted.

This property is used in the proof of the next one.

III.[20] **Property 20**

If the hypotenuse of a right-angled triangle is cut at a point into two unequal parts, [121a] and a straight line is drawn from this point to the right angle, [then] the ratio of the excess of the larger part's expansion by itself over the expansion of the drawn line by itself to the excess of the drawn line by itself over the expansion of the smaller part by itself is equal to the ratio of the expansion of the side enclosing the right angle from the direction of the large part [of the hypotenuse] by itself to the expansion of remaining side from the direction of the small part by itself.

How? Let *ABC* be a right-angled triangle with right angle *B*, and we cut the hypotenuse *AC* into unequal parts on point *D* so that part *AD* is greater than part *DC*. We join *BD*. I say that the ratio of the excess of the expansion of *AD* by itself over the expansion of *BD* by itself to the excess of the expansion of *BD* by itself over the expansion of *DC* by itself is as the ratio of the expansion of *AB* by itself to the expansion of *BC* by itself.

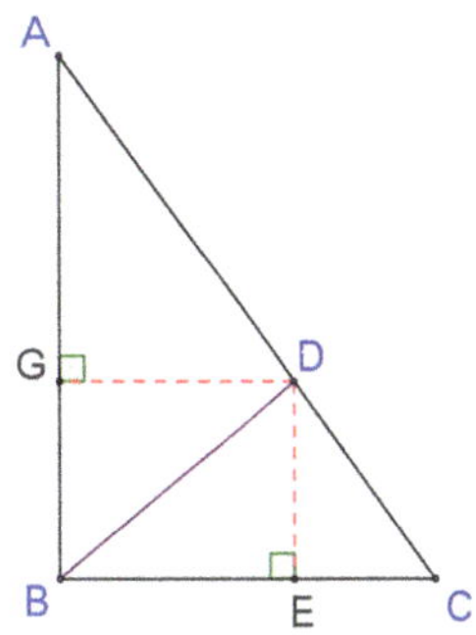

Its demonstration. From point *D* we draw perpendiculars *DG*, DE. Triangles *AGD*, *ABC*, *DEC* are similar. Therefore the ratio of the excess of the expansion of *AG* by itself over the expansion of *DE* by itself to the excess of the expansion of *GD* by itself over the expansion of *EC* by itself is as the ratio of the expansion of *AG* by itself to the expansion of *GD* by itself, which is equal to the ratio of *AB* by itself to the expansion of *BC* by itself. But the excess of the expansion of *AG* by itself over the expansion of *DE* by itself—which is equal to [the expansion of *GB*] by itself—is equal to the excess of the expansion of *AD* by itself over the expansion of

BD by itself. And the excess of the expansion of *GD* by itself—which is equal to the [expansion of] *BE* by itself—over the expansion of *EC* by itself is equal to the excess of the expansion of *BD* by itself to the expansion of DC by itself. Hence the ratio of the excess of the expansion of *AD* by itself over the expansion of *BD* by itself to the excess of the expansion of *BD* by itself over the expansion of *DC* by itself is as the ratio of the expansion of *AB* by itself to the expansion of *BC* by itself. Q.E.D.

[Corollary] And therefore if *AB*, *BC* are equal, the expansions of *AD*, *DC* each by itself [together], equal twice the expansion of *BD* by itself, as established in Proposition 19.

Mathematical Commentary

By the construction $\Delta AGD \sim \Delta ABC \sim \Delta DEC$, so

$$\frac{AG}{GD} = \frac{DE}{EC}, \frac{AG}{GD} = \frac{AB}{BC}.$$

By VI.22 and V.19:

(1) $$\frac{AG^{②} - DE^{②}}{GD^{②} - EC^{②}} = \frac{AG^{②}}{GD^{②}} = \frac{AB^{②}}{BC^{②}}$$

Now *BEDG* is a rectangle, hence $BE = DG$ and $BG = ED$. Using the Pythagorean Theorem (*Elements* I.47) for ΔAGD, ΔBGD, ΔDEB, ΔDEC we have:

(2) $$AG^{②} - DE^{②} = AG^{②} - GB^{②} = AD^{②} - BD^{②}$$

(Alfonso skips the proof that $AG^{②} - DE^{②} > 0$, which follows from *Elements* V.14). Similarly

(3) $$GD^{②} - EC^{②} = BE^{②} - EC^{②} = BD^{②} - DC^{②}$$

(again Alfonso skips the proof that $GD^{②} - EC^{②} > 0$).

Substituting (2), (3) into (1) we get $\dfrac{AG^{②} - DE^{②}}{GD^{②} - EC^{②}} = \dfrac{AD^{②} - BD^{②}}{BD^{②} - DC^{②}} = \dfrac{AB^{②}}{BC^{②}}$ as required.

If, in addition, *AB*, *BC* are equal, the last equation $\dfrac{AD^{②} - BD^{②}}{BD^{②} - DC^{②}} = \dfrac{AB^{②}}{BC^{②}}$, means (by V.16 and V.9) $AD^{②} - BD^{②} = BD^{②} - DC^{②}$, hence—using Property 19, we get $AD^{②} + DC^{②} = 2BD^{②}$.

This proposition is used in the proof of the next proposition.

III.[21] Property 21

[121b] When a line is divided into two unequal parts, the expansion of each of the parts by itself together is equal to twice their expansion by one another, [together] with the expansion of the excess of the larger part over the smaller part by itself.

How? We consider line *AB* bisected at point *C* and divided into unequal parts at point *D*, *AD* greater than *DB*. I say that the expansions of *AD DB* each by itself are equal together to twice their expansions by one another [together] with four times the expansion of *CD* by itself.

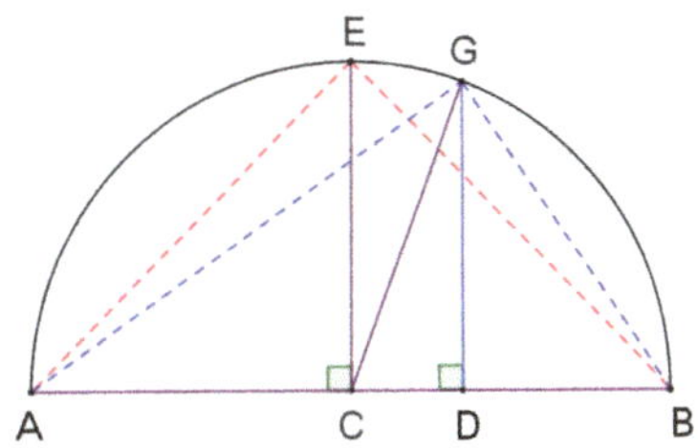

The demonstration. We describe a semicircle *AEB* about a center *C*, and erect perpendiculars *EC*, *DG* upon the diameter *AB*, and join lines *CG*, *ED*, *GB*,[510] *GA*. As has been established in Proposition 20, the expansions of *AD*, *DB* each by itself together are equal to twice the expansion of *DE* by itself, which equals twice the expansions of *EC*, *CD* each by itself, which equals twice the expansion of *DG* by itself [together] with four times the expansion of *CD* by itself. But twice the expansion of *DG* by itself is equal to twice the expansion of *AD* by *DB*. Therefore the expansions of *AD DB* each by itself are equal together to twice their expansion by one another [together] with four times the expansion of *CD* by itself, which equals the expansion of the excess of *AD* over *DB* by itself. Q.E.D.

Mathematical Commentary

ΔAGD is right-angled and isosceles, hence

(1) $$AD^{②} + DB^{②} = 2DE^{②} \quad (\text{corollary of Property 20})$$

But $DE^{②} = EC^{②} + CD^{②}$ (*Elements* I.47 for ΔECD)[511] and $EC = CG$ (radiuses), hence

$$EC^{②} + CD^{②} = CG^{②} + CD^{②} \quad (\textit{Elements} \text{ I.47 for } \Delta CDG)$$

(2) $$DE^{②} = EC^{②} + CD^{②} = DG^{②} + 2CD^{②} \quad (\text{corollary of Property 20})$$

Substituting $DG^{②}$ by $AD \otimes DB$ (*Elements*, porism of VI.8) in (2), we get:

$$DE = AD \otimes DB + 2CD^{②}$$

[510] In the text *DB*. Gluskina's correction.

[511] Theorem I.47 pertains to squares on the sides of the triangles, Alfonso uses it for expansions, i.e., half squares.

And, by substituting this in (1) we get the requested equation:

$$AD^{②} + DB^{②} = 2(AD \otimes DB) + 4CD^{②} = 2(AD \otimes DB) + (AD - DB)^{②}$$

Taking x for *AD*, y for *DB*, ($x > y$) this proposition could be algebraically interpreted as the identity $x^2 + y^2 = 2xy + (x - y)^2$, thus algebraically yielding the same result as of *Elements* II.7.[512] Geometrically the two theorems are not identical. This illustrates Alfonso's geometrical approach.

III.[22] Property 22

In every right-angled triangle whose sides which enclose the right angle are unequal, the expansion of the hypotenuse by itself is equal to twice the triangle together with the expansion of the difference between the sides enclosing the right angle by itself.

How? Let *ECD* be a right-angled triangle with right angle *C*, and [whose] side *EC* is greater than side *CD*; and we extend *CD* in the direction of *D* so that CB equals [122a] *EC*. I say that the expansion of *ED* by itself is equal to twice triangle *ECD* [together] with the expansion of *DB* by itself.

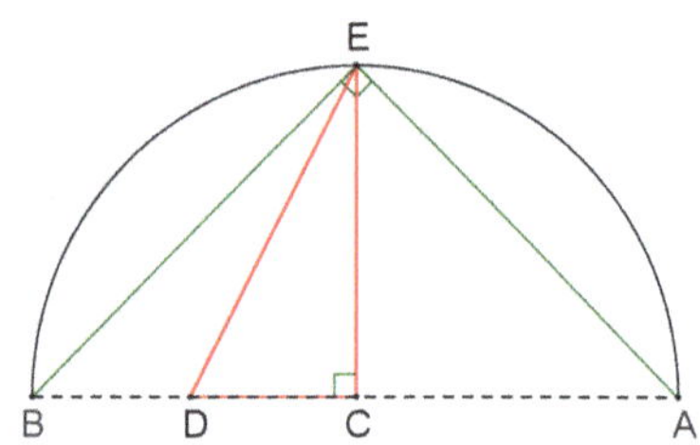

Its demonstration. We describe about point C a semicircle *AEB* whose diameter is *ACB*, and we join *EA*, *EB*. As has been established in Proposition 20, twice of the expansion of *ED* by itself equals the expansion of *BD DA*[513] each by itself together. But the expansion of *DA* by itself equals the expansion of *CD*, *CA*[514] each by itself [together] with twice the expansion of *CD* by *CA*, which equals the expansion of *ED* by itself [together] with twice triangle *ECD*. It remains that twice triangle *ECD* [together] with the expansion of *DB* by itself and the expansion of *ED* by itself are equal. Q.E.D.

Mathematical Commentary

ΔABE is right-angled and isosceles, hence

$$(1) \qquad DA^{②} + BD^{②} = 2ED^{②} \quad (\text{corollary of Property 20})$$

[512] Heath (1908/1956 I, 373, 388–9).

[513] In the text *DE*. Gluskina's correction.

[514] In the text *DA*. Gluskina's correction.

(2) $DA^{\textcircled{2}} = (CD + CA)^{\textcircled{2}} = CD^{\textcircled{2}} + CA^{\textcircled{2}} + 2(CD \otimes CA)$ (*Elements* II.4)

Now, $CA = EC$, hence

(3) $CA^{\textcircled{2}} + CD^{\textcircled{2}} = EC^{\textcircled{2}} + CD^{\textcircled{2}} = ED^{\textcircled{2}}$ (*Elements* I.47 for ΔECD)

By substituting (3) in (2) we get: $DA^{\textcircled{2}} = ED^{\textcircled{2}} + 2(CD \otimes CE)$, and by substituting the last equation in (1) we get: $2ED^{\textcircled{2}} = BD^{\textcircled{2}} + ED^{\textcircled{2}} + 2(CD \otimes CE)$.

Consequently

$$ED^{\textcircled{2}} = BD^{\textcircled{2}} + 2(CD \otimes CE)$$

The geometric presentation is different, but the algebraic presentation of this proposition is identical to that of the previous one: $x^2 + y^2 = 2xy + (x - y)^2$. Alfonso had not considered it redundant.

This is one of the few properties in which Alfonso mixes expansions with pure geometrical entities, here triangle.

Property 23: Quadrature of Lunes

III.[23] Property 23

We would like to construct a lunar figure equal to a rectilinear area.

[First quadrature] We describe a circle *ABCD* around point *E* whose diameters *AC BD* intersect at right angles, and we describe around point *A* a sector *ABGD* with a radius *AB*. Then a lunar-area *BCGD* will be formed between arcs *BCD*, *BGD*. I say it is equal to the rectilinear triangle *ABD*. For sector *ECD*[515] is a quarter of the circle, and so is sector *EBC*. And the ratio of sector *ABGD* to sector *EBC*, which is similar to it, equals the ratio of the square on line *AB* to the square on line *BE*. But, the square on line *AB* is twice the square on line *BE*; therefore sector *ABD* is twice sector *EBC* which is semicircle *BCD*. We discard the common area *BGDE*, and the remaining lunar-area *BGDC* is equal to the straight-lines triangle *ABD*. Thus, there is a moon-shaped[516] area equal to a rectilinear area.

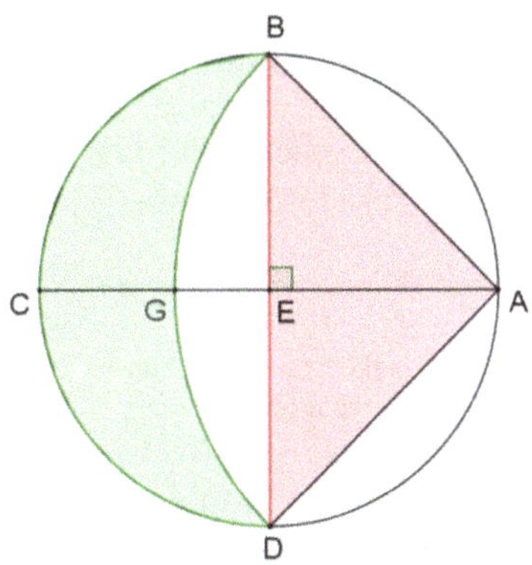

[515] In the text *ACD*. Gluskina's correction.

[516] The Hebrew word is not certain. Gluskina's reading is אחר. We suggest that it may be שהר.

[Second quadrature] We construct a straight line *AB* whose expansion by itself is equal to three times the expansion of line *BC* by itself, and we construct line AC[517] whose expansion by itself equals the expansion of line *BC* by itself together with the expansion of *AB* by *CB*. We construct from these three [122b] lines a triangle *ABC*, as was established,[518] and describe around it a circle[519] *ABCD* whose center is point *E*. And for line *AC* greater than chord *CB*, arc *ADC* would be greater than arc *CB*. We cut from it an arc *AD* equal to arc *BC*, and join lines *AD*, *DC*, *EA*, *ED*, *EC*, *EB*, *BD*. Then the sides of triangle *ABC* are equal to the sides of triangle *ADB*, respectively, and also *AC*,[520] *BD* are equal to one another. And since quadrilateral *ABCD* is inscribed in a circle, the expansion of line *AC*[521] by line *BD*, namely, the expansion of *AC* by itself, is equal to the expansion of *AD* by *CB*, which is the expansion of *CB* by itself, together with the expansion of *AB* by *CB* or *CD*. And for *CB* is equal to *CD*, then the expansion of *AB* by itself is equal to the expansions of chords *AD*, *DC*, *CB* each by itself, together. Now erect upon *AB* a triangle *ABG* similar to triangle *AED*,[522] and describe around point *G*[523] a sector *GAHB*. And therefore sector GAHB is equal to sector *EADB*, and segment *AHB* is equal to the three[524] segments *AD*, *DC*,[525] *CB*. Hence, the lunar area *ADCBH* is equal to polygon *ADCB*. Q.E.D.

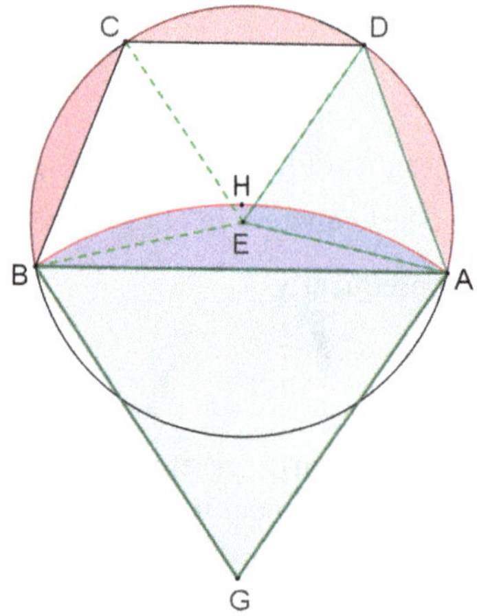

Historical Commentary

Hippocrates of Chios' quadrature of lunes is a well-known episode in the history of mathematics. Four quadratures were preserved in Simplicius' commentary on the *Physics*. Simplicius relied on Eudemus' lost book on the history of the mathematical sciences and on Alexander's lost commentary on the *Physics*.[526] As far as we know

---

[517] Missing in the text. Gluskina's correction.

[518] Alfonso skips the explanation. He presumably alludes here to his source, but unfortunately does not provide the reference.

[519] In the text "triangle." Gluskina's correction.

[520] In the text *AD*. Gluskina's correction.

[521] In the text *AB*. Gluskina's correction.

[522] In the text *AEG*. Gluskina's correction.

[523] In the text *C*. Gluskina's correction.

[524] Literally to the "triangle of segments."

[525] In the text *DB*. Gluskina's correction.

[526] Simplicius *Commentary on Aristotle's Physics*. Book I. 60.22–68.32 Diels; Heath (1921 I, 184–200); Lloyd (1987); Netz (2004). Lloyd (1987, 110) remarks that the testimonies of Themistius and Philoponus contribute little to the reconstruction of Hippocrates' argument.

Simplicius' commentary was not translated into Arabic.[527] Alexander's commentary was translated into Arabic and still circulated in the Iberian Peninsula in the end of the twelfth century.[528] Alfonso knew that the quadrature originated with Hippocrates (I.2.3 and I.2.5 below), but here Hippocrates' name is not mentioned. This is not surprising. Clagett and Langermann comment that Hippocrates is not mentioned by name in medieval texts on the quadrature of the lunes. Alfonso presents the first two quadratures following Eudemus' version quoted by Simplicius.[529] The first quadrature is simpler than the others and was probably better known. Two anonymous Hebrew variants of it are known to us.[530] Both first and second quadratures are presented in the *Hebrew Geometrical Compendium* (*HGC*).[531] Langermann has shown that Alfonso's lettering of the diagrams is identical to that of *HGC* and the proofs are similar, although Alfonso is typically briefer.[532] Did one of these Hebrew compilers draw on the other or did they use a common (probably Arabic) source?[533]

Mathematical Commentary

For some reason in the first construction Alfonso atypically uses the language of squares, in the second the language of expansions. We follow the text.

The First Construction

Sectors *ABGD, EBC* are similar (quarters of a circle), so they are in the same ratio as is the square on *AB* to the square on *BE* (*Elements* XII.2).

$$\frac{\text{sector } ABGD}{\text{sector } EBC} = \frac{AB^2}{BE^2}$$

But according to the construction $AB^2 = 2BE^2$, and by substituting it:

$$\text{sector } ABGD = 2 \cdot \text{sector } EBC = \text{semicircle } BCD$$

Now, discard the common segment *BGDE* and get:

$$\text{lune } BCDG = \Delta ABD$$

527 Sorabji (1990, 301).

528 Ibn Rushd used Alexamder's commentary intensively when he wrote his long commentary on the *Physics*. See Glasner (2009, 24–5, note 23).

529 Heath (1921, 191–3); Netz (2004, 249–50).

530 London BL MS 1013 Or. 2806, IMHM 6385 at the beginning; Paris Private collection MS of Euclid's Elements (IMHM 39116), fol. 169b.

531 See Introduction, Section 6.

532 Langermann (1996a, 33–35).

533 Langermann suggests that Alfonso drew on the Hebrew Geometrical Compendium and abridged the argument. The relation between these two texts is not easy to understand. See Introduction, Section 6 above.

The Second Construction

Alfonso considers a triangle *ABC* so that

$$(1)\quad AB^{②} = 3BC^{②}$$
$$(2)\quad AC^{②} = BC^{②} + AB \otimes BC$$

In other words, taking x for *BC*, Alfonso assumes a triangle whose sides are $AB = \sqrt{3} \cdot x$ and, substituting the first equation in the second, $AC = \sqrt{1+\sqrt{3}} \cdot x$. Alfonso skips this construction, but remarks "as was established," most likely referring to the full version (that came down to us in *HGC*[534]). It is also possible that Alfonso refers here to *Elements* I.22.

Inscribe *ABC* in a circle *E*, and from arc *ACB* cut an arc *AD* equal to arc *BC* (this is possible because $BC < AC \Rightarrow \widehat{BC} < \widehat{ADC}$).

$\sphericalangle DBA = \sphericalangle CAB$ (inscribed angles on equal arcs), $\Delta ABD \cong \Delta BAC$, hence $AC = BD$.

ABCD is a quadrilateral inscribed in a circle, so by Property 12 (Ptolemy's Theorem):

$$AC \otimes BD = AD \otimes BC + AB \otimes CD$$

But $AD = BC$ (construction), and $AC \otimes BD = AC^{②}$ (congruence of triangles), therefore:

$$(3)\quad AC^{②} = BC^{②} + AB \otimes CD$$

By comparing (2), (3) we get $BC = CD$ [$=AD$], *ABCD* is an isosceles trapezium, and (1) could be alternately formulated: (1∗) $AB^{②} = 3AD^{②}$.

Now construct on line *AB* a triangle *ABG* similar to ΔADE.

Sectors *GAHB*, *EAD* are similar, so they are in the same ratio as is the square on *AB* to the square on *AD* (*Elements* XII.2):

$$\frac{\text{sector } GAHB}{\text{sector } ADE} = \frac{AB^{②}}{AD^{②}} = 3$$
$$\Rightarrow \text{sector } GAHB = 3 \cdot \text{sector } ADE = \text{sector } ADEB$$

But the ratio of the sectors is also the ratio of their arcs and segments, so:

$$\begin{aligned} \text{segment } AHB &= 3 \cdot \text{segment } AD \\ &= \text{segment } AD + \text{segment } CD + \text{segment } BC \end{aligned}$$

and by respectively substracting them from the last equation we get:

$$\text{lune } ADCBH = \text{trapezium } ABCD$$

[534] Langermann (1996, 41–2).

Properties 24–27: Theory of Proportion

Introductory note: Euclid's theory of proportion in Arabic mathematics

Alfonso defines the aim of Part III as to provide the reader with all the knowledge needed for understanding Parts IV and V beyond Euclid's *Elements*.[535] Properties 24–7 offer additions to Euclid's theory of proportion (*Elements* V–VI). The Euclidean theory is based on the conception of ratio as a relation between magnitudes of the same kind in geometry,[536] between numbers in arithmetic.[537] Ratios can be compared, but no arithmetical operation on them can be defined.[538] The need to develop a more quantitative concept of ratio, and with it an operation of composition of ratios, was already recognized by the Greeks and was developed by Arab mathematicians.[539] Euclid himself does not define a concept of compound ratio but, nonetheless, takes such a concept for granted in Propositions VI.23 (for magnitudes) and VIII.5 (for numbers).[540] The concept became indispensable later with the developing of trigonometry and spherical geometry.

A definition of compound ratio was added to some Greek manuscripts of book VI of the *Elements*,[541] and was preserved in Ḥunayn Ibn Isḥaq's Arabic translation: "It is said that a ratio is compounded of ratios when the amounts (*aqdār*, שעורים) of the ratios multiplied together make a certain ratio." Another variant of this definition followed: "a compounded ratio is divided into ratios so that the amounts of the ratios from one another generate another ratio."[542] The definitions sound rather awkward to modern ears, but introducing the term "amount" (قدر, שעור, *quantitas*) of a ratio led the way to the understanding of the quantitative nature of ratio.[543] This trend was advanced in the Arab world and culminated in the work of Umar al-Khayyām.[544] *HGC*, The *Hebrew Geometrical Compendium* dwells at length on the notion of compound ratio.[545] In *SMA* two relatively simple Properties (24–5)

---

[535] III.0 above.

[536] *Elements* V definition 3.

[537] In the definitions of Book VII a general concept of ratio is not introduced, but three types of ratios between numbers are defined: part, parts and multiple. The word "ratio," however, is used in the propositions.

[538] Heath (1908/1956 II.132); Mueller (1981, 118, 154). Euclid recognized only compound ratios that can be reduced to a single ratio (i.e., $a{:}b{\bullet}c{:}d$ where $b = c$). See Netz (2004, 203, note 102).

[539] Vitrac (2000); Rashed and Vahabzadeh (1999, 298).

[540] These propositions were not used later in the *Elements* (Heath 1908/1956 II, 132–3, 176–8, 189–90).

[541] Vitrac (2000, 6). Heiberg and Heath agree that this definition is an interpolation (Heath 1908/1956 II, 189).

[542] For example, Judeo-Arabic ms. fol. 57a8–11; Hebrew Paris private collection ms. fol. 48a18–9. There are some differences between the mss. In the Hebrew Geometrical Compendium (Mantova ms. 2, fol. 46b) the two definitions are more neatly stated, applying multiplication and division to the ratios themselves, not mentioning "sizes of ratios."

[543] Euclid, *Elementa*, vol. II, 72, Heiberg.

[544] Rashed and Vahabzade (1999, 278); Vitrac (2000, 23–4).

[545] In Part V of this treatise. See Glasner, forthcoming.

deal with compound ratios. Alfonso's style, as always, is strictly geometrical, and the proofs follow the Euclidean pattern. Properties 24–5 draw mainly on *Elements* VI.12—the finding of a fourth proportional.

Properties 26–7 are much more sophisticated and address the difficulty that Arab mathematicians encountered with Euclid's definition of proportion. Eudoxus' definition of proportion, (*Elements* V definition 5) admired today as one of the highlights of Greek geometry,[546] was considered by medieval Arab mathematicians as one of its weaknesses.[547] Commentaries on *Elements* V, writes Vahabzadeh, were written with the aim "either to justify it by means of demonstration, or to replace it by another definition known as the antyphairetic definition of the sameness of ratios."[548] The antyphairetic definition preceded Euclid's: two ratios are equal if they are subject to the same order of divisions when carrying out the Euclidean algorithm, i.e., if they have the same expansion into continued fractions.[549] Being based on the algorithm of division, it was considered to be a more natural definition of proportion than Euclid's definition in terms of equimultiples.[550] Several mathematicians in the east, notably Al-Mahani, al-Nayrizi, and, most explicitely, Umar Khayyām (in his Commentary on the Difficulties of Certain Postulates in the Work of Euclid) found Euclid's definition problematic, and suggested the alternative approach.[551]

In the west, on the other hand, Arab and Jewish commentators accepted more willingly Euclid's definition, and some tried to improve on it.[552] The Andalusian mathematician Ibn Muʿāḏ al-Jayyānī (eleventh century) states Definition V.5 in terms of parts, rather than of equimultiples. He considers his definition to be "clear and evident to the mind" and proves its equivalence to Euclid's original V.5.[553] Centuries later the Italian mathematician Saccheri in his Defense of Euclid came back to the comparison of the two definitions, preferring Euclid's.[554] Properties 26–7 should be examined in the light of these different attitudes among Arab mathematicians to the definition of proportion. We cannot know whether Alfonso was aware of the antyphairetic interpretation of Definition V.5,[555] or whether he was

[546] For example, Heath @@, II. 120–129.

[547] Vitrac (2002, 4–8); Rashed and Vahabzade (1999, 275); Vahabzade (2013, 225); De Risi (2014, 18).

[548] Vahabzadeh (2012, 11).

[549] Much has been written on this early theory of proportions. For example, Becker (1933); Szabo (1978, Part 2); Fowler (1981, 1982); Thorup (1992).

[550] Vitrac (2002, 11).

[551] Rashed and Vahabzadeh (1999, 274–9); Vahabzadeh (2013, 225–7); Vitrac (2002, 13–17, 30). The need to correct Euclid' theory of proportion was still an issue in the modern era from the early Renaissance to the beginning of the eighteenth century (De Risi 2014, 17–24). On the persistence of the early theory of proportions in post-Euclidean Greek geometry see Knorr (1978).

[552] Ibn Hud in his *Istikmāl* follows Euclid's definition (Hognedijk 1991, 231). Gersonides, in his commentary on the fifth book of the *Elements*, provides the rigorous axiomatic basis that should underlie Euclid's definition (Carlebach 1910, 168–172).

[553] Plooij (1950, 63–4).

[554] De Risi (2014, 195–201).

[555] Alfonso had a copy of al-Nayrizi's commentary on *Elements* 1 (II.2.3 above) but we cannot be sure that his copy of al-Nayrizi included book V, in which al-Nayrizi presents the antyphairetic theory.

familiar with that of al-Jayyānī. Properties 26–7 present an alternative approach to analyzing proportion in terms of sequences, an approach that can be useful for approximating ratios.

III.[24] Property 24

Given four magnitudes such that the ratio of the first to the second is compounded of the ratio of first to the third and the ratio of the fourth to the second, and the first and the second are equal, then the third and the fourth are also equal.

How? We consider the ratio of *A* to *B* which is compounded of the ratio of *A* to *C* and the ratio of *D* to *B* (which equal it [*A*]). I say that magnitudes *C*, *D* are equal [to one another].

Its demonstration. We assume a ratio of *C* to *E* as the ratio of *D* to *B*. Hence [123a] the ratio of *A* to *E* is as the ratio of *A* to *B*, which equals *A*. And since the ratio of *C* to *E* is as the ratio of *D* to *E*, as is *D* to *B*, which equals *E*, therefore *C*, *D* are equal to one another. Q.E.D.

Terms

ערך, יחס The word יחס in the sense of ratio is not a part of the mathematical vocabulary of *SMA*. Alfonso uses (176 times) the older term ערך that was coined by Bar-Ḥiyya and commonly used in the Iberian Peninsula. The term יחס, used here in a sentence that was copied mistakenly,[556] may be due to an error of the fifteenth-century copyist.[557]

ערך מחובר The word מחובר is used five times in Parts I–II meaning "composed." In Part III ערך מחובר is used as a mathematical term for compound ratio. Following the Arabic مجمع the word מחובר was used by Abraham Ibn Ezra to denote both the result of addition and of multiplication.[558] In the Hebrew translation of Joseph Naḥmias' *The Light of the World* the word מחובר is used in exactly the same sense in a similar context, but there it is the translation of the Arabic مؤلفة.[559] As noted above the term "compound ratio" is not defined by Euclid but is used in *Elements* VI.23. In his translation of this proposition Moshe Ibn Tibbon uses the term יחס מחובר.[560]

[556] See note to the Hebrew text of Property 24.

[557] On the "errors" in Part III see Introduction, Section 7, example [iii].

[558] Sarfatti (1968, 82, 135).

[559] Naḥmias, e.g., Section A.II.10, Hebrew text, p. 196, Arabic text, p. 60.

[560] Paris private collection ms. fol. 55a7.

Mathematical Commentary

Algebraically this property is simple: if $\frac{a}{b}=\frac{a}{c}\cdot\frac{d}{b}$ and $a = b$, then $c = d$.

Given that $\frac{a}{b}=\frac{a}{c}\cdot\frac{d}{b}$ we find a magnitude e such that $(^{*})\frac{c}{e}=\frac{d}{b}$ (finding fourth proportional, *Elements* VI.12), and substitute in the given equation to get:

$$\frac{a}{b}=\frac{a}{c}\cdot\frac{d}{b}=\frac{a}{c}\cdot\frac{c}{e}=\frac{a}{e}.$$

Therefore $b = e$ (V.9).
By substituting it in (*) we get $\frac{c}{e}=\frac{d}{e}$, hence $c = d$ (*Elements* V.9).
This is the only proposition in *SMA* that includes a diagram in the text (fol. 122b). The other diagrams in Part II are marginal additions, probably by the owner Mordekhai Finzi.[561] This property is used in the proof of the next one.

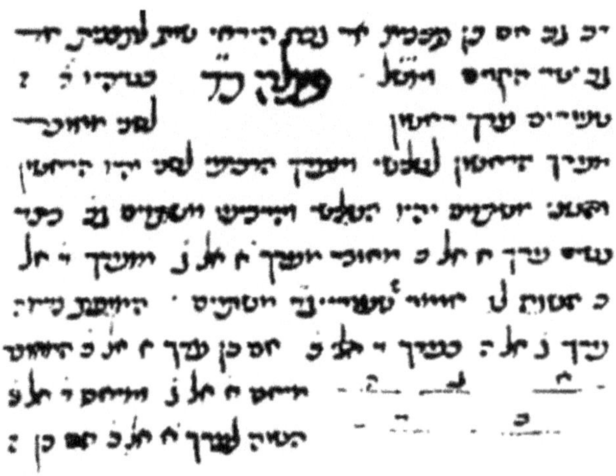

III.[25] **Property 25**

Given six magnitudes of which the ratio of the second to the first is compounded of the ratio of third to the fourth and of the ratio of the fifth to the sixth, and the first and the second are equal, then the ratio of the third to the fourth is as the ratio of the sixth to the fifth.

[561] See Introduction, Section 1.

How? We consider a ratio of A to B which is compounded of the ratio of C to D and the ratio of E to F, and A B are equal to one another. I say that the ratio of C to D is as the ration of F to E.

The demonstration. We assume a ratio of A to G as the ratio of C to D; and [also] consider a ratio of H to B equals the ratio of E to F. Therefore, as it was established in [Property] 24, G and H are equal to one another. Hence, the ratio of A to G which equals the ratio of C to D is as the ratio of B to H, which equals the ratio of F to E. Q.E.D.

Mathematical Commentary

Algebraically Property 25 is simple too: if $\frac{a}{b}=\frac{c}{d}\cdot\frac{e}{f}$ and $a = b$, then $\frac{c}{d}=\frac{f}{e}$.

For, find a magnitude g such that $(1)\frac{a}{g}=\frac{c}{d}$ and another magnitude h such that $(2)\frac{h}{b}=\frac{e}{f}$ (finding fourth proportional, *Elements* VI.12) and substitute in the given equation to get:

$$\frac{a}{b}=\frac{c}{d}\cdot\frac{e}{f}=\frac{a}{g}\cdot\frac{h}{b}$$

But $a = b$, so according to the previous property (applied for the quartet a, b, g, h), $g = h$.

Therefore $\frac{a}{g}=\frac{b}{h}$, and by (1) and by the inverse of (2) we get $\frac{c}{d}=\frac{f}{e}$.

III.[26] Property 26

When two magnitudes of the same kind are given, and when parts are taken from the first magnitude, [and] the same number of parts taken from the second magnitude, [in such a way that] each part of the former has the same ratio to its respective part of the latter, then whenever the former parts together exceed the first magnitude, [so] the latter parts together exceed the second magnitude, and whenever the former parts together fall short of the first magnitude, [so] the latter parts together fall short of the second magnitude – then the ratio of the first magnitude to the second magnitude is equal to the given ratio of one of the parts taken [of the first magnitude] to the corresponding part taken of the second magnitude.

How? We consider two magnitudes [123b] *AB*, *CD*, of the same kind. We take from magnitude *AB* parts *GN*, *NM*, *ML*, *LH*, so that each part of the former is in the same ratio to its corresponding part of the latter: *GN* to *IQ*, *NM* to *QP*, *ML* to *PO* and *LH* to *OK*, and [it is given that] whenever the parts of *GH* exceed magnitude AB the parts of *IK* exceed magnitude *CD*, and whenever the parts of *GH* fall short of magnitude *AB* the parts of IK fall short of magnitude *CD*. I say that the ratio of *AB* to *CD* is as the ratio of magnitude *GN* to magnitude *IQ*.

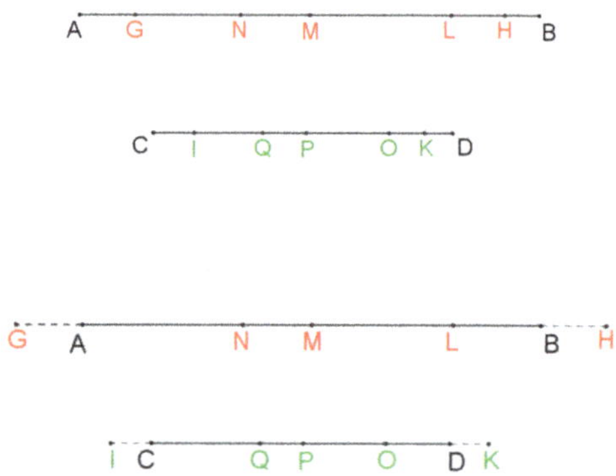

Its demonstration. For, if not, we assume the ratio of *AB* to *CD* to be greater than the ratio of *GN* to IQ, if possible, and we assume *EB* smaller than *AB*, whose ratio to CD is not smaller than the ratio of *GN* to *IQ*. And we take parts *SR*, *RF* exceeding magnitude *EB* and not exceeding magnitude *AB*, and take also parts *VU*, *UT* as many as the parts of *SR*, *RF*. And the ratio of *VU* to *SR* and of *UT* to *RF* is as the given ratio of *IQ* to *GN*. Hence, by summation [of antecedents and of consequents] the ratio of *VT* to *SF* is as the ratio of *IQ* to *GN*, and by alternation, the ratio of *SF* to *VT* is as the ratio of *GN* to *IQ*. And, since the parts of *SF* do not exceed magnitude *AB*, the parts of *VT* do not exceed magnitude *CD*. Therefore, the ratio of *SF* (which is greater than *EB*) to *VT* (which is not greater than *CD*) is as the ratio of *GN* to *IQ*. And consequently the ratio of *EB* to *CD* is smaller than the ratio of *GN* to *IQ*; but we have already assumed it is not smaller than it, and this is impossible. Hence, the ratio of *AB* to *CD* is not greater than the ratio of *GN* to *IQ*. And it was established that it is not smaller than it, and therefore it is equal to it. Q.E.D.

Terms

קבוץ Summation. Here it is the summation of the antecedents and of consequents, when a sequence of ratios is given. According to *Elements* V.12, if all the ratios = *p*, then the ratio of the sum of antecedents to the sum of consequents is also = *p*. See also comment on the term נקבץ in III.0 above.

Two notes on the diagram and interpretation

1. The diagram and lettering correspond to the case of line segments, but the argument holds for any two magnitudes of the same kind.
2. The word חלקים means parts. Properties 26 and 27 and their proofs hold whether the חלקים are [i] aliquot or [ii] aliquant parts (including in the case that the given magnitude and the parts are incommensurable). It is likely that Alfonso thought of [i] as in Euclid's Definitions V.1 and VII.3–4. Yet, since [ii] is more general, we draw the diagrams and state the commentary for the case of aliquant parts.

Mathematical Commentary

Given two magnitudes AB and CD and a ratio p,

If for any two sequences of magnitudes, taken of AB (here: $\{GN, NM, ML, LH\}$) and of CD (here: $\{IQ, QP, PO, OK\}$), so that the ratios between their respective elements is the given ratio p (here: $\frac{GN}{IQ} = \frac{NM}{QP} = \frac{ML}{PO} = \frac{LH}{OK} = p$), then it holds that when $GH = GN + NM + ML + LH < AB$ so does $IK = IQ + QP + PO + OK < CD$, and conversely, when $GH > AB$ so does $IK > CD$,

then $\frac{AB}{CD} = \frac{GN}{IQ} = p$.

Assume to the contrary that $p \neq \frac{AB}{CD}$, let's say $p < \frac{AB}{CD}$.

According to the assumption: $p = \frac{GN}{IQ} < \frac{AB}{CD}$.

We find EB such that:

(1) $EB < AB$

(2) $\frac{GN}{IQ} \leq \frac{EB}{CD}$ (by *Elements* VI.12 we can find a fourth proportional k so that $\frac{k}{CD} = \frac{GN}{IQ}$, and because $\frac{GN}{IQ} < \frac{AB}{CD}$, then $\frac{k}{CD} < \frac{AB}{CD}$, and by V.10 we have $k < AB$, so we may simply choose $EB = k$).

Then we find a magnitude SF such that

(3) $EB < SF \leq AB$,

and we divide it into some parts, say SR, RF, and find VU, UT so that $\frac{SR}{VU} = \frac{RF}{UT} =$ the given ratio $\frac{GN}{IQ}$.

(4) $\Rightarrow \frac{SR + RF}{VU + UT} = \frac{SF}{VT} = \frac{GN}{IQ}$ (*Elements* V.12).

The assumptions of the theorem apply to the two sequences $\{SR, RF\}$, $\{VU, UT\}$, and hence $SF \leq AB \Rightarrow$ (5) $VT \leq CD$.[562]

Therefore $\frac{EB}{CD} < \frac{SF}{VT}$ (by (3) $EB < SF \Rightarrow \frac{EB}{VT} < \frac{SF}{VT}$;

by (5) $CD \geq VT \Rightarrow \frac{EB}{CD} \leq \frac{EB}{VT}$ (*Elements* V.8).

[562] See below Appendix B.

But by (4) we get $\frac{EB}{CD} < \frac{SF}{VT} = \frac{GN}{IQ}$, in contradiction to (2): $\frac{a}{b} \leq \frac{E}{B}$.

Therefore it is impossible that $\frac{GN}{IQ} < \frac{AB}{CD}$.

In the same manner it could be shown that it is impossible that $\frac{GN}{IQ} > \frac{AB}{CD}$.

It follows that $\frac{AB}{CD} = \frac{GN}{IQ}$.

In the general case:

Let A, B be two magnitudes of the same kind.

If there is a certain ratio p such that for any two respectively proportional sequences of magnitudes of this ratio $\{a_1, a_2, \ldots, a_n\}$, $\{b_1, b_2, \ldots, b_n\}$ [i.e., $\frac{a_1}{b_1} = \frac{a_2}{b_2} = \cdots = \frac{a_n}{b_n} = p$] the condition

(*) If $\sum_{i=1}^{n} a_i > A$ then $\sum_{i=1}^{n} b_i > B$, and if $\sum_{i=1}^{n} a_i < A$ then $\sum_{i=1}^{n} b_i < B$ holds,

then $\frac{A}{B} = p$.[563]

Proof:

Let $\{a_1, a_2, \ldots, a_n\}$, $\{b_1, b_2, \ldots, b_n\}$ be two sequences of magnitudes such that:

$$\frac{a_1}{b_1} = \frac{a_2}{b_2} = \cdots = \frac{a_n}{b_n} = \frac{\sum_{i=1}^{n} a_i}{\sum_{i=1}^{n} b_i} = p.$$

Denoting $a = \sum_{i=1}^{n} a_i$, $b = \sum_{i=1}^{n} b_i$. We should show that $\frac{A}{B} = \frac{a}{b} = p$.

Assume to the contrary that $\frac{A}{B} \neq p$, let us say $\frac{A}{B} > p$.

We find E so that:

$$(1)\quad E < A;\ (2)\quad \frac{a}{b} \leq \frac{E}{B}$$

[563] If the condition were symmetrical, the proof would have been valid regardless of whether strong or weak inequalities were used in the proof. See Appendix B below.

Then we find a magnitude x such that

$$(3)\quad E < x \leq A,$$

and divide it into some parts: $x_1, x_2, \ldots, x_m$.

Then find $y_1, y_2, \ldots, y_m$ so that $\frac{x_1}{y_1} = \frac{x_2}{y_2} = \cdots = \frac{x_m}{y_m}$ = the given ratio $p = \frac{a}{b}$.

Let us denote $x = \sum_{i=1}^{m} x_i$, $y = \sum_{i=1}^{m} y_i$, then

$$(4)\quad \frac{x}{y} = \frac{a}{b} \quad (\textit{Elements}\ \text{V.12}),$$

Hence by assumption (*) above the inequality $x \leq$ A yields the inequality.

$$(5)\quad y \leq B.$$

Therefore $\frac{E}{B} < \frac{x}{y} \underset{(4)}{=} \frac{a}{b}$ $\left(B \geq y \Rightarrow \frac{E}{B} \leq \frac{E}{y},\ E < x \Rightarrow \frac{E}{y} < \frac{x}{y}\right)$

in contradiction to (2): $\frac{a}{b} \leq \frac{E}{B}$

Therefore it is impossible that $\frac{A}{B} > p = \frac{a}{b}$ as we assumed above.

In the same manner it can be shown that it is impossible that $\frac{A}{B} < p = \frac{a}{b}$.

Consequently $\frac{A}{B} = p$.

It should be noted that the number of terms of the two sequences, n, does not play any role in the proof. The theorems could thus be stated more simply using only the sums: Given two magnitudes of the same kind A and B and a certain ratio p, if for any two magnitudes C, D of that kind which are in this same ratio p, it holds that when $C > A$ so also $D > B$ and when $C < A$ so also $D < B$, then $A{:}B = p$. Similarly, a simplified version of Property 27 can be stated.

III.[27] Property 27

When four magnitudes are [given], and from the first and the third the same number of parts is taken, and the ratio of the first magnitude to each of the parts taken of it is [124a] equal to the ratio of the third magnitude to each of the parts taken of it, and whenever the former parts exceed the second magnitude so the latter parts exceed the fourth magnitude, and whenever the former parts fall short of the second magnitude so the latter parts fall short of the fourth magnitude—then the ratio of the first magnitude to the second magnitude is equal to the ratio of the third magnitude to the fourth magnitude.

How? We consider four magnitudes *A, B, C, D* and we take magnitudes *EG, GH, HI* of the same kind of magnitude *A*, and we also take magnitudes *LM, MN, NO* as many as the parts of *EI*]of the same kind of magnitude C[, so that the ratio of magnitude *A* to each of magnitudes *EG, GH, HI* is as the ratio of magnitude *C* to its corresponding magnitude among *LM, MN, NO*. [And it is given that] whenever magnitudes EI exceed magnitude *B*, so magnitudes *LO* exceed magnitude *D*; and whenever the magnitudes *EI* fall short of magnitude *B*, so the magnitudes *LO* fall short of magnitude *D*. I say that the ratio of *A* to *B* is as the ratio of *C* to *D*.

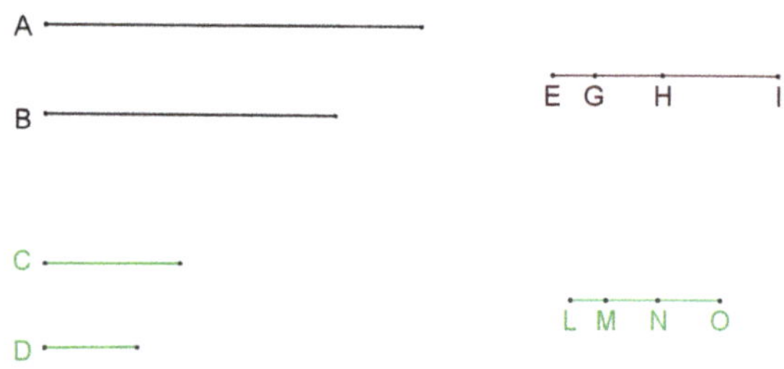

Its demonstration. For, if not, we assume the ratio of *B* to *A* to be greater than the ratio of *D* to *C*, if possible. And we consider *K* smaller than *B*, whose ratio to *A* is not smaller than the ratio of *D* to *C*. And we take parts *PQ, RQ* exceeding magnitude *K* and not exceeding magnitude *B*, and take also parts *ST, TU* as many as the parts of *PR*, so that the ratio of *A* to *PQ* is as the ratio of *C* to *ST*, and the ratio of *A* to *QR* is as the ratio of *C* to *TU*. Hence, by summation [of antecedents and of consequents] the ratio of A to the whole *PR* is as the ratio of *C* to the whole *SU*; and, since the parts of PR do not exceed magnitude *B*, the parts of *SU* also do not exceed magnitude *D*. And the parts of *PR* exceed *K* and the parts of SU do not exceed *D*. Therefore, the ratio of [124b] PR to K is greater than the ratio of *SU* to *D*, and the ratio of *K* to *A* is not smaller than the ratio of *D* to *C*, consequently the ratio of *PR* to *A* is greater than the ratio of *SU* to *C*; but they were already [assumed] to be equal to one another, and this is impossible. Hence, the ratio of *B* to *A* is not greater than the ratio of *D* to *C*. And similarly it can be shown that it is not smaller than it, and therefore it is equal to it. Q.E.D.

Mathematical Commentary

As in Property 26, the diagram and lettering corresponds to the case of line segments, but the argument holds for any two magnitudes of the same kind.

Given two pairs of magnitudes, *A, B* and *C, D*:

If for any two sequences of magnitudes of the same kind of *A* (here: $\{EG, GH, HI\}$), and of the same kind of *C* (here: $\{LM, MN, NO\}$)—whose elements are respectively proportional to *A* and *C* $\left(\frac{A}{EG}=\frac{C}{LM}, \frac{A}{GH}=\frac{C}{MN}, \frac{A}{HI}=\frac{C}{NO}\right)$ it holds that

$EG + GH + HI = EI > B$, so does $LM + MN + NO = LO > D$, and conversely, when $EI < B$, so does $LO < D$,

then $\frac{A}{B} = \frac{C}{D}$.

Assume to the contrary that $\frac{B}{A} \neq \frac{D}{C}$, let's say $\frac{B}{A} > \frac{D}{C}$.

We find k so that:

(1) $k < B$

(2) $\frac{D}{C} \leq \frac{k}{A}$ (we find k so that $\frac{k}{A} = \frac{D}{C}$,[564] and because $\frac{D}{C} < \frac{B}{A}$, then $\frac{k}{A} < \frac{B}{A}$, and by V.10 we have $k < B$, so we may simply choose $EB = k$).

Then we take magnitudes PQ, QR such that:

(3) $k < PR \leq B$

and find magnitudes ST, TU so that $\frac{A}{PQ} = \frac{C}{ST}, \frac{A}{QR} = \frac{C}{TU}$.

(4) $\Rightarrow \frac{A}{PQ + QR} = \frac{C}{ST + TU}$ or: $\frac{A}{PR} = \frac{C}{SU}$ (*Elements* V.12).

The assumptions of the theorem are applicable to the two sequences $\{PQ, QR\}$, $\{ST, TU\}$, and hence $PR \leq B \Rightarrow$ (5) $SU \leq D$.[565]

Therefore: (6) $\frac{SU}{D} < \frac{PR}{k}$ (def. V.7, $m = n = 1$); by (3) $k < PR \Rightarrow \frac{k}{A} < \frac{PR}{A}$; by (5) $SU \leq D \Rightarrow \frac{SU}{C} < \frac{D}{C}$; but according to (2) $\frac{D}{C} \leq \frac{k}{A} \Rightarrow \frac{SU}{C} < \frac{D}{C} \leq \frac{k}{A} < \frac{PR}{A}$.

But this result is in contradiction with (4): $\frac{A}{PR} = \frac{C}{SU}$.

Therefore it is impossible that $\frac{B}{A} > \frac{D}{C}$.

In the same manner it could be shown that it is impossible that $\frac{B}{A} < \frac{D}{C}$.

It follows that $\frac{B}{A} = \frac{D}{C}$.

---

[564] The finding of the fourth proportional here is not trivial because the magnitudes are not of the same kind (see Bos 2001, 105). Alfonso uses a construction of a fourth proportional in the non-homogeneous case also in III.32. Perhaps he was not aware of the difficulty or ignored it.

[565] See Appendix B below on the use of weak vs. strong inequalities.

In the general case:
Let A, B be two magnitudes of one kind, and C, D be two magnitudes of another kind.

If for any two sequences of magnitudes:

$\{a_1, a_2, \ldots, a_n\}$ of the same kind of A, and $\{c_1, c_2, \ldots, c_n\}$ of the same kind of C, whose corresponding elements are respectively proportional to A and C $\left(\text{i.e. } \frac{A}{a_i} = \frac{C}{c_i} \quad \forall i, i = 1,2,\ldots,n\right)$, the condition

(*) If $\sum_{i=1}^{n} a_i > B$, so $\sum_{i=1}^{n} b_i > D$, and if $\sum_{i=1}^{n} a_i < B$, so $\sum_{i=1}^{n} b_i < D$ holds,

then $\frac{B}{A} = \frac{D}{C}$.

Proof:
Assume to the contrary that $\frac{B}{A} \neq \frac{D}{C}$, let's say $\frac{B}{A} > \frac{D}{C}$.

We find k so that:

$$(1) \quad k < B$$

$$(2) \quad \frac{D}{C} \leq \frac{k}{A}$$

Then we find a magnitude x such that

$$(3) \quad k < x \leq B,$$

and divide it into some parts: $x_1, x_2, \ldots, x_m$.

Then find $y_1, y_2, \ldots, y_m$ so that $\frac{A}{x_1} = \frac{C}{y_1}, \frac{A}{x_2} = \frac{C}{y_2}, \ldots, \frac{A}{x_m} = \frac{C}{y_m}$.

Let us denote

$x = \sum_{i=1}^{m} x_i$, $y = \sum_{i=1}^{m} y_i$ then using V.12 we get

$$(4) \quad \frac{A}{\sum_{i=1}^{m} x_i} = \frac{C}{\sum_{i=1}^{m} y_i} \quad \text{or simply}: \quad \frac{A}{x} = \frac{C}{y}$$

Hence by assumption (*) above the inequality $x \leq B$ yields the inequality

$$(5) \quad y \leq D.$$

Therefore $\frac{y}{D} < \frac{x}{k}$ (def. V.7 for x, k, y, D, $m = n = 1$).

According to (2) $\frac{D}{C} \leq \frac{K}{A}$, and for (3) $k < x$ and (5) $y \leq D$, by V.8 we have $\frac{y}{C} < \frac{x}{A}$ in contradiction to (4): $\frac{A}{x} = \frac{C}{y}$.

Therefore it is impossible that $\frac{D}{C} < \frac{B}{A}$ as we assumed above.

In the same manner it can be shown that it is impossible that $\frac{D}{C} > \frac{B}{A}$.

Consequently $\frac{B}{A} = \frac{D}{C}$.

Appendix A. Properties 26–7 and Euclid's Definition of Proportion

According to Euclid's Definition V.5:

> "magnitudes are said to **be in the same ratio**, the first to the second and the third to the fourth, when, if any equimultiples whatever be taken of the first and third, and any equimultiples whatever of the second and fourth, the former equimultiples alike exceed, are alike equal to, or alike fall short of, the latter equimultiples respectively taken in corresponding order" (Heath' tanslation).

In a modern notation, given four magnitudes a, b, c, d, the ratios $a{:}b$, $c{:}d$ are equal if and only if for any two integers m, n:

(#) $ma > nb \Leftrightarrow mc > nd$; $ma = nb \Leftrightarrow mc = nd$; $ma < nb \Leftrightarrow mc < nd$.

Specifically, setting $n = 1$ we get:

(#1) $ma > b \Leftrightarrow mc > d$; $ma = b \Leftrightarrow mc = d$; $ma < b \Leftrightarrow mc < d$.

Properties 26–7 offer two ways to determine a proportion by means of proportional sums rather than equimultiples.

We replace ma, b, mc, d by $\sum a_i$, A, $\sum b_i$, B respectively. We get:

(1) $\sum a_i > A \Leftrightarrow \sum b_i > B$; $\sum a_i = A \Leftrightarrow \sum b_i = B$; $\sum a_i < A \Leftrightarrow \sum b_i < B$

Which is precisely the condition stated in Property 26 for the proportion $A : B = \sum a_i : \sum b_i$. If the members of the two sequences a_i and b_i are equal parts of some A', B' so that $A' : B' = p$, then theorem 26 actually determines a proportion using equal parts.

In the same manner, dividing (#) by n, we get the equivalent condition:

$$(\#2)\quad \frac{m}{n}\cdot a > b \Leftrightarrow \frac{m}{n}\cdot c > d;\ \frac{m}{n}\cdot a = b \Leftrightarrow \frac{m}{n}\cdot c = d;\ \frac{m}{n}\cdot a < b \Leftrightarrow \frac{m}{n}\cdot c < d$$

Again shifting to sums of sequences we replace $\frac{m}{n}a, b, \frac{m}{n}c, d$ by $\sum a_i$, B, $\sum c_i$, D respectively. We get:

$$(2)\quad \Sigma a_i > B \Leftrightarrow \Sigma c_i > D;\ \Sigma a_i = B \Leftrightarrow \Sigma c_i = D;\ \Sigma a_i < B \Leftrightarrow \Sigma c_i < D$$

which is the condition stated in Property 27 for the proportion $A : B = C : D$.

Since (#2) is equivalent of (#), Theorem 27 can be thought of as a variant of V.5 in which the proportion is determined using equal parts.

The difference between (1) and (2), i.e., between the two properties is probably due to the requirement of homogeneity that is made in Property 26,[566] but not in 27. In 26 there is no problem to describe the ratio between the sums ($\sum a_i : \sum b_i$), whereas in 27 the ratio has to be described indirectly, by comparing each sum to one of the given magnitudes which is of the same kind ($A : \sum a_i = C : \sum c_i$). It can easily be shown that applying Property 27 for homogeneous magnitudes yields a claim equivalent to Property 26. Therefore 27 can be viewed as a generalization of 26.

The differences between Alfonso's Properties III.26, III.27, and Euclid's Definition V.5 are briefly summarized in the table below:

	Definition V.5	Property 26	Property 27
Homogeneity	Not required	Required	Not required
Examined magnitudes	Terms of the same ratio (numerator to denominator)	Corresponding terms of the ratios (numerators apart, denominators apart)	Terms of the same ratio (numerator to denominator)
Compared magnitudes	Integer equimultiples	Parts and their sums	Parts and their sums
Comparison is applied to:	Two ratios	One ratio	Two ratios (numerators only)

If indeed Properties 26 and 27 are variants on Euclid's definition of proportion, the question comes up as to what was Alfonso's aim in including them in Part III. Several answers are possible:

[i] Alfonso preferred the determining of a ratio in terms of sums of sequences to its determination in terms of equimultiples, because he was interested in sequences

[566] The requirement of homogeneity is made in the exposition of the theorem but is not mentioned in the proof.

of ratios that can be extended indefinitely as a potentially useful mathematical tool (I.5.3, IV.1).

[ii] Alfonso was motivated by his special interest in the foundations of geometry. Postulate 5 of Book I and Definition 5 of Book V were both considered to be problematic by Arab mathematicians. In the case of the former, Alfonso was well aware of the history of the reception of Postulate 5 and of the attempts to prove it. He also attempted a "proof" of his own.[567] It is possible that he was also familiar with the history of the reception of Definition V.5 and ventured to contribute to its clarification.

[iii] Properties III.26–7 provide a practical method of approximating an unknown ratio $A{:}B$. Using Property 26, $A{:}B$ can be found to any degree of accuracy even if the two magnitudes are incommensurable:

Consider two sequences of magnitudes $\{x_i\}_{i=1}^{n}, \{y_i\}_{i=1}^{n}$, whose terms are respectively proportional: $\frac{x_1}{y_1} = \frac{x_2}{y_2} = \cdots = \frac{x_m}{y_m} = p$.

If $A < \sum_{i=1}^{n} x_i$ but $B > \sum_{i=1}^{n} y_i$ the ratio $\sum x_i : \sum y_i$ is greater than the required ratio, $A{:}B$ (since $\frac{A}{B} < \frac{A}{\Sigma y_i} < \frac{\Sigma x_i}{\Sigma y_i}$), and p should be decreased in order to approach $A{:}B$.

In the same manner, If $A > \sum_{i=1}^{n} x_i$ but $B < \sum_{i=1}^{n} y_i$ the ratio $\sum x_i : \sum y_i$ is smaller than the required ratio, $A{:}B$, and p should be increased in order to approach $A{:}B$.

Using modern terminology, applying this argument for $B = 1$ gives an approximation of any real number A by a sequence of rationals.

Appendix B. Weak and Strict Inequalities

In Euclid's *Elements* inequalities are understood to be strict. In Definition V.5 Euclid lists three conditions:

$$ma > nb \Leftrightarrow mc > nd;\ ma = nb \Leftrightarrow mc = nd;\ ma < nb \Leftrightarrow mc < nd.$$

In Properties 26–7 weak and strict inequalities are used side by side.[568] Property 26 is stated in terms of strict inequalities $\begin{matrix}(1) & \Sigma a_i > AB \Rightarrow \Sigma b_i > CD \\ (2) & \Sigma a_i < AB \Rightarrow \Sigma b_i < CD\end{matrix}$ and in the

[567] II.3 above.

[568] There are examples of the use of mixed inequalities in the sixteenth century. See, e.g., Wagner (2010, 506).

proof the same conditions are stated in terms of weak inequalities: $\sum x_i \leq AB \Rightarrow \sum y_i \leq CD$. It is not clear to what extent Alfonso was aware of the significance of the distinction between the two kinds of inequalities.

It should be noted that if the implications (1) (2) [as stated in the formulation of the theorems] were not symmetrical, then the implications $\sum a_i \leq AB \Rightarrow \sum b_i \leq CD$ (as well as the contradiction $\frac{GN}{IQ} \leq \frac{EB}{CD}$) would not be established,[569] and the proof would not be valid.[570]

If the conditions of the theorem were symmetrical (which is probably what Alfonso meant), i.e.,

$$(1)\ \ \sum a_i > AB \Leftrightarrow \sum b_i > CD$$
$$(2)\ \ \sum a_i < AB \Leftrightarrow \sum b_i < CD$$

then the argument would be valid even if weak inequalities were used in the proof.[571]

If the distinction between strict and weak inequalities were clear to Alfonso, the proof could be further simplified:

Conditions 1–2 are logically equivalent to 1*–2*:

$$(1)\ \ \sum a_i > AB \Rightarrow \sum b_i > CD$$
$$(2)\ \ \sum a_i < AB \Rightarrow \sum b_i < CD$$

$$(1^*)\ \ \sum b_i \leq CD \Rightarrow \sum a_i \leq AB$$
$$(2^*)\ \ \sum b_i \geq CD \Rightarrow \sum a_i \geq AB$$

Now, divide CD into n parts: $y_1, y_2, \ldots, y_n$, so that $\sum_{i=1}^{n} y_i = CD$, and find $x_1, x_2, \ldots, x_n$ which holds: $\frac{x_1}{y_1} = \frac{x_2}{y_2} = \cdots = \frac{x_n}{y_n} = p$.

$\sum y_i = CD \Rightarrow \sum y_i \leq CD$, and by (1*) we get $\sum x_i \leq AB$.

$\sum y_i = CD \Rightarrow \sum y_i \geq CD$, and by (2*) we get $\sum x_i \geq AB$.

The two equalities yield $\sum x_i = AB$.

---

569 The assumptions (1) $\sum a_i > AB \Rightarrow \sum b_i > CD$, (2) $\sum a_i < AB \Rightarrow \sum b_i < CD$ are not necessarily equivalent to $\sum a_i \leq AB \Rightarrow \sum b_i \leq CD$, $\sum a_i \geq AB \Rightarrow \sum b_i \geq CD$, so it is possible that $\sum a_i \leq AB$ but $\sum b_i > CD$; In that case $EB: CD \geq EB: \sum b_i$. But this inequality cannot lead (together with the inequality $EB: \sum b_i \leq \sum a_i: \sum b_i$) to the conclusion $EB: CD < p$, so the desired contradiction cannot be achieved, and the theorem cannot be proved.

570 The theorem can be proved also under these assumptions, but in a slightly different way.

571 In this case $\sum a_i = AB \Leftrightarrow \sum b_i = CD$, and the conclusion $EB : CD < p$ is valid.

But by the assumption above $\frac{x_1}{y_1} = \frac{x_2}{y_2} = \cdots = \frac{x_n}{y_n} = p$, therefore $\frac{\sum x_i}{\sum y_i} = p$ (V.12), and $\frac{AB}{CD} = p$, as required.

Property 28: Spherical Trigonometry

III.[28] Property 28

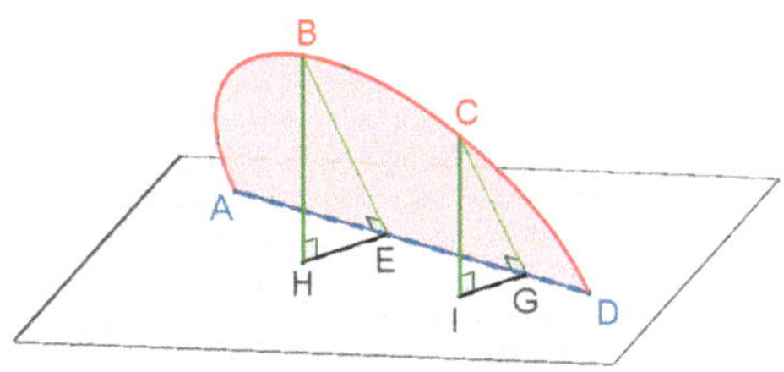

When the surface of a semicircle is inclined over a plane surface, and perpendiculars are drawn from the circumference of the [semi-]circle to the plane surface, then the ratio of the perpendiculars to one another is as the ratio of the sines of the arcs, from whose extremities the perpendiculars are drawn.

How? We consider a surface of a semicircle *ABCD* whose diameter is *AD*, inclined on a plane surface. From points *B*, *C* two perpendiculars *BH CI* are drawn on the plane, and the [corresponding] sines *BE CG* are drawn on the diameter *AD*. I say that the ratio of perpendicular *BH* to perpendicular *CI* is as the ratio of sine *BE* to sine *CG*.

Its demonstration. We join lines *EH GI*. The two perpendiculars *BH*, *CI* are parallel; and so the sines *BE CG* are parallel, and these [two pairs of] lines are not in the same plane. Hence angle *HBE* is equal to angle *ICG*, as was proved by Euclid, *H* and *I* being right angles. Therefore triangles BEH CGI are similar, and the ratio of *BE* to *CG* is as the ratio of *BH* to *CI*. Q.E.D.

Terms

משען See Section III.0 above.

Mathematical Commentary

Given a plane and a semicircle whose diameter is on the plane, it should be demonstrated that the perpendiculars to the diameter from any two points on the circumference are proportional to the perpendiculars from these points to the plane.

In the diagram above, it should be demonstrated that $\frac{BH}{CI} = \frac{BE}{CG}$.

CI, *BH* are perpendiculars to the given plane, therefore $BH \parallel CI$ (*Elements* XI.6).

CG, *BE* are perpendiculars to the diameter *AD* in the circle's plane, therefore $CG \parallel BE$ (I.27).

BE, *BH*, *CG*, *CI* are two pairs of intersecting lines not in the same plane whose sides are parallel, therefore $\sphericalangle ICG = \sphericalangle HBE$ (*Elements* XI.10).

But $\sphericalangle H = \sphericalangle I$ [=90°], so $\triangle BEH \cong \triangle CGI$ (*Elements* VI.4), and it follows that $\frac{BH}{CI} = \frac{BE}{CG}$.

Without the requirement that *BE CG* be perpendicular to the diameter *AD*, the theorem could be extended to any arc of a circle *ABCD* (not necessarily semicircle).

Historical Commentary

This lemma was proved by Thabit Ibn Qurra as a preliminary lemma to his proof of Menelaus theorem,[572] and is copied as a marginal note in *HGC*, the Hebrew Geometrical Compendium.[573] Alfonso's formulation is rather different from that of *HGC*.

Properties 29–32: The Conchoid and Its Applications[574]

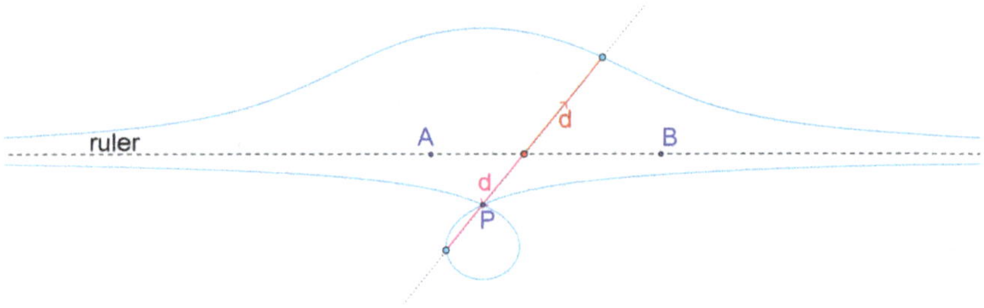

Introductory remark

Properties 29–32 form an independent unit, dealing with a special curve—the Conchoid. Being located towards the end of the third part of SMA, it can be considered as one of its pinnacles. The conchoid, like the quadratrix, the cissoid, and the spiral of Archimedes, is a special curve that Greek mathematicians used in order to deal with the famous geometrical problems unsolvable by the means of ruler and compass or by conic sections.[575]

Given a straight line (the "ruler" or "canon" *AB*), a point outside it (the "pole" *P*), and a distance (*d*), the conchoid is the locus of all points lying at the given distance *d* from the ruler *AB* along the segment that connects them to the pole *P*.[576]

---

[572] Van Brummelen (2009, 177).

[573] Mantova ms fol. 212 B. We are grateful to Akiva Kadari who drew our attention to this marginal note.

[574] Avinoam Baraness had already published his interpretation of these four properties in Katz (2016).

[575] See Heath (1921, vol. I, 225–232, 264–266).

[576] It can also be viewed as the trace of a fixed point on a straight-line through the pole, moving along the ruler. See commentary on Property 29 below.

[577] Its Cartesian equation is little more complicated: $(y - a)^2 (x^2 + y^2) = d^2y^2$.

If P is the origin and AB is the line $y = a$, then the curve is defined by the polar equation $r = \frac{a}{\sin\theta} + d$.[577] The curve has two branches on the two opposite sides of the ruler, to which both are asymptotic. The branch passing on the side of the pole has three different distinct forms, depending on the ratio between a and d: If $a < d$, it has a loop (as is in the diagram); if $a = d$, P is a cusp point; and if $a > d$, the curve is smooth. The other branch has one topological form.

It is assumed that the conchoid was introduced by the Greek mathematician Nicomedes in the second half of the third century BC.[578] His treatise On Conchoid Lines was not preserved and we know about it *from secondary sources*.[579] Geminus, in his famous classifications of lines, mentions a line that first "forms a [closed] figure, and for the rest, extend without limit,"[580] presumably the conchoid that forms a loop and then proceeds indefinitely.[581] Pappus in his classification of lines describes the conchoid as a "high order" curve, belonging to the category of curves "which have a more varied and forced genesis, because they are generated out of less structured surfaces, and out of twisted motions."[582] Later it has been shown that the conchoid can be used for solving any problem involving equations of third and fourth degree. For this reason, Newton suggested that it would be treated as a "standard" curve.[583]

Nicomedes supposedly described the generation of the curve, its classification into types, some of its properties, and a mechanical device for drawing it. Nicomedes also applied the curve to solve two of the classical problems of antiquity: the trisection of an angle and the doubling of the cube. The latter problem was reduced to that of finding two mean proportionals between two given straight lines.[584] The two solutions depend on an auxiliary construction that can be implemented with a conchoid, due to its fundamental characteristic neusis: the length of the segment intercepting between the curve and the ruler along any line through the pole is constant. Toomer refers to the following statement as the "Lemma of Nicomedes"[585]:

---

[578] Nothing is known of his life; the dates were estimated by references to his work. See Heath (1921, 238); Toomer (2008).

[579] Pappus' *Collection*, Eutocius' *Commentary on Archimedes' Sphere and Cylinder*, Proclus' *Commentary on the First Book of Euclid's Elements*. See Toomer (2008).

[580] Proclus, *Commentary* 138–9 Morrow translation (176–7 Friedline).

[581] Heath (1908/1956, vol. I, 160–1).

[582] Sefrin-Weis (2010, 144, see also pp. 144–6, 271–3); Heath (1921, vol. ii, 362).

[583] http://xahlee.info/SpecialPlaneCurves_dir/ConchoidOfNicomedes_dir/conchoidOfNicomedes.html.

[584] In case the greater line is double of the less. The reduction was probably established in the fifth century BCE by Hippocrates of Chios (Heath 1981, 244–246).

[585] Toomer (2008).

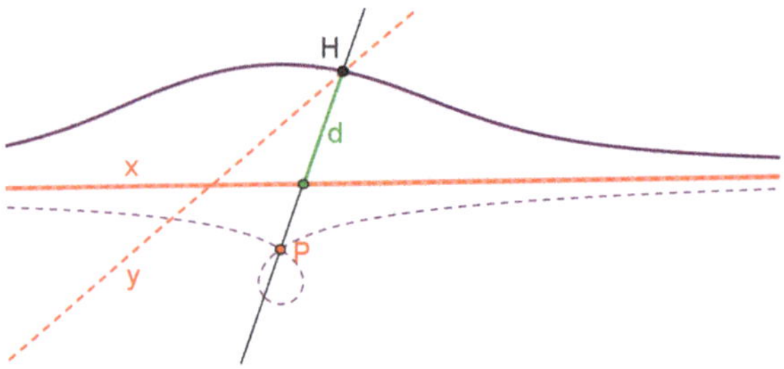

> Given two straight lines, *x*, *y*, meeting at a given angle, and a point *P* outside the angle, it is possible to draw a line through *P* cutting *x* and *y* so that a segment of given length, *d*, is intercepted between **x** and ***y***.

This is achieved by constructing a conchoid with canon *x*, pole *P* and distance *d*; the line through *P* and through the intersection point of this conchoid and *Y* meets the demands.

It is usually assumed that all the applications of the conchoid made in antiquity were developed by Nicomedes himself, and that the interest in it was revived in the late sixteenth century.[586] It has been shown by Rashed that in the ninth century Banu Musa offered a solution to the trisection of the angle problem using another type of conchoid, a circular conchoid.[587] Alfonso's text presented below thus provides a rare evidence that the linear conchoid was known and used in the West in the fourteenth century. In Property 29 Alfonso presents the construction of the conchoid, and in the three following properties (30–32)—its uses in trisecting an angle, finding two mean proportionals, and constructing a parallelepiped of the same volume as a given parallelepiped which is also similar to another given parallelepiped. Properties 29–33 are thus of a great historical and mathematical interest. The method that he uses in these properties is not known to us from other sources. As is often the case, we do not know whether Alfonso follows an unknown Arabic text based on Pappus' Collection, or whether Propositions 30–32 render his own elaboration, based on an oral tradition.

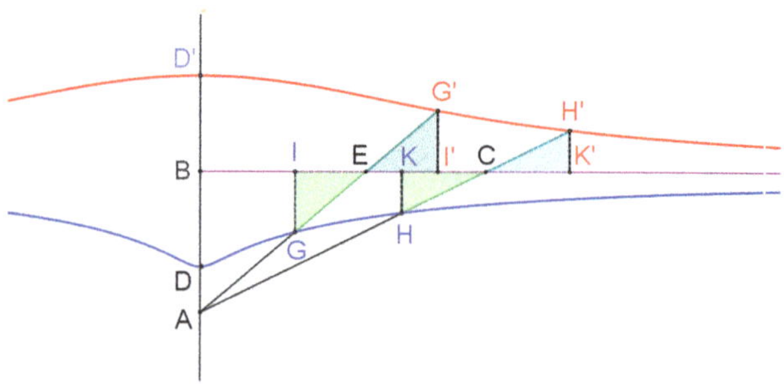

[586] Toomer (2008).

[587] Rashed (2012, 66–9); see also Heath, The Works of Archimedes, c-cii and p. 310 note.

III.[29] Property 29

We wish to find two lines separating [from each other] (יציאת שני קוים) the one straight and the other curved, so that there is an initial distance between them, and as long as the two lines are continued, the distance between their extremities decreases and one of them approaches to the other, but they do not intersect even if produced indefinitely.

How? We consider two lines *AB BC* enclosing a right angle *B*, and we pick a point *D* on the line *AB* at a certain distance, whether between [points] *A* and *B* whether outside [point] *B*. Then we move [125a] point *B* of line *BD* onto the line *BC* in the direction of point *C*, so that the line *BD* is moved with it in such a manner that the point *D* is opposite to point *A*, namely the three points *A, D, B* are always collinear. By this motion point *D* describes a segment of a curved line *DGH*, which we call the split line [הקו הפרוץ]. I say that as long as lines *BC DGH* are continued in the direction of *CH*, the distance between their extremities decreases, and they never meet.

The demonstration. We draw the two straight lines *AHC AGE*. The three lines *BD, GE, HC* are equal to one another. We draw from points *G H* two perpendiculars *GI HK* onto line *BC*. Since (i) the square of line *GE* is greater than the square of *GI*, (ii) the two triangles *GIE KHC* are right-angled, (iii) the hypotenuses *GE HC* are equal, and (iv) angle *GEI* is greater than angle *KCH*,[588] it follows that perpendicular *GI* is greater than perpendicular *HK*. Similarly it can be shown that of all the perpendiculars drawn from the split line *DGH* to the straight line *BC*, the closer to line *AB* is greater than that more distant from it. Hence, as long as the two lines *BC DGH* are extended in the direction of *CH*, the distance between their extremities decreases, and they never meet. Q.E.D.

And the instrument designed to draw the split line is common among artisans, and is very useful in this discipline.

Terms

קו פרוץ The term "conchoid" follows the Greek and Latin word for mussel shell.[589] The use of the term *qaw paruṣ* for conchoid may be unique in Hebrew. The word *paruṣ* means cracked or broken into, and in his extant Hebrew book *Theshuva la-Meḥaref* Alfonso uses this word in this sense, referring to the wall of Jerusalem.[590] Perhaps Alfonso thought of the conchoid as a broken line because of the "crack" between the two branches that keeps them apart. It is also possible that he had in

---

588 In the text *KHC*. Gluskina's correction.

589 The curve was called "conchoid" by Proclus, but Pappus has called it "cochloid" (Heath 1921, 238).

590 Ms. Parma, fol. 45b, Following Nehemiah 2:13.

591 This result can be justified by simple geometric (or trigonometric) considerations. Alfonso—as he often does—skips the details. It is possible that the subject was discussed in the beginning of chapter three that is missing in the manuscript.

mind the different but phonetically similar Arabic فرج in the sense of "split," "cleave," "diverge," "spread apart." In this vein we translate *ha-qaw ha-paruṣ* as "the split line."

Mathematical Commentary

The picking of point D between A and B or outside B determines which of the two branches of the conchoid is generated. According to the construction $GE = HC$, and ΔEIG, ΔCKH are right-angled triangles with equal hypotenuses. $\sphericalangle GEI$ is an exterior angle of ΔACE; therefore $\sphericalangle GEI > \sphericalangle HCK$ (*Elements* I.16), and so $GI > KH$.[591] It is clear that the perpendiculars from the curve to BC (in the direction away from line AB) become smaller, namely the curved line DGH approaches line BC. The construction implies that the lines will not meet.

This argument is valid for both branches of the conchoid (referring to the plain letters or letters marked by apostrophe in the diagram respectively). Alfonso apparently intended to construct both of them, as he uses both in the following propositions: the external one in Proposition 30, and the internal one in Proposition 31.[592]

The conchoid can be defined in terms of the neusis property as the locus of all the points on any line through the pole whose distance from the ruler [along the line] is constant. It can also be defined in terms of motion as the trace of a fixed point on a straight line through the pole, moving along the ruler. It is not sure how Nicomedes thought about his curve.[593] It is likely that Alfonso, who encouraged the use of motion in geometry and who mentions the drawing device (in Property 30), thought of it in terms of the second definition. The last sentence indicates that Alfonso took for granted the use of a special instrument for drawing the conchoid. This is confirmed also in I.2.25, and used in the next property.

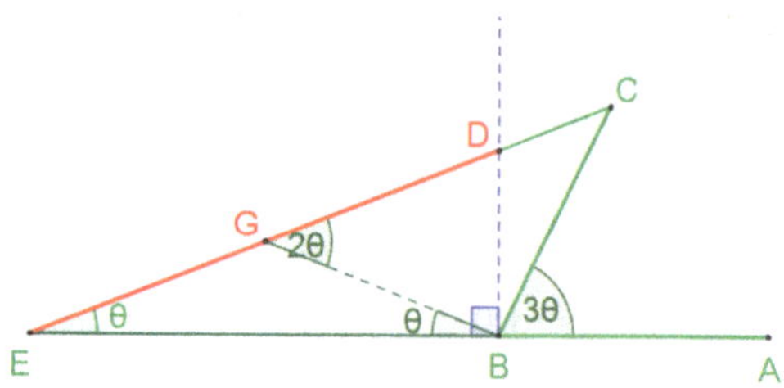

III.**[30] Property 30**

To divide any rectilinear angle into three equal parts.

[592] Alfonso mentions *en passant* that the constructions can be done by means of the conchoid, not explaining the *neusis* property of the conchoid on which the demonstrations of Propositions 30 and 31 rely.

[593] See Knorr (1987, 303); Sefrin-Weis (2010, 243–4 and 244, note 3).

[594] In the text *DHB*. Gluskina's correction.

[595] In the text *BDE*. Gluskina's correction.

How? We consider a rectilinear angle *ABC* and erect a perpendicular *BD* upon *AB*, we assume the magnitude of *BC* as we wish, and we extend *AB* indefinitely in the direction of *B*. We put the plotting device (משרטטת) on point *C* in such a manner that it meets the two lines *AB BD* in two points *D E*, so that line *DE* is twice line *BC*. This can be done by drawing [125b] the split line [conchoid] as mentioned above. Then angle *DEB*[594] would be a third of the given angle *ABC*.

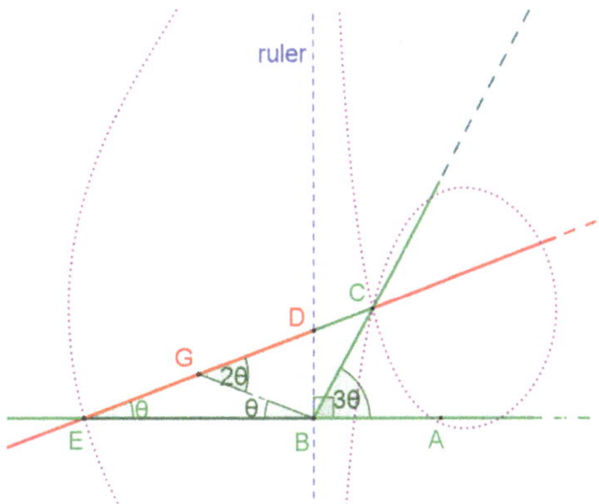

The demonstration. Since line *DE* of triangle BDE is the hypotenuse of the right angle *DBE*,[595] and when we cut it at the [mid]point *G* and join *BG*, [the line] *EG* becomes equal to *GB*, which is equal to *DG*, hence angle BGC, which is equal to angle *BCG*, is equal to angle *GBE* together with angle *GEB*; and angle *DEB* is a third of the two angles *GEB*, *GCB* together, which are equal to the given angle *ABC*. Q.E.D.

Mathematical Commentary

In order to trisect $\sphericalangle ABC$, Alfonso finds a point *E* on line *AB* (beyond *B*) so that $ED = 2BC$. This is done using a conchoid with pole *C*, ruler *BD* and distance $2BC$.[596]

BG is the median to the hypotenuse *ED* of the right-angled ΔEBD, hence $BG - EG - GD$.

Note. $\sphericalangle E = \theta$, then $\sphericalangle BGD = 2\theta$, being an exterior angle to ΔGEB.

According to the construction $BC = BG$, so ΔBGC is also an isosceles, and $\sphericalangle BCD = 2\theta$. Therefore $\sphericalangle ABC = 3\theta$ [being an exterior angle to ΔBCE], and $\sphericalangle E = \frac{1}{3}\sphericalangle ABC$.

This procedure is intended for trisection of an acute angle. A right angle can be easily trisected using ruler and compass,[597] and an obtuse angle—by dividing it into a right angle and an acute angle, and trisecting them separately.[598] Alfonso does not discuss these cases.

---

596 As noted in the previous property, Alfonso most probably took for granted the use of a special instrument to draw the conchoid.

597 Relying on *Elements* I.1 an angle of 60° is constructed, which can then be bisected.

598 In Pappus. See Sefrin-Weis (2010, 149).

599 Heath (1921, 235); Sefrin-Weis (2010, 148–149).

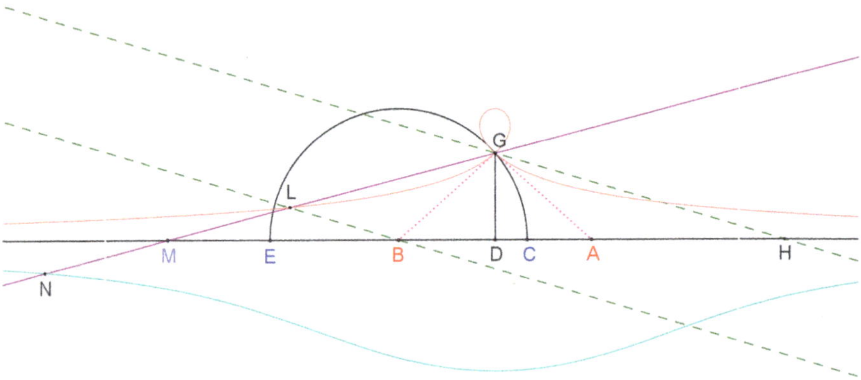

The angle trisection is achieved in a similar manner to that attributed to Nicomedes,[599] but the position of the perpendicular (relative to the angle's side) is different, hence the required trisection is constructed outside the given angle, rather than inside. Of the three applications of the conchoid that Alfonso presents, only this one resembles one of those known from the Greek tradition.

III.[31] The 31st Proposition

We wish to find two straight lines mean proportionals between two other straight lines that are unequal to one another.

How? We consider *AB* the smaller line and *CE* the larger, and we describe a semicircle whose diameter is *CE* the larger [line] and whose center is *B*, and *D*, the midpoint of *AB*, falls on the diameter [*CE*]. We draw a perpendicular *DG* on the diameter [and extend it] to the circumference of the circle, and we extend *AD* in the direction of *H* until *AH* equals *AB*. We join *HG*, and draw from point *B* a line *BL* parallel to *HG* and to indefinitely. We draw from point *G* a line *GM* towards line *CEM*, meeting line *BL* at point *L* in such a manner that line *LM* is half the diameter. This can be done by drawing the split line [conchoid]. I say that the ratio of *AB* to *GL* is as the ratio of *GL*[600] to *BM* and as the ratio of *BM* to *CE*[601]

Its demonstration. We extend *GM* until *LN* is equal to the diameter. Since the excess of the expansion of *MG* by itself over the expansion of *BG* by itself is as the expansion of *MB* by itself and by twice *BD*,[602] which is also[603] as the

600 In the text *GM*. Gluskina's correction.

601 In the text *NE*. Gluskina's correction.

602 In the text *BL*. Gluskina's correction.

603 In the text "with." Gluskina's correction.

604 In the text *GN*. Gluskina's correction.

605 In the text *AL*. Disagreeing with Gluskina here.

606 In the text *F* instead of *G* and the order is reversed: "*MB* to *LF*."

607 If $AB = CE$ the proposition becomes trivial.

expansion of *GL* by itself and by twice *LM*, then the expansion of *AM* by *MB* is as the expansion of *NG* by *GL*. Moreover, the ratio of *AM* to *NG* is as the ratio of *GL*[604] to MB, and the ratio of *AB*[605] to GL is as the ratio of half *MB* to *LM*, which is equal to the ratio of *MB* to *LN*. And it was already established that the ratio of *AB* to *GL* is as the ratio of *AM* to *NG*, which is equal to the ratio of *LG* to *MB*.[606] Hence, the ratio of *AB* to *GL* [126a] is as the ratio of GL to *MB* and to the ratio of *MB* to *LN*, which is equal to the diameter.

Mathematical Commentary

Alfonso constructs two mean proportionals between two given line segments $AB < CE$.[607]

The basic construction is: semicircle on *CE* as diameter, *AB* copied on *CE* so that *B* coincides with the center of the circle. A perpendicular *DG* is drawn from the mid-point of *AB* to *CE* and a point *H* on line *CE* is selected so that $AB = AH$. *GH* is joined and a line parallel to it is drawn through *B*. A conchoid is constructed with pole *G*, ruler *CE* and distance $BE = \frac{1}{2}CE$. By means of it one can find a line through *G*, cutting the parallel through *B* at point *L* and the line *CE* at point *M*, so that $LM = BE$. The point *L* is actually the intersecting point of the conchoid with the parallel through *B*. Then *LM* is extended in the direction of *M* to *N* so that $MN = BC =$ radius of the circle on *CE*.

By I.47 (*the Pythagorean theorem*):

(1) $$MG^2 = GD^2 + MD^2 \quad (\Delta MDG)$$

(2) $$BG^2 = GD^2 + BD^2 \; (\Delta BDG)$$

By subtracting we have $MG^2 - BG^2 = MD^2 - BD^2$.

(3) $$\begin{aligned} MG^2 - BG^2 &= (MG + BG) \otimes (MG - BG) = ((ML + LG) + LM) \otimes (MG - LM) \\ &= (LG + 2LM) \otimes GL = (LG + LN) \otimes GL = GN \otimes GL \end{aligned}$$

(4) $$\begin{aligned} MD^2 - BD^2 &= (MD + BD) \otimes (MD - BD) = ((MB + BD) + BD) \otimes MB \\ &= (MB + 2BD) \otimes MB = (MB + AB) \otimes MB = AM \otimes MB \end{aligned}$$

According to the construction $AB = 2BD$, and by comparing (3) and (4) we get $AM \otimes MB = NG \otimes GL$, hence:

(5) $$\frac{AM}{NG} = \frac{GL}{MB}.$$

According to the construction $LB \parallel GH$. Looking at ΔMGH and by *Elements* VI.2:

$$\frac{HB}{GL}=\frac{MB}{LM} \Rightarrow \frac{\frac{1}{2}HB}{GL}=\frac{\frac{1}{2}MB}{LM} \Rightarrow \frac{AB}{GL}=\frac{MB}{2LM}=\frac{MB}{LN}$$

so:

(6)
$$\frac{AB}{GL}=\frac{MB}{LN}$$

Applying *Elements* V.12 to the last equation and comparing to (5) we get:

(7)
$$\frac{AB}{GL}=\frac{AB+MB}{GL+LN}=\frac{AM}{GN}\underset{(5)}{=}\frac{GL}{MB}$$

Finally, combine Eqs. (6) and (7) to obtain

$$\frac{AB}{GL}=\frac{GL}{MB}=\frac{MB}{LN} \quad \text{as required.}$$

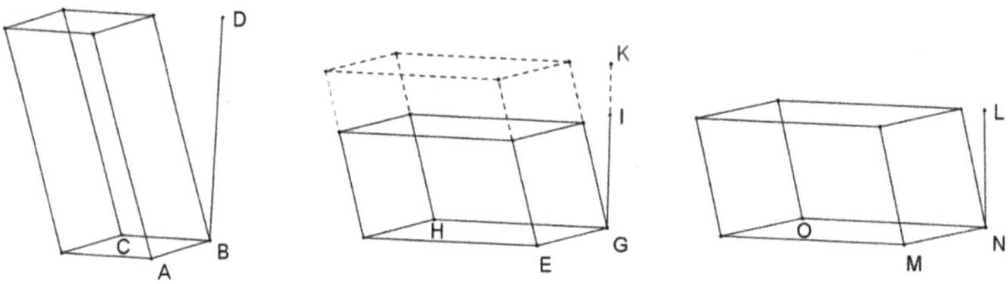

Historical Commentary

This impressive construction of two mean proportionals neither resembles nor alludes to any Greek or Arabic solutions known to us.[608] Its uniqueness is in using the *internal* branch of the conchoid for achieving the *neusis*. Nicomedes' construction of the two mean proportionals is carried out using the external branch of the conchoid, and the two given lines are set up perpendicularly.[609]

This property is used in the proof of the next one.

---

608 See Knorr (1989, pp. 251–319); Heath (1981, pp. 244–268); Clagett (1964–1984, I pp. 335–345, 658–665, III pp. 27–30, 849–854, 1163–1179); Rashed (2011–2013, pp. 60–69, 103–107).

609 See Heath, pp. 260–262.

610 In the text *EG*, our correction.

611 In the text *HG*, our correction.

612 In the text *MI*, Gluskina's correction.

III.[32] The 32nd Proposition

We wish to construct a polyhedron ("מושגם מיושר הקווים") which is equal to a given polyhedron, and which is similar to a second given polyhedron.

How? We consider a polyhedron whose base is area *AC* and whose height is *DB*, and a second polyhedron whose base is area *EH*[610] and whose height is *IG*. We wish to find a third polyhedron that is equal to the first and is similar to the second. We erect upon area *EH* another polyhedron whose height is *GK*, which is equal to the first polyhedron. We draw between *GK* and *GI*; two mean proportionals, *P*, *LN*, so that the ratio of *GK* to *P* is as the ratio of *P* to *LN* and as the ratio of *LN* to *IG*.

And we consider the ratio of LN which is known, to *IG* which is known to be as the ratio of *NM* which is unknown to *EG*[611] which is known. And we erect on line *NM* an area *MO*[612] which is similar to area *EH*. It [area *MO*] would be the base of the required third polyhedron and *LN* would be its height.

Its demonstration. Since the ratio of *GK* to *LN* is as the duplicate ratio of *LN* to *IG*, which equals the duplicate ratio of *NM* to *EG*, which equals the ratio of area *MO* to area *EH*, then the expansion of *GK* by area *EH* is as the expansion of *LN* by area *MO*, and the two polyhedra are equal. Therefore the third polyhedron is equal to the first. And since the ratio of *LN* to *IG* is as the ratio of *NM* to *EG*, the third polyhedron is similar to the second.

Terms

ערך שנוי This expression means duplicate ratio. This concept is defined in *Elements* V.9. Moshe Ibn Tibbon uses the expression נשנה בהשנות.

Mathematical Commentary

Though the general word for "polyhedron" is used, it seems that Alfonso actually had in mind parallelepipeds or simple boxes. A generalization of the proposition for prisms is possible, since no assumption on the basis is made or used in the proof. Property 32 is the only one in which Alfonso uses the concept of three-dimensional expansion. It is possible that he thought of "expansion of a straight line by a rectilinear area" as defined in III.0 definition [e], namely as a pyramid. It is more likely that he had in mind the "product" of a straight line by a rectilinear area as defined in III.0 definition [e*], namely a prism. Both interpretations are valid, and we offer here a general interpretation in terms of parallelepipeds.

Given two parallelepipeds P_1 and P_2, it is required to construct a new parallelepiped P_3 of the same volume as P_1 and which is similar to P_2. *AC* is the base of P_1, *BD* its

[613] This is not the simple classic problem of finding a fourth proportional, because not all the magnitudes are of the same kind. In such a case neither the existence nor the constructability of the fourth proportional is obvious. See Bos (2001, 125). Alfonso does not give any details about the way this construction should be carried out (unless he did it in one of the missing Properties 2–9).

height, EH the base of the P_2, and GI its height. Alfonso constructs a new parallelepiped $P' = P_1$, whose base is EH (the base of P_2) and whose height is GK.[613]

Relying on the previous proposition he constructs the two mean proportionals between GK and IG: P and LN.

$$\frac{GK}{P} = \frac{P}{LN} = \frac{LN}{IG} \tag{1}$$

He also constructs NM such that:

$$\frac{NM}{EG} = \frac{LN}{IG} \quad (\textit{Elements}\ \text{VI.12}). \tag{2}$$

Now the required parallelepiped P_3 can be constructed: its base area MO (whose side is NM) is similar to EH (the base of P_2), and its height is LN.

P_3 meets the demands, for:

$$\frac{GK}{LN} \underset{(1)}{=} \frac{GK}{P} \cdot \frac{P}{LN} \underset{(1)}{=} \left(\frac{LN}{IG}\right)^2 \underset{(2)}{=} \left(\frac{NM}{EG}\right)^2 = \frac{\text{area}\,MO}{\text{area}\,EH} \quad \left(\textit{Elements}\ \text{VI.20}\right).$$

Hence $GK \otimes \text{area}\,EH = LN \otimes \text{area}\,MO$, which means that the volume of P_3 is equal to the volume of P_1. The similarity of P_3 to P_2 follows from the construction of line NM (2) and of area MO.

Historical Commentary

Property 32 can be "contextualized" in more than one way:

It can be viewed as a three-dimensional version of *Elements* VI.25: "To construct a figure similar to one given rectilinear figure and equal to another." A theorem similar to Property 32 presented as a generalization of VI.25 appears in Stevin's *Problematum Geometricorum* from 1583.[614]

It can be viewed as an extension of *Elements* XI.27: "To describe a parallelepipedal solid similar and similarly situated to a given parallelepipedal solid on a given straight line."

It can also be viewed as a generalization of the problem of doubling the cube [the Delian problem], where P_1 is any parallelepiped of volume of two cubic units and P_2 is a cube whose side is one unit. The use of two mean proportionals to solve the Delian problem may have originated in Hippocrates and since then, remarks Heath, "the problem seems to have been attacked in the latter form exclusively."[615] It is thus of special interest that Alfonso "attacked" the original problem.

---

[614] See Bos (2001, 79–81).

[615] Heath (1921, 246); Knorr (1986, 23).

Property 33: The Ṭūsī Couple

III.[33] Property 33

Introductory comment

Property 33 is not a purely geometrical proposition like the rest of Part III. It is taken from astronomy and focuses on a debated question in Aristotelian physics: whether there must be a temporary rest at the two endpoints of an oscillatory linear motion. Aristotle's affirmative answer in *Physics* VIII.8 supports his thesis that only circular motion can be infinite. Alfonso disproves it by using the famous construction that later became known as the Ṭūsī couple. Naṣīr al-Dīn al-Ṭūsī, the Persian thirteenth century astronomer, had ingeniously shown that in planetary motion, when the parameters are chosen so that the radius of the epicycle is half that of the deferent and its angular velocity twice that of the deferent in the opposite direction, then the resulting motion of the planet is linear, back and forth along the diameter of the deferent.

Alfonso most probably knew about al-Ṭūsī's construction, but perhaps was not familiar with the details. Al-Ṭūsī uses an auxiliary construction that is not included in Alfonso's version. In 1996 Langermann commented that "we have yet no clues as to how these constructions may have reached Alfonso."[616] Twenty years later, we cannot say anything definite about the route of transmission, but we are fairly convinced that Alfonso had some information on the mathematical activity that was carried out in al-Ṭūsī's school.[617] The facts that he was familiar with the outline of al-Tūsī's construction but not with the details may suggest that the transmission was oral.

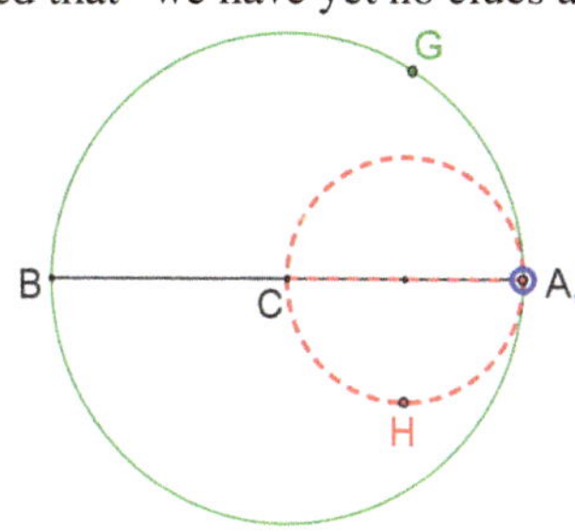

The text of Property 33 is confused. Perhaps while working on his proof Alfonso made some corrections in the lettering, which the copyist could not (or did not try to) understand. Parts of the proof are muddled and incomprehensible. Property 33 actually comprises two arguments, the first being included in the "how" part of the property, the second in the "its demonstration" part.

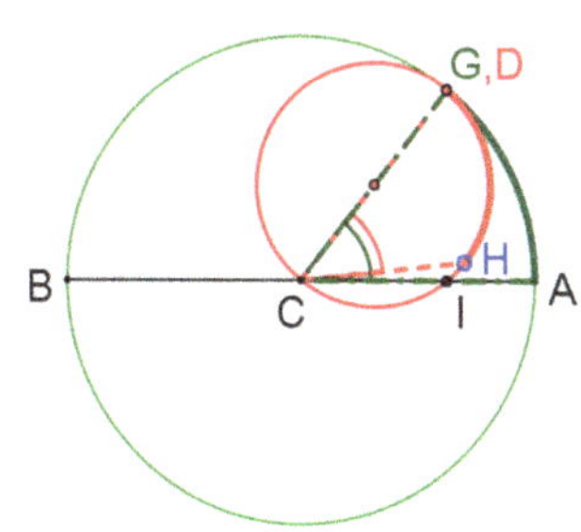

[616] Langermann (1996a, 34).

[617] See Introduction, Section 6.

[618] In the text: *E*.

[619] In the text: *E*. Several readings are possible. Seems that "*D*" agrees better with the rest of Property 33.

[620] In the text: the motion of *E* is twice that of *D*.

[621] In the text *DH*, Gluskina's corrects to *AG*.

[126b] We wish to describe a continuous perpetual linear motion forth and back on a finite line without a state of rest between forth motion and back motion.

[Argument I] **How**? We construct a straight line AB and describe a circle AGB about center *C*, and we describe on line *AC* another circle *DHC* touching circle *AGB*. We assign three letters *AED* to the point of touch and we assume that the center of the planet *D*[618] rotates on the circle DHC from point *D* to point H in a uniform motion. And point *E*, which is the touching point of the two circles, rotates from point *A* to point *G* in an opposite motion to that of *D*,[619] as also the motion of *D* is twice that of *E*,[620] namely when *D* completes [its rotation] about circle *DHC* twice, point E completes [its rotation] about circle *AGB* once.

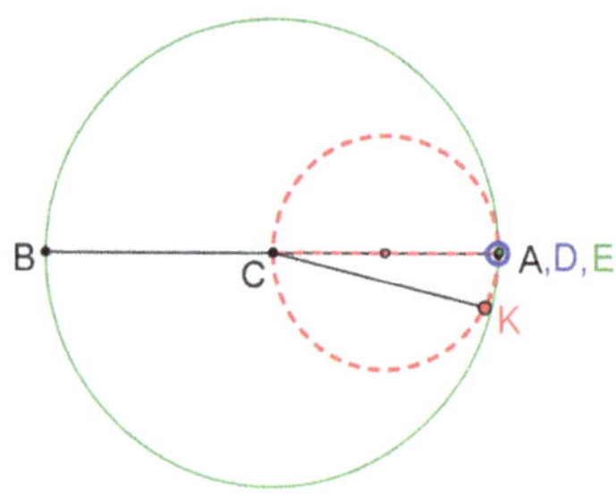

We join the straight lines *CG CH*. Since arc *DH* is similar to twice arc *AG*, and angle *DCH* at the circumference of the [inner] circle subtends arc *DH* [of the small circle], and angle [*ACG*] at the center of the [big] circle subtends arc *AD*[621] [of the inner circle], then angle *DCH* equals angle *ACG*. And therefore when line *DC* is superposed on line *CG* then line *HC*[622] is superposed on line *AC*,[623] and point *H*, which is the place of the center of the planet on line *AC* would fall on point I, it follows that the planet moved from point *A* to point *I*. Similarly it will be established that when the planet completes its circle once and point *A* arrives at point *B* and the planet will be at point *B*, and also it will return from point *B* to point *A* without deviating from line *AB* to one of the sides. I say that the planet will not rest at any time, neither at point *A* nor at point *B*.

[**Argument II**] Its demonstration. If it were at rest the planet would cover at this time no [127a] arc, since it always moves on the circle. We assume that it moves from point *D* to point *K*.[624] We join line *KC*. If point *K* were falling on point *A*, line *KC* would have equaled line *AC*, which is false. Therefore the planet is not at rest at point *A*. We thus found [that] the aforementioned linear motion [continues] without rest between going forth and back.

[**Corollary**] This implies that the cause of the rest that exists between two opposite motions is not the actual extremity and end of the distance along which the motion is, as Aristotle and his followers thought, because the

[622] Gluskina reads *AC*.

[623] Gluskina correct to *CI*.

[624] In the text: *C*.

[625] The word is not certain.

[626] Barker and Heidarzadeh (2016, 21–2).

extremity and end is not an act of the moved object so that it could come to a rest by its [the extremity's] effect, but, rather, it is the effect of the distance. The moved object can traverse it in its motion without coming to a rest at it. This is definitely the case when the motion of the moved body is subsequent upon another circular and continuous motion, or upon a force[625] that maintains continuity, as is the case in the two spheres of Venus and Mercury. And Abu 'Ali Ibn Ridwan has written that what had been claimed that between opposite motions there is a [state of] rest is not always true and that [a state of] rest between continuous motions does not occur unless the moving force loosens its grip.

Historical Commentary

The astronomical and mathematical activity that took place in the thirteenth and beginning of the fourteenth century around the observatory of Maragha did not find its way to the west fast.[626] The only known possible reference in fourteenth century to the Tūsī couple is somewhat dubious: in his *Questiones de spera* Nicole Oresme suggests: "it is possible for some planet to be moved perpetually according to its own nature in a rectilinear motion composed of several circular motions."[627] This may or may not testify to an acquaintance with the construction of the Tūsī couple.[628] Claudia Kren conjectures that "some account of the Eastern astronomer's work may have been available in Europe during the first half of the fourteenth century."[629] A lively argument has been going on recently on the question whether it is likely or not that knowledge that originated at the Maragha observatory reached Copernicus. The objectors claim to be skeptical "until the missing link is found."[630] Tzvi Langermann cautiously points out a possible such link through Alfonso and Mordekhai Finzi, who read *SMA* in Mantua in the 1470s.[631] It is also a possibility (though equally speculative) that Nicole Oresme's sentence quoted above could be based on oral transmission line that goes back to Alfosno.[632]

Mathematical Commentary

Argument I. The Planet Moves on the Diameter of the Outer Circle

The construction is of two touching circles, the radius of the inner circle is half that of the outer circle and its angular velocity twice that of the outer circle in the

---

[627] *Questiones de spera*, Droppers edition English 285–9, Latin 286–290; Kren (1971, 490–1), Kren's translation.

[628] Goddu (2010, 478–484).

[629] Kren (1971, 496).

[630] For example, Goddu (2010, 476). Blåsjö (2014) are skeptical as to eastern influence on Copernicus. Barker and Heidarzadeh (2016) adduce strong arguments to support the possibility of such influence.

[631] Langermann (1996, 34–5). See also Ragep (2017, 190).

[632] Oresme's proof, as Kren (1971, 492) and others noted, is confused.

opposite direction. Intuitively, with the ratio between the velocities being 1:2 the outer circle pulls the planet in the clockwise direction the same amount that the inner circle pulls it to the counterclockwise direction.

Diagram [I] describes an imaginary starting point of the two circles system. Alfonso assigns to each point a letter depending on whether it is static, moving on the outer circle, or moving on the inner circle. The big circle is denoted by *AGB*, with diameter *AB* and center *C*. Points *A*, *C*, *B*, are static, at rest relative to the fixed stars. The small circle is denoted by *DHC* with diameter *DC* which coincides with the radius of the big circle, so that the small circle always passes through the center *C*.

Points *A*, *D*, and *E* coincide at the beginning. *A* is the edge point of the diameter of the big circle, *D* is a point on the inner circle that represents the position of the planet on the inner circle. *E* represents the touching point of the two circles. The positions of *G* and *H* are not specified.

Diagram [II] describes the state of the system after the two circles had rotated a certain distance. $\widehat{DH}$ is the arc described by the motion of a point (described by the center of a planet) on the inner circle. $\widehat{EG}$ should have been the arc described by the motion of a point on the outer circle, but the letter *E* is no longer used, and it seems that $\widehat{AG}$ is taken to describe the motion of the outer circle. It seems that *G* represents the tangent point of the two circles considered as a point of the outer circle, *D* is the tangent point considered as a point of the inner circle. *H* is the point on the small circle carrying the planet. It has to be shown that *H* is on line *AB*, i.e., the planet moves on the diameter of the outer circle.

The outline of the proof is: $\widehat{AG}$ measures the clockwise motion of the big circle, $\widehat{DH}$ measures the counterclockwise motion of the small circle. Since the angular velocity of the small circle is twice that of the big circle, then angularly $\widehat{DH} = 2\widehat{AG}$. But $\sphericalangle DCH$ subtends $\widehat{DH}$ as inscribed angle of the small circle, whereas $\sphericalangle GCA$ subtends $\widehat{AG}$ as central angle of the big circle. Therefore $\sphericalangle DCH =$ half of twice $\sphericalangle GCA$ (*Elements* III.20 for similar arcs), hence $\sphericalangle GCH = \sphericalangle GCA$.

After having established the equality of the two angles, Alfonso uses superposition to show that the two triangles actually coincide: the vertex *C* and side *GC* coincide, the other sides *CA* and *CH* are in the same direction, the angles *GCH* and *GCA* are equal, hence *CH* must coincide with *CA*. Since *A*, *C*, and *H* are on the same straight line, by their definitions *H* and *A* must coincide. This means that *H* stays stable on the *AB* axis "without deviating from line *AB* to one of the sides."[633]

[633] Nice animations are available in the internet. For example, http://en.wikipedia.org/wiki/Tusi_couple.

Argument II. The Planet Does Not Rest at the Endpoints of Its Oscillation

Alfonso uses al-Tūsī's construction in order to consider the question whether or not the planet is at rest at the endpoints (a question that al-Tūsī himself had not addressed). On the one hand, the planet can be said to be at rest because it changes direction and therefore its instantaneous velocity at the endpoints is zero. When the planet passes through point A the two components of its motion are perpendicular to the diameter AB and in opposite directions.[634] On the other hand, the planet can be said not to be at rest because, being a projection of a uniform circular motion, its motion is continuous.

From the perspective of modern mathematics, the velocity is a vector and the question is rather meaningless. From the perspective of Aristotelian philosophy, however, the question is meaningful. Motion, according to Aristotle, is from a *terminus ex quo* to a *terminus ad quem*.[635] Arriving at the terminus and turning back "necessarily implies coming to a stand."[636] Aristotle's thesis was a subject of debate among Arab scholars. Rashed shows that the first to object to Aristotle's view was al-Qūhī in the ninth century. Al-Qūhī's approach, like Alfonso's, was to present the back and forth linear motion as a projection of a continuous motion.[637] The argument was renewed in the Avicennian tradition.[638]

Alfonso uses al-Tūsī's construction to show that the planet must not come to a rest at the two endpoints. His mathematical argument consists of one sentence. Since the argument is fragmentary and the reading of the letters (of the diagram) cannot be correct as is, no reliable reconstruction of the proof is possible. A possible suggestion (with two letter corrections) is: Had the planet stopped at A for a time interval Δt and the inner circle maintained its continuous motion, then after time Δt the point of the outer circle would still be at A, while the motion of the inner circle would describe an arc $\widehat{AK}$. We join lines AC and KC. Line AC and KC cannot be equal because AC is a radius of the outer circle and K is inside the outer circle. Therefore $KC < AC$ contrary to the assumption that the planet stayed at A during the time interval Δt.

Corollary. Alfonso's proof is rather vague (probably due to his tendency to be brief and to errors of the copyist who did not understand the argument), but his conclusion is clear-cut and offers a strong argument against Aristotle: it is not the endpoints that govern the oscillatory motion of the

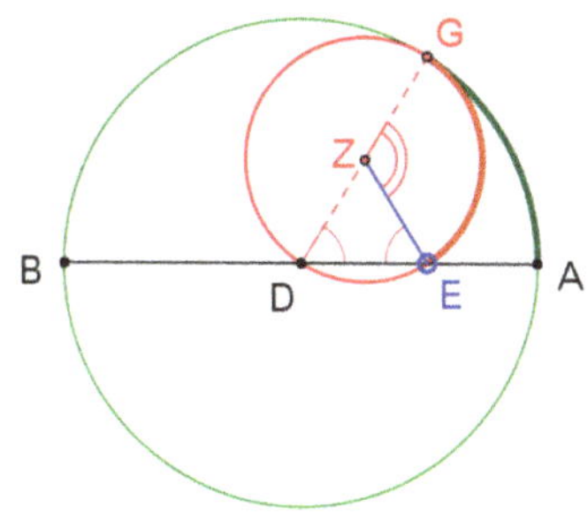

[634] The angular velocity of the inner circle is twice of that of the outer circle but the tangential velocities are equal and therefore the velocity of the planet at A is zero.

[635] *Physics* V.1, mainly 224b1–225a12.

[636] *Physics* VIII.8, 262a14.

[637] Rashed (2005, Appendix Section II). I am grateful to Prof. Rashed for the reference and for sending me the appendix.

[638] Langermann (2009).

planet, but rather the distance, namely the amount of motion that was traversed. Place itself has no effect on the motion.[639] The motion of the planet is induced either [i] by a continuous circular motion (as in the case of the epicycle described above) or [ii] by a force, such as the force that links the motions of Venus and Mercury (whose motions are affected by that of the sun[640]).

The importance of Property 33 to Alfonso's agenda is hard to exaggerate. It implies that linear motion and circular motion are not two distinct types of motion, as Aristotle contends. A linear motion can be generated by circular motions and can be continuous.

Appendix. Ṭūsī's Argument (From the Tadhkira)[641]

In order to prove that the point does not deviate at all from the line let the larger be circle *ABG*, its diameter *AB*, its center *D*, and the smaller be circle *GED*, its diameter *GD*, its center *Z*; the given point is *E*. Initially let diameter *GD* coincide with line *AD* and point *G* with point *A*, and let *E* be there with the two [points]. Then let circle *GED* move in the direction *GE* and let point *E* be moved along due to its motion until it has described, let us say, arc *GE*. And let circle *AGB* move [concurrently] in the direction *AG* with half the former's motion, and let the endpoint of diameter *DG* be moved along until it has described be moved along until it has described arc *AG*, which will then be equal [lit., similar] to half of arc *GE*. 10 We connect *E* to *Z* and *E* to *D*. Then angle *GZE* is twice angle *GDA* on account of the wo motions, and it is also twice this [angle] because it is an exterior [angle] of triangle *EZD* and equal to the two interior [angles] *ZED* and *ZDE*, which are equal because the two legs *ZE* and *ZD* are equal. Thus the two angles *GDE* and *GDA* are equal, and line *DE* coincides with line *DA*. Therefore point *E* is on diameter *BA*, not deviating from it, and the same will be the case in all other positions. Hence point *E* will continually oscillate between the endpoints of line *AB* and will not deviate from it.

[639] The natural places as causes of the natural movements play a major role in Aristotle's natural philosophy, yet, Aristotle himself is ambiguous on the question whether place can be considered a cause. Physics IV.1209a 18–22, 208b8–25. See Glasner (2015, 23, especially note 21).

[640] Pedersen Jones (2010, 295–8).

Part IV to Explain How a Body is Divided into Surfaces, and the Surface into Lines, as Plato Taught, and the Way of Evaluating and Measuring Them by One Another, and in it 16 Principles

Commentary

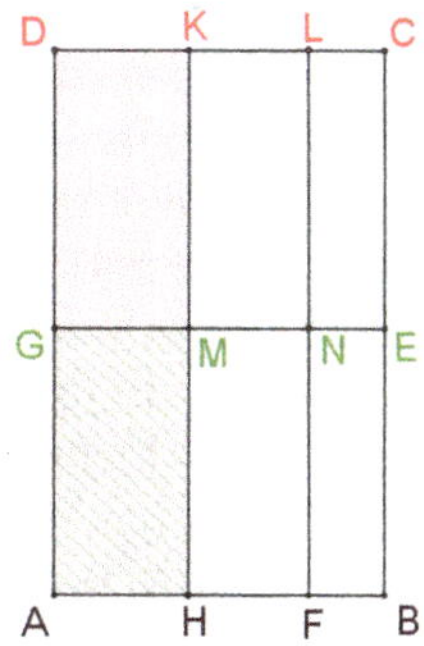

Part IV, according to the outline, endeavors to offer a precise mathematical account of Plato's geometrical reduction. Unfortunately only the first 32 lines are extant. The outline at the beginning of *SMA* and the recapitulated outline at the beginning of Part IV are not identical. The difference is significant. The first outline reads: "The fourth [part] explains how a body is divided into surfaces, and the surface into lines and the line into points, as Plato taught, and the way of evaluating and measuring them by one another."[642] This annunciation is misleading. Plato never carried out his analysis to the level of points. His analysis was terminated with the triangles.[643] The thesis quoted by Alfonso was nowhere stated by Plato nor ascribed to him by Aristotle or by one of the commentators. It calls to mind several Pythagorean theories, but is not identical to any of them. In the second outline Alfonso omits the words "and the line into points," thus partially correcting the first. Even with this correction the statement that Alfonso ascribes to Plato seriously conflicts with a basic tenet of Greek geometry, as stated by Aristotle and Euclid, namely that a magnitude is divisible only into magnitudes of its kind.[644] Point, line, plane and solid are entities of different types. Lower-dimensional entities are boundaries, but not parts of higher-dimensional entities.[645] Trying to rely on the authority of Plato, and in sharp contrast to Aristotle, Alfonso allows division of magnitudes into lower-dimensional magnitudes. This scheme was presented in I.4.2, and should have been put to use in Part IV. Perhaps what Alfonso intended to so in Part IV was to show how two nonhomogeneous two-dimensional magnitudes that cannot be equated qua areas can be equated by decomposing them into one-dimensional parts.

---

641 Copied from Ragep (1993, vol. I, English translation 198, Arabic text 199).

642 Outline, Section 0 above.

643 On the significance of this correction see commentary to I.3.2, thesis ii.

644 *Phys.* VI.1, 231a21–b18; *Meta.* V.131020a11–14.

645 *Elements* I, definitions 1–6, XI, definitions 1–2.

[4.1] The First Principle

When there are two areas with four straight sides and four right angles [i.e., rectangles] [127b] on one basis or on two equal bases, [then] the ratio of one area to the other is as the ratio of that which falls inside the first quadrilateral of the infinitely many heights drawn on the basis and divide it and the two quadrilaterals into equal parts—namely the heights described on it by the linear equally divided motion of the height (as I mentioned in Part I)—to all that which falls inside the second quadrilateral. This is so also with the similar cylinders and similar prisms with the same height, that the ratio of the larger to the smaller is as the ratio of the infinitely many parallel planes that cut the large magnitude into equal parts to all of them [the parallel planes] that fall in the small magnitude. This is also so with the circular cylinders described by the right-angled areas when they are moved [i.e., rotated] about their axis. The ratio of one [cylinder] to the other is as the ratio of the squares of all the perpendiculars drawn on the first area on its axis to the squares on all the perpendiculars drawn in the second are on its axis.

How? We consider two right-angled four-sided areas [i.e., rectangles] *ABCD* and *ABEG* on basis AB. I say that the ratio of *ABCD* [to *ABEG* is as the ratio] of the infinitely many perpendiculars described by the equally divided linear motion in area *ABCD* to that which falls of these perpendiculars in area ABEG.

Its demonstration. We divide basis *AB* into any number of equal parts by lines *HMK*, *FNL* and their likes. Thus the ratio of area *AK* to area *AM* is as the ratio of lines *AD HK* together to lines *AG HM*[646] Similarly the ratio of area *HL* to *HN* is as the ratio of line *FL* to line *FN*. And the ratio of area [text terminated].

Terms

מוצרף The word appears only here and no further information is provided. We assume that it means "prism." The term used by Moshe Ibn Tibbon for prism is תמונה מוגשמת מגוררת (Definition XI.13) or simply מגורר (for instance, in Proposition XII.7).

חץ Axis. The word is mentioned also in Section I.5.

שטח בעל ד' צלעות ישרות נצב הד' זויות See commentary on Section III.0.

Commentary

AB is divided into equal segments $a_1, a_2, \ldots, a_n$, which are bases of the smaller rectangles $s_1, s_2, \ldots, s_n$ with equal heights $h_1, h_2, \ldots, h_n$, and of the bigger rectangles $S_1, S_2, \ldots, S_n$, with equal heights $H_1, H_2, \ldots, H_n$. The statement is that the ratio between the big rectangle and the small rectangle is equal to the ratio between what falls in them of "the infinitely many heights on the basis." What remains of the proof is:

$\frac{S_1}{s_1} = \frac{H_1 + H_2}{h_1 + h_2}$ (*Elements* VI.1 and V.12). Similarly $\frac{S_2}{s_2} = \frac{H_2}{h_2}$ (Elements VI.1).

[646] In the text: *HG*.

At this point the mathematical proof is discontinued, but from the setting out we learn that the argument should lead to the conclusion that the ratio between the big rectangle and the small rectangle is equal to the ratio between *all* the infinitely many heights that fall in the big rectangle and *all* the infinitely many heights that fall in the small rectangle. How can the geometer consider all these heights? Alfonso's answer is that they are generated by the motion of the side along the basis. So the first step was arguing that presumably by increasing n, the number of small rectangles of which $ABCD$ and $ABEG$ are composed, indefinitely. His scheme could be:

$$\frac{S_1}{s_1}=\frac{H_1+H_2}{h_1+h_2},\quad \frac{S_i}{s_i}=\frac{H_i}{h_i}\quad i=1,\ldots,n$$

$$\frac{S(ABCD)}{s(ABGH)}=\frac{\sum_{i=1}^{n}S_i}{\sum_{i=1}^{n}s_i}=\frac{\sum_{i=1}^{n}H_i}{\sum_{i=1}^{n}h_i}\quad (\textit{Elements}\ \text{V.12})$$

In I.4.1 Alfonso contended that motion can divide a magnitude into infinitely many parts. By moving the side of the rectangle, maintaining a right angle with the base, all the infinitely many heights would be instantiated. Consequently

$$\frac{S(ABCD)}{s(ABGH)}=\frac{\sum_{\text{all } i}H_i}{\sum_{\text{all } i}h_i}$$

In this simple example all the H_i are equal to one another and so also all the h_i and the complex method presented in Principle 1 is not needed. Perhaps Alfonso will try to use this method for computing the ratio between a circle and the square on its diameter by dividing them into small rectangles.

It is interesting to compare Alfonso's idea with Cavalieri's:

Suppose that we have a plane figure and we draw a straight line inside it, and suppose, furthermore, that we then draw all the possible lines inside the figure that are parallel to the first. "In that case," he writes, "I call the lines so drawn 'all the lines' of that plane figure. Similarly, given a three dimensional solid, all the possible planes inside a solid that are parallel to a given plane are 'all the planes' of that solid." Is it permissible, he inquires of Galileo, to equate the plane figure with "all the lines" of the figure and the solid with "all the planes" of the solid? Furthermore, if there are two figures, is it permissible to compare "all the lines" of one with "all the lines" of the other, or "all the planes" of one with "all the planes" of the other.[647]

[647] Alexander (2014, 138).

Chapter 3
Hebrew Text

R. Glasner, A. Baraness, *Alfonso's Rectifying the Curved*,
Sources and Studies in the History of Mathematics and Physical Sciences,
https://doi.org/10.1007/978-3-319-77303-2_3

[0. תוכן החיבור]

[293ב] הספר הזה יש בו חמישה שערים: **הראשון** בהקדמת הספר ולברור שציור התנועה השוה בחלקיה הוא ראוי להיות מן ההתחלות לחכמת השיעור.[1] **השני** במקרים המתחברים אל התנועה המונעים את ציורה מלהיות התחלה לחכמת השעור, ואיך השתבשו בם רבים מהחכמים הקודמים. **השלישי** בסגולות השעורים ישרי הקווים והשטחים המועילות בזו החוכמה. **הרביעי** לבאר דרך החלק הגוף אל השטחים והשטח אל הקווים והקווים אל הנקודות, כמו שאמר אפלטון, ודרך שעוריהן ומדידתם אלה באלה.[2] **החמישי** בקצת סגולות חלוקי הגופים אל השטחים והשטחים אל הקווים והקווים אל הנקודות הצריכות להישרת העגלה, ובמדידת הכידור והאסטונא המעוגלת והמוצק העגול והטבעת הכידורית והטבעת הרבועה והעגולה ושבריהם והקווים והשטחים המקיפים בם, והשבת קצתם לדמות קצתם, והשויתם לישרי הקווים, ובחלוקי הזויות לכל החלקים השווים שנרצה.

[1] המילה "שעור" כתובה כאן ולהלן בפיסקה I.2.2 בכתיב מלא, אך בשאר המקרים – בכתיב חסר.

[2] בכתב היד: בלה.

השער הראשון בהקדמת הספר

[1.0 שיר פתיחה]

אמר אלפונשו המחבר, אני הוא המדבר.

ברוך יי׳ לעולם אמן ואמן, אשר כונן לבי וזמן.

אמרת חכמים להאמן: יגעתי ומצאתי האמן,

יגעתי ולא מצאתי אל תאמן.

כי לבי התמיד ועיונו לא סר, ולהג הרבה יגיעת בשר.

ולא יגע לתהו ואפס, למצוא דברי חפץ,

לקנות דעה צלולה [94א] מרוח הקדש אצולה.

להראותי את כבודו ואת גדלו, לחזותי בנועם יי׳ לבקר בהיכלו.

ולעשות לו שם וגדלה, ואפי׳ במשיחת העגלה,

שהיתה מנעורי ועד זקוני אחת[1] שאלתי מאת יי׳,

שהכל ברא לכבודו והכל הדרו והודו מלא כל הארץ כבודו.

[1. הצגת הבעיה המתמטית: האם ייתכן שוויון בין גדלים ישרים ומעגליים]

[1.1 הצגת שתי העמדות] רצוני לומר שבקשתי לדעת אם אפשר לימצא שטח ישר הקוים שוה לשטח העגלה כפי האמת, לא בקירוב הנמצא לחכמים הקודמים, ולא כפי מחשבות בעלות[2] הספקות שעלו על לבות קצתם, ככתוב בדברי ארסטו ושאר הנמשכים לדעתו. רק[3] בדרך מופת נחתך אמתי בלי שום קירוב ולא שום זיוף. זה שנים רבות שהיתה תשוקתי למצאו כי לא[4] נכון בעיני מה שכתב בן רשד בזה שהוא נמנע להמצא לו מופת אמתי בשום פנים, וכי נמנע שיהיה קו ישר שוה לקו מעוגל, לפי שלא ירכבו חלקי הקו האחד על חלקי הקו האחר, וכן שאי אפשר שישוה שטח ישר הקוים לשטח העגול לטעם זה בעצמו,

---

[1] בכתב היד: אחר.

[2] אולי צ״ל: בעלי.

[3] בכתב היד: רך.

[4] בכתב היד: לו.

כי לא ירכבו חלקי השטח הישר על כל חלקי העגול. וכן החכם בן סינא היה סובר שהקו או השטח הקשתיים לא יחזרו לעולם להיות ישרים.

[1.2 **ראיות מאוקלידס וארכימדס**] ונ״ל שאין טעם בדבריהם כי אנו מוצאים עגול אחד שוה לשני עגולים קטנים כמבואר מאקלידיס, ואע״פ שלא ירכבו שני העגולים יחד על העגול האחד בכללו ולא כל חלקיהם על כל חלקיו. וכן השטח[1] המקיף בכדור שוה לשטח עגולה ולשטח אסטונא עגולה [94ב] ולשטח המצק שלה כמו שביאר ארשמידס, ולא ירכב שום חלק מהאחד על שום חלק מהאחר. ולזה לא נתיאשתי מחקור בדבר כי חשבתי אולי יש אפשרות ברבוע העגול אעפ״י שלא ירכבו כל חלקי המרובע על כל חלקי העגול.

[1.3 **ראיה מבריסון**] ונ״ל כי לזה אמר החכם פרוס המשער כי לפי שהיה העגול גדול מכל שטח ישר הקווים עשוי בתוכו וקטן מכל שטח ישר הקוים עשוי חוצה לו, והעשוי בין שני אלו קטן מהחיצון וגדול מהפנימי, יהיה העגול שוה לזה השטח האמצעי בלי ספק. כי זה החכם ידע באמת שהעגול והשטח ישר הקוים הם תחת סוג השטח ואינם מתחלפים רק במקיפים המגבילים, וכי על כן היה שטח ישר הקוים גדול או קטן מהעגול בלי ספק. כי לא יתכן לומר גדול ולא קטן בשני שעורים שלא יהיו תחת סוג אחד קרוב, כמו שלא יאמר שההר גדול ולא קטן מהיום. ולזה כל שטח שאינו גדול מהעגול ולא קטן ממנו יהיה בהכרח שוה לו, והוא התבנית שקראו החכם פרוס אמצעי בין השטח החיצון ובין הפנימי. ובאלו[2] אמר הקדמה תנאית מדובקת, והיא:

> שאם נמצא שטח ישר אמצעי בין כל שטח שאפשר לעשות תוך העגול ובין כל שטח שאפשר ליעשות חוצה לו, ישוה בלי ספק אל העגול.

וזו הקדמה אמתית ראויה להיות התחלה לחכמת השעור כשאר התחלות האמתיות שלה. וביחוד ראויה להיות התחלה והקדמה גדולה לרבוע העגול. [95א] ונשאר על בעלי חכמת השעור לחקור על מציאות ההקדמה הקטנה והיא קיום המוקדם בהקדמה זו שהיא גזרה תנאי מדובקת. כלומר למצא שטח ישר הקוים שיהיה אמצעי קטן מכל שטח ישר הקוים שיעשה חוצה לו וגדול מכל שטח ישר הקוים שיעשה בתוכו. ואז תחוייב תולדת ההקש והוא קיום הנמשך בהקדמה התנאית המדובקת.

---

[1] מילים מיותרות ומחוקות: השוה בכדור מקיף.

[2] בכתב היד: וכאלו.

[1.4 ראיות נוספות מארכימדס] ועל הקדמה זו בנה ארשמידס מופתיו לבאר שהעגול שוה לרבוע חצי קטרו בחצי מקיפו. כלומר שאם נמצא שטח ישר שאינו גדול ולא קטן מרבוע חצי קטר העגול בקו שוה לחצי מקיפו, הוא יהיה השטח הישר השוה לשטח העגול. אלא שהוא לא מצא לעולם קו ישר שוה לחצי מקיף העגול, אלא בקירוב. וארשמידס הוסיף בראיתו[1] על דברי פרוס כי בהמצא שטח אמצעי כמו אותו שאמר פרוס שוה לעגול, יתחייב להמצא עוד קו ישר שוה למקיפו, ובהפך זה, כי בהמצא קו ישר שוה למקיף העגול יחויב להמצא אותו שטח אמצעי שאמר פרוס. אלא שהוא לא מצא מעולם ההקדמה הקטנה לכך, כי לא מצא שטח אמצעי כמו זה ולא קו ישר שוה למקיף העגול.

ועל כמות אותה ההקדמה הגדולה בנה ארשמידס עוד מופתיו למצא עגלה ישרה שוה למקיף הכדור ושוה למקיף האסטוונא[2] ושוה למקיף המוצק שלה,

כלומר שאם נמצא עגול ישר אמצעי שאינו לא גדול ולא קטן מכל אחד מהמקיפים הנזכרים הוא יהיה שוה לו. [95ב] אלא שביאר עוד ההקדמה הקטנה בהקש התנאי, כי מצא העגול האמצעי הנזכ׳.

[2. הצעת המושג הרכבה שכלית]

[2.1 שתי דוגמאות להרכבה שאינה איזומטרית] ויחויב מזה שאין הרכבת השעורים קצתם על קצתם באי זה דרך שיהיה היא ההרכבה העצמית להודיע השויתם קצתם בקצתם, כמו שחשב בן רשד, כיון שאותם שעורים השוים זה לזה אי אפשר שירכבו זה על זה, ועם שנמצא הרכבת שעורים מתחלפים בכמותם רוכבים זה על זה.

[דוגמה ראשונה] כגון חצי גלגל המזלות, שהוא ק״פ מעלות, וירכב כלו על קשת הרחב בגלגל חצי היום שהוא מ״ח מעלות, ופוגשת כל נקודה מהקשת הגדולה בנקדה שהיא חברתה מהקשת הקטנה, ולא תיסוף הגדולה על הקטנה.

[דוגמה שניה] ובדרך זו בתנועת הגלגול היה אפשר כמו שנבאר [i] שיגע שטח אסטונא כלו את שטח ישר גדול מאד ממנה, ויגע גם כן את שטח ישר קטן מאד ממנה, [ii] ויגע הקו המקיף בעגלה את קו ישר ארוך מאד ואת קו ישר קטן קצר מאד, [iii] ותרכב זוית חדה על זוית נרוחת ולא יוסיפו ולא יגרעו זה מזה,

[1] בכתב היד: בראתיו; תיקון גלוסקינה: באותו.

[2] בד״כ כתוב ב-ו׳ אחת, כאן בשתיים.

עד שרבים מחכמי הטבע וכ״ש שאר אנשים יתמהו מאוד בזה ולא יאמינו בו עד שיראוהו בעיניהם. ועד כשיראוהו יהיו במבוכה גדולה כי יאמרו: איך אפשר שיגע בשוה כל חלק משעור גדול את חלק חברו משעור קטן ולא יוסיף עליו כלל, כמו שיוסיף עליו בהרכבה השוקטת. ולא ידעו באי זה צד תחייב התנועה זה.

[2.2 תיקון לאקסיומת ההרכבה של אוקלידס] ולזה אומר שצריך להוסיף תנאי בהרכבה שזכר אקלידס בתחלת ספרו שלא תהיה בתנועה וחלק אחר חלק אלא כלה [96א] בבת אחת וברגע אחד. כי אין ספק שהשעורין שיצטייר הרכבתן זה על זה בבת אחת (ולא קצתם אחר קצת כמו שהוא כך בתנועה) יהיו שוים זה לזה; כמו שהשיעורים השוין זה לזה יתכן שיצוייר הרכבתן זה על זה בשום ענין אלא [...].[1] כי לפי שאותן שעורים שזכרנו השוים זה לזה [בסעיף I.1.2] לא יצוייר בם שירכבו זה על זה בבת אחת בזו ההרכבה שזכרנו לציירה [האוקלידית], יתחייב לומר בהכרח שיש שם הרכבה אחרת עליונה במעלה על הרכבה זו שאנו רגילין לציירה, היא הסבה העצמיית להשואת השעורים אלו באלו.

[2.3 הרכבה שכלית אצל היפוקרטס] ואני חושב כי בהרכבה זו העליונה רמז אבקרט המשער באמרו שאם תחתך[2] עגולה לתבניות ירחיים השוים לשטחים ישרים הקוים, תהיה העגלה שוה לאותם שטחים הישרי הקוים יחד. כי אין ספק שאם זה אבקרט היה יודע דבר בזו החכמה שהיה מפורסם כחכם בה, לא היה אומר שהעגולה תחלק לתבניות ירחיים כדי שירכבו כלם יחד על העגולה בהרכבה זו מפורסמת שהרגלנו לציירה כדי שיהיו שוים לה. אבל ר״ל שיש ערך בין השטח העגול בקצותיו ובין השטח הישר בקצותיו, שהתחיב מכח היות עגולה אחת שוה לשתי עגולות קטנות או שלש. ולא היה זה אלא בשביל זו ההרכבה השכלית והעליונה.

אלא שנודע לנו דרך השויתן מדרך ראיה הנקרא בערב דליל, לא מדרך מופת העצמי הנקרא בערב ברהאן. כי בהרכבה זו השכלית לבד תהיה העגולה נחלקת לתבניות שוים לשתי עגולות [96ב] קטנות או שלש. ומשם התחייב בעצם היות התבנית ירחי שוה לשטח ישר הקוים. וכמו כן יש לחשב מזה שאולי ימצא בכך היות העגולה שוה לשטח ישר הקוים. ולפיכך[3] אמר אבקרט הקדמה תנאית ובלשון חששא. ואינו

[1] כנראה שסוף המשפט חסר.

[2] בכתב היד: תחתוך.

[3] בכתב היד: לפי, ללא גרש.

מתנאי ההקדמה התנאית היות המוקדם שלה צודק בהכרח, שאלו כן לא היה צריך לחזרו בהקש פעם שנייה ולהיות לו כהקדמה קטנה.

[2.4 הרכבה שכלית אצל ארכימדס] אך נרצה שעל ציור זו ההרכבה השכלית בנה ארשמידס מופתו על שווי העגולה לרבוע[1] חצי קטרה בחצי מקיפה, כי הוא בארו במופת החלוף הנקרא ברהן אלכלף, ולקח בו הקדמה תנאית מתפרקת והיא היות הקו הישר גדול או קטן או שוה לקשת ממקיף העגולה. וכשלא יהיה גדול ולא קטן יחויב שיהיה שוה לה. ולא יהיה קו ישר שוה לקשת ממקיף העגולה אלא מכח הרכבה זו השכלית שזכרנו. וכמו כן בזה הצד בנה ארשמידס מופתיו על ההרכבה השכלית כשבאר שווי שטח העגלה לשטח האסטוונא המעוגלת, וכן לשטח מוצק העגול או לשטח הכידור. והוא היה יכול להסכים בכך מכח התבנית הירחי המחייב להבין שיש שם הרכבה שכלית נותנת ערך בין השעורים המשתנים בתבניותיהם.

[2.5 הערה מתודולוגית] ובכאלו הודיע אבקרט כי שווי העגולה לשטח ישר הקוים הוא דבר טבעי ונודע אל השכל, אלא שיש ספק אם יתכן לבעל חכמת השעור להגביל אותו שטח ישר הקוים. ויש עליהם לחקור עליו, כי אין אותה ידיעה הכללית של השכל מספקת להם לפי דרכם, אעפ״י [97א] שתהיה מספקת לחכמת השימע[2] וחה שאחר הטבע. כמו שהוא נודע לשכל על אי זה מקום יגיע הקו הישר השוה בקו ישר אחר מונח בהיות התחלתו מנקדה מונחת, אלא שלא יספיק זה לבעל חכמת השעור, אבל יחקור על הגבלת אותו קו ומציאותו בפעל אצלו, כמו שעשה זה אקלידס בתחלת ספרו. וזה יהיה להם בכלים מוכנים לכך כמחוגה והמשרטטת או כלים אחרים. וכל שיהיה השעור צריך לכלים פחות ויותר קלים להמצא, תהיה הידיעה בו יותר משובחת ויותר קרובה למעשה.

[3. הפניה לאפלטון]

[3.1 עקרון פרשני ״מדרך השכל והמוסר״] ואני הייתי ימים רבים מחשב על זו ההרכבה השכלית לדעת מה היא ואיך אוכל לציירה. ויהי בשוטטי ברחובות העיון בספרי העיון ארסטו והנמשכים לדעתו, וראיתי את מיני התשובות והתפיסות שכתב על אפלטון ואבן דקליס ושאר הקודמים, ולא ישרו בעיני וזה מפני הקדמה אחת נכונה מדרך השכל ומדרך המוסר יחד שהכניס הב״ה בלבי, לא ראיתי ארסטו

[1] בכתב היד: לרביע.

[2] בכתב היד: השעור.

משמש בה ברוב דבריו. היא היתה פתח גדל לי ומבא לקצת תעלומות. וזו היא ההקדמה אשר אני רואה: שאין ראוי לייחס אל החכמים שהתעסקו כל ימיהם במחקרי החכמות את הטעיות שיטעו בם שאר אנשים שלא התעסקו בחכמה כלל. וכ״ש מהטעיות הברורות מאד שלא יטעה אחד מהב״ח בלתי מדברים לא יתכן ליחסם לחכמים שהתעסקו כל ימיהם בחכמה ובדרישה ובחקירה וחבורי ספרים בהם. אלא אם נמצא שום דבר [97ב] להם שיש בו כנראה אחד מטעיות אלו, יהיה ראוי בלי ספק לבקש לו התנצלות ופנים שיהיו כפי האמת מתיישבים בשכלו.

[3.2 שלוש תזות מיוחסות לאפלטון: הצגה ראשונה]

[שלוש תזות מיוחסות לאפלטון] ומצאתי באותן תפיסות שהשיב ארסטו לאפלטון במה שאמר [i] שיש שם מונח מקבל את הכל והוא ההיולי, וכי [ii] הגופים נחלקים ושבים אל המשלשים ואל הקוים, וכי [iii] סבת הכובד לגופים הכבדים היא רבוי הגרמים הראשונים שמהם הורכב הגוף וסבת הקלות היא מיעוט הגרמים ההם.

[חמש שאלות מיוחסות לאריסטו] ותמה ארסטו מאד בזה ואמר [a] איך יתכן שיהיה ההיולי שטחים? [b] וגם אם יהיו שטחים איך יהיו מוצקים, כי המוצקים הם גופים מחודדים מלמעלה ורחבים מלמטה[1] ואינם שטחים? [c] ועוד, אם ההיולי שטחים איך יתחלקו השטחים וישובו למשולשים?[2] שאם כן היו המשולשים ראויים להיות היולי יותר מהשטחים. [d] ועוד, אם היה ההיולי משולשים איך יתכן שיהיה קוים? [e] ועוד, שאם הגרמים שהורכב הגוף מהם לדעתו היו שטחים איך יתכן שיהיה רבויים סבה אל הכובד, כי לא יתכן לו שהשטח יהיה יותר כבד משטח אחר אפילו היה גדול ממנו?

[3.3 תיקון ההצגה הראשונה] אז אמרתי מדרך אותה הקדמה שזכרתי, שאין ראוי ליחס איש חכם כאפלטון טעיות ברורים כמו אלו שהיה ארסטו מיחס אליו. אלא שצריך לבקש פנים לדבריו שיהיו מתישבים אל השכל. ואמרתי כי כל אשר היו אותן קושיות שעשה ארסטו על אפלטון יותר חזקות ויותר מורות על סכלות אפלטון, כך הם אצלי יותר מורות על [98א] אמתת דבריו ועל גודל מעלתם. וכאשר מצאתי עם זה מאמר כתוב בשם אפלטון על פתח מדרשו – כל מי שלא יהיה בקי בהגיון ובחכמת השעור

---

[1] בכתב היד: מלמעלה.

[2] הכתיב של המילה אינו עקיב. ברוב המקרים הכתיב חסר, אולם כאן ובמקומות אחדים נוספים הכתיב מלא.

אל יכנס לכאן – אז אמרתי שמא היה אפלטון מדבר באותם מאמרים בשני צדדין: האחד מצד הטבע שרוב דבריו הנמצאים היו בו, והשני מצד השעור שהיה חפץ בו. והתחזקה אצלי מחשבה זו במצאי שלא [אמר] אפלטון שהשטחים והמוצקים יהיו היולי אל הגופים כמו שחשב ארסטו עליו, ולא שרבוי הגופים[1] המשלשים יהיו סבה אל הכובד. אך אמר שהגופים מתחלקים ושבים אל המשלשים ואל הקוים, וכי הגוף הכבד הוא המורכב מחלקים[2] יותר והקל הוא המורכב מחלקים פחות.

[3.4 שלוש תזות מיוחסות לאפלטון: הצגה שנייה]

[i] ואמרתי שחלקי הגוף אצלו שני מינים. **האחד** כפי הטבע הוא הנקרא אצלו היולי וגרמים. והוא שאמר עליו שהוא מונח אל הכל ושחיצונו פעל ותיכונו כח, ושאלו היו כל הדברים נקראים בשמו, היה בתכלית האמת, כמו שאלו היו הענקים והטבעות מהזהב נקראים בשם זהב היה בתכלית האמת.

[iii] ואמר כי כשהיו בגוף אחד גרמים יותר ממה שיהיו בגוף אחר שוה לו בגדלו וכמות שעורו, תהיה צורת אותו גוף בטבעה מבקשת המקום הראוי לה כפי כמותו, והוא המקום הקטן בטבעו. אבל צורת הגוף אשר לו חלקים פחות, כי יהיו נרחבין וצריכין מקום רחב וגדול, תדרוש בטבעה המקום הגדול הראוי לה. והמקומות הטבעיים הם מוגבלים[3] אל הגופים כפי קרבם ורחקם [98ב] ממקיף הגלגל. ולכן כל שהיה מהם יותר קטן יהיה קרוב למרכז העולם, וכל שהיה מהם יותר גדול יקרב יותר למקיף הגלגל. ועם היות מקום החלק והכל אחד בגופים הטבעיים, כמו שאמר ארסטו', ויהיה החלק מבקש המקום הראוי אל הכל. וזו היא סבת הקלות והכובד בכלל אצל אפלטון ודימריטיש ולוקיש ושאר הקודמים.

[ii] ואין כונתי עתה לדבר בחכמת הטבע ולא לקיים מציאות אותם גרמים הטבעיים ולא לבטלם. אלא שאמרתי זה אגב גררא עבור המין **השני** של חלקי הגוף אצל אפלטון, והוא מן החלקים הנאמרין כפי השעור והמדידה שכונתי בו לדבר בו. ומבואר באקלידס שהשטח המוגבל יחלק למשלשים וכן שהגוף יחלק למוצקים. וכמו כן אפלאטון[4] אמר זה. ונשאר עלי לחקור איך יחלק הגוף וישוב אל השטחים, ואיך יחלק השטח וישוב אל הקוים, כמו שאמר אפלטון, וזה מצד השעור והמדידה.

---

[1] רוח גדול בין המילים. ייתכן שכתב היד שעמד בפני המעתיק לא היה חד משמעי, והמעתיק לא ידע באיזו משתי המילים לבחור.

[2] בכתב היד: מתחלקים.

[3] שתי המילים האחרונות כתובות פעמים.

[4] הכתיב של השם אינו עקיב.

[3.5 המבוכה בדבר החלקים הבלתי מתחלקים] וחשבתי שמא היה חושב אפלטון שהגוף מחובר משטחים והשטח מן קוים והקו מנקדות וזה כמאמר אנטיפרס ברבוע העגולה. כי כשעשה תבנית ישר הקוים בתוך עגולה ואחר חילק כל קשת שאצל כל צלע מצלעי התבנית לחציין חצאין, והדביק יתרים ישרים לכל אותן קשתות, ו[לא][1] סר מעשות כך בכל קשת, עד שחשב שיגיע החלוק לחלקים שמהם הורכב הקו הישר ומהם הורכב מקיף העגולה. והנה זה היה, כמאמר ארסטו׳, הפך סברתם [99א] של חכמי השעור והוא שאין הקו מחובר מנקדות ושאין השעורים מתחלקים לאין תכלית.

ואז הייתי במבוכה גדולה עם מה שיחוייב לי מן העיון כמו שזכרתי, שצריך להיות שם הרכבה שכלית עליונה מהרכבה זו שהרגלנו לציירה, כי כשלא תהיה מצד חזרת הסובבים[2] אל החלקים שאינם מתחלקים, איך יתכן לציירה? וחזרתי לעיין בדברי אנטיפרש,[3] כי תמהתי מאד איך בא לסתור התחלות החכמה שהיה חכם בה. ואמרתי שאולי לא ר״ל שיהיה השעור המדובק מחובר מהחלקים בלתי מתחלקים,[4] אלא כפי מה שאפלטון אמר כן, כלומר מצד השעור והמדידה לא מצד החבור הטבעי. כי לא אמר שיהיו שם חלוקים אין תכלית למספרם מצואים בפעל, עד שיסתור התחלות החכמה. וידעתי עם זה שאם יהיו החלקים[5] שאין תכלית למספרם מצואים בכח בלבד, לא יהיו משערים את המרחקים כלל.

[4. הפניה לאריסטו]

[4.1 מציאות אמצעית בין כח לפעל] ואז עלה בדעתי לומר שאולי יש שם מציאות אחר לא בכח ולא בפעל שהמרחקים יחלקו בו לחלקים אין תכלית למספרם. ולא זה המציאות, אלא המציאות שהיא בתנועה, שהיא אמצעית בין הכח והפעל. וכשראיתי עוד בהרכבה אשר בתנועה שהיא אמצעית שעושה חלוקים לאין תכלית למספרם בזמן התנועה, והוא[6] המציאות[7] האמצעי בין הכח והפעל.[8] כי מצד היות בקו הקטן חלוקים אין תכלית למספרם כמו שהם בקו הגדול היה אפשר שירכבו עליו בתנועה כל קו

[1] תיקון לפי גלוסקינה.

[2] מעל המילה יש נקודות. גלוסקינה מתקנת: שעורים .

[3] הכתיב של השם אינו עקיב.

[4] בכתב היד: מתחלפים.

[5] בכתב היד: החלוקים.

[6] כנראה צ״ל ״והיא״.

[7] אולי צ״ל במציאות.

[8] המשפט אינו מנוסח טוב מבחינה תחבירית.

אחר גדול [99ב] ממנו באי זה שעור שהיה (כמו שזכרתי בחצי גלגל המזלות עם קשת הרחב שבגלגל חצי היום). וכל אותן חלוקים שאין תכלית למספרם יהיו בזמן בעל תכלית לאותו המציאות האמצעי בין הכח והפעל.

[4.2 הגדרת ההרכבה השכלית] מאז [שהבנתי זאת] התחייב לי לומר שציור התנועה המשתוה בחלקיה היא ההרכבה השכלית שאמרתי, המשערת את המרחקים קצתם בשעורי קצתם. כי בזה יהיו נמצאים חלוקים אין תכלית למספרם לא בכח ולא בפעל אלא ביניהם. ויהיו אותם החלוקים הולכים על סדר שוה לפי שהתנועה ההיא שוה בחלקיה, ובזה יהיו משערים את המרחק על ערך שוה. וזה יסכים עם המבואר בשמע טבעי שהמרחק והזמן והתנועה מתערכים שלשתם זה בזה בערך שוה. כי לא יהיה זה[1] הערך אל התנועה בהיותה מתחלפת בחלקיה [אלא בהיותה ?[2]] ומשתוה בחלקיה, כמו שהזמן והמרחק הם משתוים בחלקיהם. וכבר יחזור המרחק אל החלקים בלתי מתחלקים במציאות האמצעי בין הכח והפעל, שהשכל לוקח אותו כאלו הוא בבת אחת ומשער קצתו בקצתו, כמו שלוקח את הקודם והאחור בציור הזמן כאלו הם בבת אחת ומשער ערך ההצטרפות ביניהם.

וידעתי כי בכך יהיה ציור הרכבת המרחקים קצתם על קצתם בבת אחת הראוי להכיר השווייתם אלו באלו (הן[3] שיתדמו בתבניותיהם הן שיהיו בלתי מתדמים), כלומר בציור התנועה הפשוטה ומתדמה בחלקיה, [100א] כמו שהמרחק השוקט אשר התנועה עליו הוא פשוט ומתדמה בחלקיו. מצד התדמות חלקי התנועה אלו לאלו יהיו חלקי השוקט משוערים אלו לאלו, ומצד שהתנועה[4] עם זה פשוטה ולא מורכבת מתנועות מתחלפות בשעורי מהירותן, יהיו חלקי המתנועע שוים לחלקי השוקט, לא כמו קשת חצי גלגל המזלות עם קשת הרחב.

[4.3 זיהוי הגרמים האפלטונים עם החלקים המינימלים האריסטוטלים] וזו היא הסברא הנכונה הראויה ליחס לאפלטון ואנטיפרס ושאר האומרים בחלקים בלתי מתחלקים אין תכלית למספרם: הגרמים שהם מסכימים בהם אינם בלא תכלית למספרם, ועם שאינם בלתי מתחלקים בשעורם ולא בלתי מתחלקים בצורותם. והוא כמו החלק שהוא יותר קטן בכל מין ממיני הגופים הטבעיים שהסכים

---

[1] מילה מחוקה: הָעל.

[2] קשה להבין את הטקסט בלי התיקון המוצע. ו׳ החיבור לפני המילה ״משתוה״ מעידה על כך שנפלה עוד מילה. אולי ״מתדבקת״?

[3] בכתב היד: הוא.

[4] מילה לא קריאה; תיקון לפי גלוסקינה.

בו ארסטו׳ בשמע טבעי, כך הגרם אצל אלו הוא החלק שהוא יותר קטן בגוף הכולל לכלל העולם. כי כמו שהגוף לא יגיע בכמות שעורו לגדל בלי תכלית כך לא יגיע בקוטן בלי תכלית, אך הגודל והקטן הם הם מגבלים[1] ומונחים אל הטבע.

אעפ״י שרוב הפלוסופים היו מיחסים לאפלטון מציאות חלקים בלתו מתחלקים אין תכלית למספרם בגוף המוגבל ועם ריקות ביניהם, וזה מן הנמנע, בלתי ראוי ליחסו לשום חכם אלא מאהבת הנצוח. והתכבד בקלונו[2] ומדרך נשוא פנים אל האהובים החולקים עליו.

[4.4 הרחבה על שאלת הריק] וכבר ראיתי מי שהיה מקיים החלק הגוף לחלקים בלתי מתחלקים שאין תכלית למספרם, ושיתנועע כל אחד מהם על מרחק מוגבל בלא זמן, [100ב] וזה כדי שלא יתחייב לו משם מציאות הריקות. שאינו [? ?].[3] וזה יהיה בהפרד גוף קשה מאצל גוף קשה שהיה תחלה נוגע אותו, כי מיד יתחייב[4] גוף אחר רך כגון אויר או מים להכנס ביניהם. ואמר כי לא ימלט אותו אויר הנכנס [א] מלהיות בו חלק או חלקים שכל אחד מהם נכנס יחד (לא [ב] קצתו קודם קצת ואז היה נמצא רקות אצל האמצע שבין שני הגופים המתפרדים מראש עת התפרדם זה מזה עד סוף עת כניסת אותו חלק מהאויר ביניהם). או [ב] שלא יהיה באותו אויר חלק נכנס יחד אלא קצתו אחר קצתו. ולא יהיה זה [הסבר א] אלא בהחלק גוף האויר ושובו אל השטחים שנרכב מהם, ושיתנועע כל אחד מהם על אותו מרחק בלא זמן. והסכים בחלוק זה ובתנועה זו מפני מניעת הרקות אצלו. וכמה יש בזה מן השגעון. ואין כאן מקום לדבר על מציאות הרקות. ונשוב למה שהינו בו .

[5. הוספת שני פוסטולטים של הרכבה לגאומטריה]

[5.1 תוספת ראשונה: פוסטולט ציור התנועה]

[פוסטולט *] שידוע הוא כי ציור התנועה הוא מההתחלות לחכמת השעור. שהרי ציור הרכבת המשלשים זה על זה שעשה אקלידס לא יהיה רק בציור תנועת המשלש האחד ממקומו עד מקום המשלש האחר, ולא ציור התנועה ישרה לבד הוא מההתחלות לחכמת השעור אלא גם ציור התנועה הסבובית. כי לא

[1] בכתב היד: מגבילים.

[2] תלמוד ירושלמי, חגיגה ב.א.

[3] גלוסקינה: אוהב אותו.

[4] מילה מטושטשת.

יצוייר היות כל הקוים הישרים היוצאין ממרכז העגול למקיפו שוין אלא בציור תנועתן סביב המרכז והרכבתם אלו[1] על אלו או אחד מהם על כלם. כמו שלא יצוייר שווי כל הקוין [101א] היוצאים ממרכז הכדור למקיף אלא בציור רשומו בתנועת חצי העגלה הגדולה סביב קטרה השוקט. וכן מוצק האצטולא[2] המעוגלת יהיה בסבוב המשלש נצב הזוית סביב אחת משתי צלעות הזוית[3] הנצבה השוקטת. כמו שרישום האסטונא הנצבת יהיה בסבוב השטח נצב הזויות סביב אחת מצלעותיו או בתנועת תושבתה על יושר בזויות נצבות על חצה. וכמו כן היות שתי הזויות[4] שעל התושבת מהמשלש השוה השוקים שוות זו לזו לא התבאר לאקלידס באמת, אלא בציור תנועת המשלש בסבוב עד שירכב על עצמו.

אלא שאקלידס העלים זה בצורה המכרחת, וכשניה[5] לה בדברים אחרים, אם שלא מדעתו, כי לא הרגיש איך נכלל אותו ציור התנועה בכח דבריו [באמרו] ולרכוב השטח על עצמו, אם כדי להעלימו מהתלמידים כדי שלא יקשה עליהם היות ציור התנועה מההתחלות לחכמת השעור. והיה מעלימו כפי היכלת, כי בהרכבה לא זכר תנועתה[6] ואם היא שם בכח. כי לכך הקדים לה חבור קו ישר לנקדה מונחת שוה בקו ישר מונח במקום אחר, כדי שיצוייר בכך הרכבת קוי המשולש האחד על קוי המשולש האחר מבלי שיזכיר תנועה. וסמך על היות הקוים היוצאין ממרכז העגול למקיפו שוין, שגם זה לא יחויב אלא מן התנועה.

[5.2 פוסטולט ציור התנועה מחליף את פוסטולט המקבילים] ונ״ל כי לפי שלא רצה אקלידס לגלות שהתנועה מההתחלות לחכמת השעור והעלים זאת ההרכבה כפי היכולת, הוצרך לקחת הקדמה אחת [101ב] שרב החכמה בנוייה עליה כמוסכמת, כאלו היתה מבוארת בעצמה, והיא אינה כן, כי צריך שתתבאר במופת כשאר הדברים המתבהרים שם במופת. והיא ״כשיפול קו ישר על שני קוים ישרים והיו שתי הזויות יחד אשר בצד אחד פחות משתי נצבות, כשימשכו שני קוים אלו באותו צד יפגשו בהכרח״. ואין זה מבואר בעצמו שיפגשו שני קוים אלו בהכרח, כי לא בעד שיתקרבו זה לזה כל אשר ימשכו יחוייב שיפגשו. שהרי מצאנו שני קוים אחרים נמשכין לצד אחד וכל אשר ימשכו יתקרבו זה לזה ולא יתכן שיפגשו לעולם, אפילו יצאו לאין תכלית. אבל תתבאר אותה הקדמה על דעת רבי האחרונים

---

[1] בכתב היד: אלא.

[2] הכתיב שגוי. על הכתיב של מילה זו ראו הערה למבוא סעיף 7.

[3] בכתב ביד: הזויות.

[4] בכתב היד: היות.

[5] מלה שאינה ברורה.

[6] בכתב היד: תנועת.

בציור התנועה. אלא שאני דחיתי כל הבאורים ההם לא מפני התנועה, אלא מפני מקרים אחדים שחברו לה, כמו שאבאר בשער השני. ואבאר אז אותה הקדמה בדרך ההרכבה המוסכמת להם. ויתחייב לומר מן הנזכר שציור התנועה הוא מההתחלות הצריכות מאד לחכמת השעור, אבל שהוא התחלה להתחלותיה המוסכמות לאותם החכמים הקודמים.

[5.3 פוסטולט המדידה] ואחר שציור התנועה במה שהיא תנועה היא התחלה לחכמת השעור, יתכן שיהיה גם כן שעור התנועה השוה בחלקיה מן ההתחלות לה, ובלבד בערכי השעורים ומדידתם אלו באלו, כי לפי שדרך בעל חכמת השעור לידע ערך השעורים ומדידתם אלו באלו. ולא יהיה לו זה אלא בחתכו את השעורים המדובקים לחלקים שוים קצתם לקצתם, כעניין שיהיה השעור המדובק [102א] אצלו מתפרק ושוה בחלקיו עם שהוא מדובק בעצמו.[1] כי מצד זה הוא נקרא חכמתו תשבורת, בעד השברים שעושה בה. היה ראוי לקחת זה ממה שהוא כך בטבעו, ולא נמצא זה אלא בתנועה השוה בחלקיה[2] ובזמן המונה והמשער אותה. ולא שנדע[?][3] איך יהיו הערכים נגבלים ונודעים בשעורי המרחקים מצד התנועה השוה בחלקיה, ומאין תהיה לנו תנועה כזאת לכך, ואיך יסתייע בעל חכמת השעור בזה הציור.

ואמרתי כי זה יהיה לו לבעל חכמת השעור בחלוק הרחקים לחלקים משתוים אלו באלו כי בכך ידמה המרחק בכח אל הזמן ואל התנועה השוה בחלקיה וישוער בשעורים. כי כשיהיו השעורים מתרבים והולכים על ערך שוה יהיו תמיד על אותו ערך, ואפילו הגיעו לאין תכלית. וזה כמו שבאר אקלדס בתבנית י״ג ממאמר ה׳, והוא כפי מה שיאמרו חכמי הטבע שערך סבובי השמש בגלגלו לסבובי הירח בגלגלו הוא תמיד בערך אחד לשלשה עשר בקרוב, עם הסכמתן בתמידות העולם. וזה ממעלות השכל[4] בהשיגו והקיפו הדברים שהם בלי תכלית כאלו הם בעלי תכלית לפניו, וכמה יקשה זה למי שלא ידע מענין השכל, רק השם לבדו.[5] **[פוסטולט **]** וזו היא התחלה לזה העיון, ר״ל בציור חלוקים שוים בעלי מספר, הולכים על ערך אחד עד אין מספר, כפי מה שהם נמצאים בתנועה המשתוה בחלקיה.

---

[1] המילים כתובות פעמיים בשינוי נוסח קל: מתפרק ומשתוה בחלקיו, עם שהוא מדובק בעצמו.

[2] בכתב היד: בחלקיו.

[3] בכתב היד: שצע; הקריאה ״שנדע״ בעקבות גלוסקינה.

[4] בכתב היד: הספר.

[5] מלה מקושטת שאינה ברורה.

[5.4 הערה מסכמת על מטרת הספר] וכשהגעתי להכיר את ציור ההרכבה השכלית הזה ואיך הוא התחלה לחכמת השעור [102ב] בעצמה ואיך ישתמשו בה בעלי החכמה להגביל את השעורים בפועל, אז חזרתי לחקור במה שהיתה כונתי בספר זה והוא [i] למצא קו ישר מוגבל שיהיה שוה אל הקו המקיף בעגלה ולכל אחד מהקוים הפתלתלים מתחלפי המינים אין תכלית למספרים, [ii] ולמצא שטח ישר הקוים מוגבל שוה לשטח העגלה ולשטח המקיף בכדור ולשטח המקיף באסטונא המעוגלת ולשטח המקיף במוצק האסטונא <ואל המקיף בעגול>[1] ולשאר שטחים עקומים מתחלפי הסוגים אין תכלית למספרים, [iii] ולמצא גוף בעל שטחים ישרים מוגבל שוה לגוף הכדור ולגוף האסטונא ולטבעת העגולה[2] ולשאר גופים בעלי שטחים עקומים מתחלפי המינים אין תכלית למספרים, והשבת כל מינין אלו קצתם לדמות קצתם, וכל זה באמצעות המישור המשתף לכלם, [iv] ולהמציא דרך חלוקת כל זוית ישרת הקוים או מקשתות עגולות על רקיע כידור לכל חלקים שוים שנרצה ובאי זה ערך שנרצה מדובר או אלם, להיות זה מן הצרך להשבת מיני השעורים אלו באלו ובאמצעות המישור המשותף להם. ולכן קראתי ספר זה מיישר עקוב כי זה עקר הכונה בו והוא מועיל מאד בחכמת הלמודים.

[1] המלים כנראה השתרבבו ממקום אחר.

[2] בכתב היד : ולעגול. זו טעות כי עגול הוא צורה דו ממדית. התיקון שאנו מציינים תואם את תוכן העניינים בתחילת הספר.

השער השני במקרים המתחברים אל התנועה המונעין את ציורה מהיות התחלה לחכמת השעור ואיך השתבשו בה רבים מהחכמים הקודמים

[פתיחה]

אמר אלפונשו יש מבני אדם יחברו לתנועה מקרים זרים שימנעו ציורה מהיות התחלה לחכמת [103א] השעור והם שני מינים. האחד מקרים נמנעים בעצמם אצל הציור השכלי, ואשר יהיו ככה אינם ראויים מאותו צד להיות התחלה לשום חכמה מהחכמות. והשני מקרים אחרונים לחכמת השעור כי אין אלו ראויים להיות התחלה לה מאותו צד שהם אחרונים.

[1. תנועה בגיאומטריה: טעויות עקרוניות]

[1.1. טעות ראשונה: סיבוב הקרן (גרסה חלשה)] אמנם המין הראשון שהם המקרים הנמנעים בעצמם אצל הציור השכלי, הוא כגון הסכמתם בציורם תנועה סבובית מגבלת קצה ותכלית לכל קו ישר. וזה שאמרו שאם היינו מניחים קו ג״ד בלי תכלית מצד ג׳ והיינו מניעין[1] את קו א״ב בעגלתו לצד ג׳ עד שיהיה נכחי לו, יהיה זה אפשר בלי ספק. ואז נניעהו מן הנכחות כדי שיתקרב אליו ב[2] יהיה תחלה על יושר נקדה בקו ג״ד ואחר כך יהיה על יושר שאר הנקודות, נקדה אחר נקדה, עד שיחזור להיות נכחי בצד האחר. וזה שקר, כי נטייתו אצלו בלי שיהיה על יושר הנקודות הוא שקר. והיותו על יושר הנקודות גם הוא שקר בקו שאין לו תכלית, מפני היותו תחילה על יושר נקודה אחת ראשון ואחר כך על יושר האחרות, ואין בקו אשר הונח בלי תכלית נקדה ראשונה. כי כל נקדה שהסכמנו[3] היות קו א״ב על ישרה יש נקדות[4] אחרות קודם לה אין תכלית למספרם שהיה על יושר להם. ולכן ראוי בהכרח לצייר לו נקדה ראשונה ויהיה בעל תכלית בלי ספק. ואמרו כי זה מופת חותך מחכמת השעור למניעת מרחק בלי תכלית הן שהונח במלא או הונח ברקות.

---

[1] בכתב היד: מגיעין.

[2] האות אינה ברורה.

[3] בכתב היד: שהכמנו.

[4] בכתב היד: נקודת.

[**1.2 הסבר הטעות**] ונ״ל שאין זה מופת כלל אך הוא בטל לגמרי כי הוא בנוי על הקדמה התבאר בטולה [103ב] בשמע טבעי, והיא מציאות חלק ראשון אל התנועה מצד שהיא תנועה. כי לפי שהיתה התנועה כמחוברת ממה שממנו יתנועע וממה שאליו יתנועע, וכשלא[1] היו שם שני דברים אלו יחד לא תהיה שם תנועה, ולא נקדה שהסכמנו אליה התנועע, [אלא] יש נקדות אחרות קודם לה שתתנועע[2] להן, כי בין כל שתי נקודות קו מדובק מקבל נקודות אין תכלית למספרן. ולזה לא תמצא נקודה ראשונה שאליה תהיה[3] התנועה. וכן יתבאר שלא תמצא נקדה אחרונה שממנה תהיה התנועה. איך יחוייב שתהיה בקו אשר הונח בלי תכלית נקדה הראשונה שאליה תהיה התנועה או נקדה אחרונה שממנה תהיה התנועה? והם נפלו בטעות זה לפי שהיו מחברים אל התנועה מה שהוא נמנע בציור השכלי והוא מציאות חלק ראשון שאליו התנועה או חלק אחרון שממנו התנועע.

[**1.3 טעות שניה: *הזזת הקרן לארכה***] וכמו כן טעו בדרך אחר כי אמרו שאלו היה אפשר להסכים במציאות קו בלי תכלית, נשים אותו קו א״ב בלי תכלית מצד ב׳ ונרשום נקדת ד׳ באי זה מקום שנרצה ממנו. ואם היה ד״ב בלי תכלית, כשנצייר הרכבת ד״ב על א״ב [i] יהיו ד״ב וא״ב רוכבים זה על זה והיה הקטן שוה לגדול והוא שקר. או [ii] יחסר ממנו לצד ב׳ בכדי א״ד הוא ג״ב[4]. ויתנועע קו ד״ג[5] על קו א״ב וכשנוסיף עליו ג״ב א״ד שהם בעלי תכליות יהיה הכל בעל תכלית והוא קו א״ב. וכבר הונח בלא תכלית זה שקר. אם כן א״ב בעל תכלית בהכרח.

[**1.4 הסבר הטעות**] ונ״ל שגם זה טעות לפי שחברו אל התנועה דבר נמנע [104א][6] בציור השכלי והוא היות המתנועע בלא תכלית. וזה לפי שהתנועה היא ממקום למקום, והמקום שממנו תהיה התנועה אל[7] הקו היא נקדה נגבלת, ואיננה מצואה אל הקו שהונח בלא תכלית. וכ״ש לדעת ארסטו שאומר שהמקום הוא המקיף, ואין המקיף אלא קצה וגבול אל המוקף בו.

---

[1] הכ׳ תוספת מעל השורה.

[2] בכתב היד: שתנועע.

[3] מילה מיותרת: הנקדה.

[4] מילה מיותרת: כן. אולי מחוקה.

[5] בכתב היד: ד״ב.

[6] מילה שגויה: בשעור. בכתיבה החוזרת בתחילת הדף הבא: בציור.

[7] ייתכן שצריך להיות ״שלי״.

[1.5 טעות שלישית: הזזת קו אינסופי במקביל לעצמו (הניסון של משה מסיביליה ואחרים להוכיח את פוסטולט המקבילים)] וכמו כן בדרך זו טעו אחרים ור׳ משה הלוי ממדינת אשבלייא[1] בכללם, שהתפארו לבאר ההקדמה הלקוחה בראש ספר אקלידס והיא ״כשיפול קו ישר על שני קוים ישרים והיו שתי הזויות יחד שחדש עמהם בצד אחד פחות משתי נצבות, כשימשכו שני הקוים ההם באותו צד יפגשו בהכרח״. ואמרו שנסכים ראשונה מה שהוא ידוע בעצמו :

[ההנחה של משה מסביליה ואחרים] והוא כשיחתכו שני הקוים שאין תכלית להם[2] על שטח כגון קוי א״ב ג״ד, ואחר כן נדמה אחד משני קוים אלו מתנועע אל השטח שהם בו, לא יתכן שיפרד אחד מהם מהאחר, אלא יהיו תמיד מתחתכים או ירכב האחד על האחר. וזה כי לא יצוייר שיפרד ממנו כי אם בהגיעו אל קצהו קודם החתוך,[3] וכבר הונח כל אחד מהם בלא תכלית, ולכן לא יתכן שיפרד אחד מהם מהאחר. ואמרו אחר זה נסכים קו אב״ג נופל על שני קוי א״ה ב״ז הישרים ונשים שתי זויות הא״ב זב״א פחות יחד משתי נצבות - נאמר שאם ימשכו שני קוי א״ה ב״ז לצד ה ז יפגשו בהכרח באותו צד.

והמופת ששתי זויות זב״ג זב״א יחד כשתי נצבות ושתי זויות זב״א הא״ב פחות [משתי] נצבות, אם כן זוית[4] זב״ג גדולה מזוית הא״ב. ונשים זוית בא״ח שוה לזוית זב״ג ונדמה ג׳ קוי א״ה א״ח[5] [104ב] ב״ז כל אחד בלא תכלית. ונדמה א״ח מתנועע בענין שהתנועע נקדת א׳ ממנו על קו א״ב וישמור א״ח תמיד את שעור זוית חא״ב. ונדמה א״ה שוקט. ולפי שקוי א״ח א״ה ב״ז כל אחד מהם בלא תכלית יהיה א״ח בלתי מתפרד מקו א״ה. ולפי שקו א״ח שומר תמיד את שעור זוית חא״ב לא יתכן שירכב א״ח על א״ה. ולזה יחתוך א״ח תמיד את א״ה. ולכן בהגיע א׳ לנקדת ב׳ בתנועת א״ח ירכב א״ח על ב״ז כי זוית גב״ז שוה לזוית גא״ח, וקו א״ח שומר תמיד לשעור זוית[6] חא״ג. אם כן כשירכב א״ב על ב״ז וקו א״ה חותך את קו א״ח,[7] יחתוך כמו כן את ב״ז. אם כן קוי א״ה ב״ז נפגשים בהכרח.

[1.6 ההצדקה של משה מסיביליה לשימוש בתנועה בגיאומטריה] ואמר ר׳ משה הנזכר שצריך לדעת שאינו חוץ מהראוי לצייר תנועת הקוים בדברים כאלה אע״פ שהוא חוץ מהראוי במעשה. וזה לפי שבעל

---

[1] בכתב היד : אשכליים, בשוליים האות ה׳.

[2] בכתב היד : לאחד.

[3] בכתב היד : הפרוד.

[4] בכתב היד : זויות.

[5] בכתב היד : א״ז.

[6] מילה מחוקה : הא״ח.

[7] אות מיותרת : ת׳.

חכמת השעורים לוקח הקוים והשעורים בכלל ומפשיטם מהחמר, עם שלא ימצאו כך חוץ מן הצייור[1] לא בעצם ולא במקרה. והקו לא ימצא מחוץ לציור מתנועע אלא במקרה, כלומר מצד שהוא במתנועע. והואיל שיכול בעל חכמת השעור לקחת בציורו את הקוים בענין שאין נמצאין כך חוץ לציור לא בעצם ולא במקרה והוא היותם מופשטים ונקיים מחמר; כ״ש שיוכל לקחת אותם בענין הנמצא להם במקרה מחוץ לציור והוא היותם מתנועעים.

אמר וכבר נמצא רבים מחכמי השעור שרגילין בחכמתן להשתמש בתנועת הקוים, וכלם משתמשים במה שדומה לתנועה וזה בהרכיבם הקוים והשטחים קצתם על קצתם, וכמו שנמצא ארסטו׳ משתמש בתנועת [105א] הקוים במופתיו לחכמת הטבע.

[1.7 הסבר הטעות] ונ״ל שזה המופת שזכר ר׳ משה בשם הקודמים ושהוא נכון גם כן בעיניו אינו נכון כלל, לא מפני שאין התנועה ראויה לציירה מצד שהיא תנועה, אלא מפני שחברו לה מציאות קו בלא תכלית, אשר נמנע להמצא קו בלא תכלית בפעל הן מתנועע הן בלתי מתנועע. וא״ת שאין לבעל חכמת השעור לדחות מציאות קו בלא תכלית לארכו[2] שהרי אקלידס הסכים לציור קו בלא תכלית, אשיבך שלא יכון בעל חכמת השעור למציאות קו בלא תכלית בפעל שזה שקר גמור, כמו שאבאר זה בשער זה, כי לא ראיתי לו קודם זה ביאור מספיק ולפי שהוא מועיל מאד בזה העיון שאנו בו.

אבל ירצה בעל חכמת השעור לומר שימצא קו בלא תכלית מוגבל, כלומר שתהיה לו תכלית אלא שלא תהיה תכלית מוגבלת וקצובה בפעל כך[3] שלא יצטייר תוספת עליו. אבל כל אשר יהיה הקו גדול אפשר לצייר תוספת, ועל התוספת תוספת, על מנת שיהיה תמיד עם כל התוספות בעל תכלית בפעל. וזה כמו שאומ׳[4] שהקו יחלק לחלקים אין תכלית למספרם. ואין כונתם בזה שיגיע החלוק לאין מספר בפעל, כי זה ג״כ שקר גמור, אך ר״ל שיחלק הקו בכח לחלקים אין תכלית למספרם, ויהיה בפעל בעל חלקים בעלי מספר קצוב. וכן לא ימשך הקו עד[5] שיהיה בפעל בלי תכלית אלא שימשך אל לא תכלית [? ?][6] על מנת שיהיה תמיד בעל תכלית בפעל.

---

[1] כתיב חריג.

[2] המילה אינה ברורה.

[3] בכתב היד: כדי.

[4] שתי המילים: תוספת בשוליים.

[5] בכתב היד: עוד.

[6] שתי מילים לא ברורות.

והענין[1] שוה בגדל ובקטן חלקיו, אינו אלא [i] מצד המגביל אשר בפעל, או [ii] מצד ה המיוחדת אל החמר הנושא והמקבל את [105ב] הגבול בפועל, או [i+ii] מצד שניהם יחדו, או [iii] מצד היות הכמות הראשונה בלתי מקרה בגוף הראשון. לא מצד חלוק האפשרות שמדרך בעל חכמת השעור לציירו. כי הוא לא יקח מרחק בעל חמר ונושא מיוחד, שא״כ לא היתה חכמתו כללית על כל החמרים והנושאים. ומה שהוא משולח אין לו תכלית קצובה לא בגודל ולא בקוטן. וזה היה הטעם בהסכמת בעל חכמת השעור להמשיך הקו בעל תכלית עד שיהיה בלא תכלית ולהדביק קו ישר בין כל שני נקדות. ולכן ר׳ משה הנז׳ [טעה] במה שהמשיך במציאות קו א״ח בלא תכלית בפעל והוא בהיותו על ב״ז, וכ״ש בהסכימו מתנועע ובלא תכלית.

[2. תנועה בגיאומטריה: טעויות מקריות]

והמין השני מן המקרים הזרים המונעים ציור התנועה מהיות התחלה לחכמת השעור ואינם ראויים[2] להיות התחלה לה מאותו צד שהם אחרונים לה.

[2.1 הנסיון של אבן אל-הייתם ואחרים להוכיח את פוסטולט המקבילים] כגון מה שכתב בן הייתם ואחרים רבים בבאור הקדמת אקלידס בנפילת קו ישר בין שני קוים ישרים שזכרתי. כי אמר : נשים קו א״ב הישר ונקים עליו קו א״ג בזוית נצבת ונדמה שיתנועע קו א״ג בענין שישמור תמיד זוית א׳ הנצבה. ונסכים שיתנועע נקדת א׳ על קו א״ב ויתנועע קו א״ג עד שיהא כמו ב״ד. ותרשום נקדת ג׳ קו ג״ד הישר שוה לקו א״ב. ונדביק א״ד הישר. ולפי ששלשת קוי משלש אב״ד שוה כל אחד מהם לחבירו מקוי משלש אג״ד יהיה המשלש שוה אל המשלש וזויותיו לזויותיו כל אחת לחברתה. אם כן זוית ג׳ נצבה שוה לזוית ב׳ נצבה וזוית אד״ב[3] שוה לזוית דא״ג וזוית בא״ד שוה לזוית גד״א. ולפיכך יהיו שתי זויות בד״א [106א] גד״א שוות לשתי זויות דא״ג בא״ד יחד שהן כנצבה אחת. ולפיכך יהיה תבנית אבד״ג נצב ארבע הזויות וצלעותיו הנכחיות שוות זו לזו. ושאר המופת שלו בנוי על הקדמה זו שתהיה נקדת ג׳ רושמת בתנועה את קו ג״ד הישר שוה לקו א״ב שכנגדו.

---

[1] המילה אינה ברורה; הקריאה לפי גלוסקינה.

[2] שגיאה תחבירית במשפט : ציור התנועה הוא האמור להיות התחלה לחכמת השעור ולא המקרים.

[3] בכתב היד : אד״ג.

[2.2 הסבר הטעות] ולא יהיה [זה] אלא ששני קוים המדביקים בין קצוות שני הקוים הישרים והנכחיים שוין זה לזה. כי בכך יהיו כל הנקודות שבקו המתנועע תנועה ישרה רושמות קוים ישרים ושוין זה לזה. והקדמה זו לא התבארה באקלידס כי אם בתבנית ל״ד ממאמר א׳ אחרי הסכמתו בהקדמת נפילת הקו על שני קוים ישרים. ולזה היה זה הקש הסבוב שהוא בטל בהגיון.

[2.3 נסיון של אל-נייריזי להוכיח את פוסטולט המקבילים] ולפי שראה המשער אלניריזי וכפי מה שכתב סנביליקיוס שלא היה נכון לבאר הקדמה זו של נפילת הקו על שני קוים ישרים בציור התנועה, ביארה הוא בהקש אחר לפי מחשבתו. והקדים לכך הקדמות **[הקדמה ראשונה]** והוא שאמר כשהיו שני קוים ישרים ונכחיים יהיה המרחק היותר קצר בינהם הוא עמוד על כל אחד מהם. **כיצד**? נשים שני קוי א״ב ג״ד נכחיים ונשים המרחק היותר קצר בינהם קו ה״ז. אומר שקו ה״ז עמוד על כל אחד מקוי א״ב ג״ד. **מופתו**, שאם לא היה א״ז עמוד על כל אחד מקוי א״ב ג״ד נשים שתי הזויות שאצל נקדה ה׳ בלתי נצבות והחדה זוית אה״ז.[1] ונוציא מן ז׳ עמוד ז״ח על א״ב ויפול בצד א׳. אם כן יהיה ז״ה ממאמר אקלידס גדול מן ז״ח [ה]קצר שבכל הקוים המדביקים בין קוי א״ב ג״ד. [106ב] וכן יתבאר שהקו היותר קצר המדביק בין שני קוים ידועים בלתי נכחיים הוא עמוד על אחד מהם. **[הקדמה שניה]** ואמר עוד נשים שני קוי א״ב ג״ד ויפול עליהם קו ה״ז, על כל אחד מהם בזוית נצבה. אומר שקוי א״ב ג״ד נכחיים וכי ה״ז הוא המרחק היותר קצר בין קוי א״ב ג״ד. **מופתו**, שאם לא היה ג״ד נכחי לא״ב נעביר על נקדת ז׳ קו ח״ט נכחי לקו א״ב, וה״ז הוא המרחק היותר קצר בין קוי א״ב ח״ט. אם כן זוית הז״ח נצבה על פי תבנית י״ט הנזכרת, וכבר הונח זוית הז״ג נצבה. זה שקר. אם כן קוי א״ב ג״ד נכחיים וה״ז הוא המרחק יותר קצר ביניהם.

[2.4 הסבר הטעות] ועל שתי אלו הקדמות בנה אלניריזי הנשאר מההקש שלו בבאור הקדמת אקלידיס הנזכרת. וכשתעיין היטב תמצא שלא התבאר לו שבהיות כל קו עמוד על אחד משני קוים נכחיים יהיה כמו כן עמוד על האחר, ולא שיהיה הוא המרחק היותר קצר ביניהם. כי לא יסכים המקשה שזה העמוד על שניהם יצא מכל נקדה שתהיה מביניהם [או] שתהיה באחד משניהם, ולא שהמרחק ישוה תמיד בין שני הקוים הנכחיים. אלא [יסכים] שלא יפגשו באחד משני הצדדין לבד. כי זה בלבד הוא גדר הנכחיים, ר״ל שלא יפגשו באחד משני הצדדין, לא שהמרחק ישוה תמיד ביניהן עד שיתבאר אחר כן.

---

[1] בכתב היד: אח״ז.

ולפי שהונח ה״ז בהקדמה זו השנית עמוד על כל אחד מקוי א״ב ג״ד, ויצא ביניהם קו ח״ט [דרך ז׳ נכחי לא״ב], לא[1] יהיה [ה״ז] עמוד על ח״ט ונשאר עמוד על א״ב. ולכך יהיה הוא המרחק היותר קצר היוצא מן ז׳ אל קו א״ה, לא המרחק היותר קצר היוצא [107א] מן ה׳ אל קו ח״ט, ולא מן זולת ה׳ אל קו ח״ט[2]. כי אולי יאמר אומר שיצא מן ה׳ אל קו ח״ט קו יותר קצר מן ה״ז שיהיה עמוד על ח״ט לצד ט, ושלא יפגשו קוי א״ב ח״ט בעבור שהונחו נכחיים, ושלא יפגשו כמו כן קוי א״ב ג״ד כי ה״ז מקיף עמהם בזויות נצבות. וכשלא יהיה ה״ז המרחק היותר קצר היוצא מן ה׳ אל קו ח״ט לא תהיה זוית הז״ח נצבה אע״פ שיהיה ג״ד נכחי אל א״ב. ולפי שהנשאר מההקש של אלנריזי על אותה הקדמה הנזכרת בנוי על זה שזכרתי ממנו שיש בו ספקות אלו אינו צריך לכתבו כאן ועם שהוא ארוך.

[3. מוסף: ״הוכחת״ פוסטולט המקבילים]

[3.0] אך אזכיר ההקש המופתי שחדשתי על ההקדמה שהוא אצלי[3] נקי מכל הספקות. ואומר שהוא ידוע למי שקרא אקלידס שהוא לא השתמש באותה הקדמה אלא אחר כ״ח תבניות כפי נסח אלחג׳אג׳. ולפיכך כשנבאר ההקדמה ממה שהתבאר באותם תבניות קודמים יהיה הבאור נכון לא סבובי. והנני מבארה בששה תבניות אחרים.

[3.1] התבנית הראשון. כשתחתוך צלע שבמשולש ישר הקוים לשני חצאין על נקדה, ויצא קו מהנקדה ההיא על הזוית שכנגדה ונמצא חוצה למשלש עד שיהיה מה שיפול ממנו חוץ למשלש שוה אל מה שיפול ממנו בתוך המשולש, ויצאו שני קוים ישרים מראשו לשתי הזויות הנשארות מהמשלש, יחדשו שני קוים אלו עם המשלש שטח בעל ד׳ צלעות וכל שתי זוית וכל שתי צלעות מתנגדות בו שוות זו לזו.

כיצד? נשים משלש אב״ג ישר הקוים, ונחתוך צלע ב״ג לשני חצאין בנקדת ה׳ ונדביק קו א״ה [107ב] ונמשיכהו חוץ המשולש עד שיהיה קו ה״ד ממנו שוה לקו א״ה, ונדביק קוי ד״ב ד״ג. אומר שקו ב״ד שוה לקו א״ג שכנגדו ושקו ג״ד שוה לקו א״ב שכנגדו, וכן שזוית בא״ג שוה לזוית בד״ג שכנגדה.

---

[1] האות השניה אינה ברורה.

[2] בטקסט ג״ד. להנמקת התיקון ראו הערה בתרגום.

[3] בכתב היד: אצל.

מופתו לפי ששני קוי א״ד ב״ג מתחתכים יהיו הזויות המתנגדות שוות זו לזו. אם כן שני קוי ה״ב ה״ד במשולש הב״ד שוין לשני קוי ה״ג ה״א במשלש אה״ג כל אחד לחברו, ה״ב אל ה״ג וה״ד אל ה״א, וזוית בה״ד שוה לזוית אה״ג המתנגדת לה. אם כן יהיה ב״ד שוה אל א״ג וזוית הד״ב שוה לזוית הא״ג וזוית הב״ד לזוית הג״א. וכן יתבאר שקו ג״ד שוה לקו א״ב שכנגדו ושזוית הד״ג שוה לזוית הא״ב ושזוית הג״ד שוה לזוית הב״א. ולפי שיתבאר שזוית הב״ד שוה לזוית הג״א ושזוית הב״א שוה לזוית הג״ד תהיה כל זוית אב״ד שוה לכל זוית אג״ד המתנגדת לה וכן יתבאר שזוית בא״ג שוה לזוית בד״ג[1]. ומש״ל.

[3.2] התבנית השני כשיחתוך יתר זוית הנצבה במשלש נצב הזוית לשני חצאין בנקדה ויצא קו ישר מאותה נקדה אל הזוית הנצבה, ונמשיך אותו קו חוץ למשלש עד שיהא מה שיפול ממנו חוץ למשלש שוה למה שיפול ממנו תוך המשלש, ויצאו מראשו שני קוים ישרים לשתי הזויות הנשארות שבמשלש, יחדשו שני קוים אלו עם המשלש תבנית בעל ד׳ צלעות נצב ארבע זויות. כיצד? נשים משלש [108א] אב״ג ישר הקוים ונצב זוית א׳ ונחתוך קו ב״ג לשני חציין בנקדת ה׳, ונדביק קו א״ה הישר ונמשיכהו לחוץ עד שישוה ה״ד אל ה״א, ונדביק קוי ד״ב ד״ג הישרים. אומר שכל [אחת] מזויות אב״ד אג״ד נצבה.

מופתו [בניה] נמשיך קו א״ג לצד א׳ עד שישוה א״ז אל א״ג, ונקים מנקדת ז׳ עמוד על ז״ח מעל קו גא״ז שוה לכל אחד מקוי א״ב ג״ד השוין, וזה כמעשה תבנית י״א ונדביק ח״ג ונקים על משלש זח״ג[2] הנצב זוית ז׳ שטח בעל ד׳ צלעות, כמו שעשינו על משלש אב״ג, והוא שטח זחכ״ג. אם כן תפול נקדת כ׳ על נקדת ד׳ או לא תפול עליה. ונמשיך ג״ד עד אין תכלית כמו שהסכים[3] אקלידס בהתחלותיו[4] לעשות כן, ובארתי דבריו אני. וכן נמשיך ב״ד לצד ד׳ עד אין תכלית. אם כן כשלא תפול נקדת כ׳ על נקדת ד׳ תפול בהכרח [I] על נקדה אחרת בקו ג״ד או [II] על נקדה אחרת בקו ב״ד או [III] על זולת שני קוי ב״ד ד״ג, ולא ימלט מזה.

[בדיקת מקרים]

[1] בכתב היד בד״א. תיקון לפי גלוסקינה.

[2] בכתב היד: זה״ג. תיקון גלוסקינה.

[3] בכתב היד: החכים.

[4] חזרה מיותרת: נקדת...בהתחלותיו.

[II] אם תפול [נקדת כ׳] בנקדה אחרת בקו ב״ד כמו בצורה ראשונה, אם כן יפול עוד קו ח״כ על נקדת כ׳ ועל נקדת ב״ד,[1] ואז היה משלש גכ״ד שוה שוקי ג״ד ג״כ, וכל אחת משתי זויות גד״כ גכ״ד נצבה כי זוית גכ״ח נצבה שוה לזוית ז׳ שכנגדה וזוית בד״ג נצבה שוה לזוית בא״ג שכנגדה כמו שבארתי. והוא שקר למשלש אחד שתי זויות נצבות מי״ג[2] ממאמר א.

[III] או לא יפול קו ח״כ על נקדת [108ב] ב אלא בחוץ כגון קו חמכ״ס. וזה או **[III.א]** יהיה שוה, ואז היתה זוית גד״כ הנצבה שוה לזוית גכ״ד שהיא גדולה מזוית גכ״ס הנצבה והוא שקר. או **[III.ב]** יהיה אחר ד׳, ותהיה זוית גד״כ הנצבה שוה לזוית גכ״ד שהיא קטנה מזוית גכ״ח[3] הנצבה וגם הוא שקר. או יפול קו ח״כ תחת נקדת ב׳ כגון חנכ״ס. וזה או **[III.ג]** יהיה לצד ב׳, ואז היתה זוית גד״כ הנצבה שוה לזוית גכ״ד שהיא קטנה מזויות גכ״ס הנצבה וזה שקר, או **[III.ד]** יהיה אחר ד׳, והיתה זוית גד״כ הנצבה שוה לזוית גכ״ד שהיא גדולה מזוית גד״כ הנצבה וזה שקר.

[I] ואם תפול נקדת כ׳ על נקדה אחרת בקו ג״ד כמו בצורה שנית היה ג״ד שוה לג״כ הקטן לגדול זה שקר.

['III][4] ואם תפול נקדת כ׳ על זולת קוי ב״ד ג״ד ונדביק קו ד״כ אם כן היתה זוית גד״כ שהיא גדולה מנצבה שוה לזוית גכ״ד שהיא קטנה מנצבה. [...][5] שוה לזוית גכ״ד שהיא גדולה מנצבה והכל שקר.

[מסקנה] ולזה תפול נקדת כ׳ על נקדת ד׳ בהכרח, וירכב ג״כ על ג״ד השוה לו, ותרכב זוית גכ״ח הנצבה על זוית גד״ב הנצבה, וירכב קו ח״ב על קו ב״ד ויהיו בקו אחד. ויהיו שתי זויות אב״ח אב״ד שוות לשתי נצבות מי״ב[6] ממאמר א׳. ולפי שהיתה זוית אב״ד שוה לזוית אג״ד שכנגדה וכן זוית זח״ד שוה לזוית אג״ד שכנגדה תהיה זוית זח״ד הנצבה שוה לזוית אב״ד.

---

[1] כנראה הכוונה לנקודה על קו ב״ד.

[2] לפי המספור של הייברג ההפניה היא למשפט 17.

[3] כתב היד: גח״כ. דתיקון לפי גלוסקינה.

[4] סעיף זה נראה שייך לשכבה שונה של הטקסט.

[5] ככל הנראה נפל כאן משפט.

[6] לפי המספור של הייברג ההפניה היא למשפט 13.

ולזה שנרכיב שטח זחב״א על שטח אבג״ד[1] בכדי שתפול נקדת א׳ משטח זחב״א על נקדת ג׳ משטח אבג״ד,[2] [109א] [1] ותפול נקדת ז׳ משטח זחב״א על נקדת א׳ משטח אבג״ד כי קוי א״ז א״ג שוין זה לזה, [2] ותרכב זוית חז״א הנצבה על זוית בא״ג הנצבה, וירכב קו ז״ח על קו א״ב השוה לו, [3] ותרכב זוית זח״ב על זוית אב״ד[3] השוה לה, ויפול קו ח״ב[4] על קו ב״ד[5] כי שתי זויות זח״ב[6] אב״ד[7] שוות,[8] [4] ותפל נקדת ב׳ על נקדת ד׳ כי קו ח״ב[9] שוה אל ב״ד[10], וירכב קו ח״ב[11] על קו ב״ד, כי זוית[12] זח״ב שוה לזוית[13] אב״ד, **אז** תרכב זוית אב״ח על זוית בד״ג הנצבה ותהיה נצבה, ותשאר זוית אב״ד נצבה, ותהיה זוית אג״ד שכנגדה גם כן נצבה, וכבר הונחו שתי זויות בא״ג[14] בד״ג נצבות. אם כן שטח אבג״ד בעל ד׳ זויות נצבות. ומש״ל.

[3.3] התבנית השלישי שתי זוית החדות שבמשלש נצב הזויות שוות יחד לנצבה אחת. **כיצד?** נשים זוית א׳ נצבה במשלש אב״ג ישר הקוים. אומר ששתי זויות ב׳ ג׳ הנשארות שוות יחד לנצבה אחת. **מופתו** נחתוך יתר ב״ג לשני[15] חצאין על נקדת ה׳ ונדביק א״ה ונמשיכהו על יושר עד שיהיה ה״ד [109ב] שוה אל א״ה. ונדביק ד״ב ד״ג. אם כן מתבנית השנית היה זוית אב״ד נצבה ומתבנית הראשון תהיה זוית דב״ג שוה לזוית אג״ב המתחלפת לה, ונשתף להם זוית אב״ג ויהיו שתי זויות אב״ג אג״ב שוות יחד לזוית אב״ד הנצבה. ומש״ל.**[3.4] התבנית הרביעי** כשיקומו שני עמודין שוין זה לזה מעל קו ישר, ונדבק בין ראשיהם קו ישר, יהיה אותו הקו המדביק שוה לקו ישר שכנגדו ויחדש בשני העמודים שתי זויות נצבות. **כיצד?** נקים שני עמודי א״ב ג״ד השוין זה לזה מעל קו א״ג המונח, ונדביק קו ב״ד. אומר כי ב״ד שוה

---

[1] הסדר הנכון כאן ובהמשך צ״ל אבד״ג.

[2] טקסט שככל הנראה מיותר (מופיע אחר כך בשלב 7) : ותפול נקדת ב׳ משטח אבג״ד.

[3] מילה שאינה ברורה, וכנראה מחוקה.

[4] בכתב היד : א״ב.

[5] בכתב היד : ג״ד. אחר כך שתי מילים מחוקות : ״ויפול קו״.

[6] בכתב היד : זא״ב.

[7] בכתב היד : אג״ד.

[8] בכתב היד : נצבות.

[9] בכתב היד : א״ב.

[10] בכתב היד ג״ד.

[11] בכתב היד : א״ב.

[12] בכתב היד : זויות.

[13] בכתב היד : לזויות.

[14] בכתב היד : בא״ד.

[15] אות מיותרת : א׳.

אל א״ג ושתי זויות ב׳ ד׳[1] נצבות. **מופתו** נשים ה״ח שוה אל א״ג ונקים מעליו עמוד ה״ז שוה לכל אחד משני עמודי א״ב ג״ד, ונדביק ז״ח ונחתכהו לחציין על נקדת כ׳. ונדביק ה״כ ונמשיכהו לחוץ עד היות כ״ט שוה [אל] כ״ה. ונדביק ט״ז ט״ח. אם כן מהתבנית השני ישוה ט״ז לקו ה״ח שכנגדו השוה לקו א״ג, ושכל אחד מזויות הז״ט חט״ז[2] נצבה. אם כן משנרכיב שטח החט״ז על שטח אגד״ב בכדי שתפול נקודת ח׳ על נקדת ג׳ ונקדת ה׳ על נקדת א׳, לפי שקוי א״ג ה״ח שוין זה לזה, ותרכב זוית זה״ח הנצבה על זוית בא״ג הנצבה, ותרכב נקדת ז׳ על נקדת ב׳ וכן נקדת ט׳ על נקדת ד׳ וירכב קו ז״ט השוה לקו א״ג[3] על קו ב״ד.[4] וישוה ב״ד אל א״ג, ותרכב זוית הז״ט הנצבה על זוית אב״ד ותהיה נצבה וכן תרכב זוית חט״ז הנצבה על זוית גד״ב[5] ותהיה נצבה. ומש״ל.

[3.5] *התבנית החמישי* כשיחתוך יתר הזוית[6] הנצבה שבמשלש ישר הקוים [ניצב הזוית] לשני חצאין בנקדה ויצא מאותה נקדה עמוד על אחד משני הקוים המקיפים בנצבה, היה אותו עמוד חותך את הקו שיפול עליו לשני חציין. **כיצד?** נשים משולש אב״ג נצב זוית בא״ג ונחלק יתר ב״ג לשני חצאין בנקדת ד׳, ונוציא מן ד׳ עמוד ד״ה על א״ג. אומר שקו א״ה[7] שוה לקו ה״ג. **מופתו** לפי שזוית בג״א חדה וזוית בא״ג נצבה לא יפול עמוד ה״ד חוץ למשולש לצד ג׳ ולא לצד א׳ כדי שלא יהיה במשלש אחד שתי זויות נצבות או יותר משתי נצבות. ונחתוך בקו א״ב קו א״ז שוה לקו ה״ד, ונדביק ד״ז. אם כן מתבנית הרביעי ישוה ז״ד אל א״ה וכל אחד מזויות הד״ז אז״ד נצבה. וזוית הד״ג חדה והיא עם זוית הד״ב כשתי נצבות. ולכן תהיה זוית הד״ג עם זוית זד״ב כנצבה אחת. ויפול לפיכ׳ קו ד״ה בתוך משלש אב״ג. ולפי שהיו שתי זויות הד״ג זד״ב יחד כנצבה אחת ושתי זויות זד״ב זב״ד שבמשלש זב״ד שוות יחד לנצבה אחת מתבנית השלישי, תהיה זוית זב״ד שוה לזוית הד״ג. וכן יתבאר שזוית זד״ב שוה לזוית הג״ב. אם כן משנרכיב משלש זב״ד על משלש הד״ג בכדי שתפול נקדת ד׳ מקו ב״ד על נקדת ג׳ מקו ד״ג ונקדת ב׳ על נקדת ד׳. ולפי שקוי ב״ד ד״ג שוים ותרכב זוית זב״ד על זוית הד״ג השוה לה ותרכב זוית זד״ב על זוית הג״ד

[1] בכתב היד: ג׳. תוקן מעל השורה.

[2] בכתב היד: הט״ז.

[3] בכתב היד: א״ד. תיקון לפי גלוסקינה.

[4] בכתב היד: ב״ג. תוקן מעל השורה.

[5] בכתב היד: גב״ד. תיקון לפי גלוסקינה.

[6] בכתב היד: הזויות.

[7] המילה חסרה בכתב היד; מעל השורה נכתב: א׳.

השוה. ולזה תרכב נקדת ז׳ על נקדת ה׳ וקו ז״ד על קו ה״ג. וכבר התבאר שקו ז״ד שוה אל א״ה [110ב] אם כן א״ה שוה אל ה״ג. ומש״ל.

[3.6] התבנית השישי כשיפול קו ישר על שני קוים ישרים וחדש עמהם מצדו האחד שתי זויות שהן יחד פחות משתי נצבות, אם ימשכו אותם שני קוים באותו צד לאין תכלית יפגשו בהכרח. **כיצד?** נשים קו א״ג הישר נופל על שני קוי א״ב ג״ד הישרים וחדש עמהם שתי זויות אג״ד גא״ב שהן יחד פחות משתי נצבות. אומר שאם ימשכו שני קוי א״ב ג״ד בצד ב׳ וד׳ יפגשו. **מופתו**, שאחת משתי זויות אג״ד גא״ב חדה בהכרח. ונוציא אותה זוית אג״ד. ונוציא מן א׳ עמוד א״ה על קו ג״ד. ועמוד א״ה יפול לצד ד׳, שאם לא כן היה מחדש עם א״ג משלש ששתי זויותיו יחד יותר משתי נצבות, וזה לא יתכן על פי הנזכר. ועל [פי] המבואר בתבנית שלישי כי יהייו שני זויות הא״ג הג״א שוות יחד לנצבה אחת, ותשאר זוית זא״ח[1] חדה. ונרשום נקדת ז׳ על קו א״ב באי זה מקום שנרצה ממנו ונוציא ממנה עמוד ז״ח על א״ה[2] והוא יפול לצד ה׳ לפי שזוית זא״ה[3] חדה. ונחתוך ז״ט מן א״ב שוה אל א״ז ונוציא מן ט׳ עמוד ט״כ על א״ה עוד. ולפי שמשלש אט״כ נצב זוית כ׳ ויצא מן ז׳ שבחצי יתר א״ט עמוד ז״ח על א״כ יחתכהו על חציו מתבנית החמישי. אם כן א״ח שוה אל ח״כ. ונחתוך עוד ט״ל מן א״ב שוה אל א״ט ונוציא מן ל׳ עמוד ל״מ על א״ה. ויתבאר כמו שביארנו שקו כ״מ שוה אל כ״א. אם כן א״כ כפל א״ח וא״מ כפל א״כ.

ומשנעשה כן מן הכפלים פעמים רבות נגיע בהכרח לקו שיהיה [111א] גדול מן א״ה[4] ונשים אותו א״מ. וזה על פי אחת מהתחלות המוסכמות בחכמה זו זכרה אקלדס בהקדמת מאמר חמישי והיא שהשעורים אשר ביניהם ערך הם אשר הם מסוג אחד ואשר כשיכפול אחד מהם פעמים רבות יהיה גדול מן האחר. ואם א״ב ג״ד לא היו נפגשים לעולם היה קו ל״מ בהכרח חותך את קו ג״ד לפי ששתי נקדות ל״מ הם משני צדי קו ג״ד ויחתכהו על נקדת נ׳ אם אפשר. ואם כן יהיו שתי זויות ה׳ מ׳ במשלש הנ״מ נצבות, וזה שקר. אם כן שני קוי א״ב ג״ד יפגשו בהכרח. ומש״ל.

[3.7 הערת סיום] והנה ששה תבניות שזכרתי ראויים להכנס באי זה מקום שתרצה מסוף תבנית י״ו עד סוף תבנית כ״ו ממאמר א׳ מאקלידס מעקרי חכמת השעור, כי הם מבוארים על פי דרכם בלי ציור תנועה אחרת מאשר היה אפשר להטיל בו ספק.

---

[1] בכתב היד זא״ח.

[2] בכתב היד א״ב. תיקון גלוסקינה.

[3] בכתב היד זא״ח.

[4] בכתב היד: א״כ. תיקון לפי גלוסקינה.

[4. תנועה בגיאומטריה: טעויות עקרוניות (המשך)]

[4.1 הפרכת אפשרות קו אינסופי בפועל] ואחר זה יתבאר מה שיעדתי שאי אפשר להמצא קו בלא תכלית בפועל. **מופתו**: נשים א״ב הישר בלי תכלית בפעל אם אפשר, ונקים מעליו עמוד א״ג ונוציא מנקדת ג׳ קו ג״ד נכחי לקו א״ב כמעשה תבנית ל״א. ולכן תהיה זוית אג״ד נצבה כמו זוית בא״ג. ולפי ששני קוי א״ב ג״ד נכחיים יהיה המרחק שבינהם שוה תמיד א״ג ב״ד כמו[1] שביאר אקלידס בתבנית ל״ג. ונסובב קו ג״א סביב מרכז ג׳ עד שירכב על קו ג״ד. ונדמה נקדה מתנועע על קו א״ב[2] [111ב] שהיא תמיד על יושר ג״א המתנועע. כי כשימשך ג״א המתנועע על יושר יפגוש תמיד את קו א״ב קודם שירכב על ג״ד, כמו שבארתי מנפילת הקו הישר על שני קוים ישרים.

ולפי שאותה נקדה המתנועעת היתה בקו א״ב ואחר כן היתה בקו ג״ד, ואין שני הקוים נפגשים זה בזה אלא המרחק שמור ביניהם תמיד להיותם נכחיים, יתחייב שלא תעבור מקו א״ב אל קו ג״ד אלא בעתה,[3] לפי שכל תנועה היא בזמן מצד שהמתנועע בחצי המרחק הוא טרם היותו בסופו, כלומר שישוער בה קודם ואחר. אם כן בתחלת אותו זמן היתה הנקדה המתנועעת בקו א״ב, ולא היתה בו אחר כן בכל אותו זמן, והיתה בסוף אותו זמן על קו ג״ד. ואלו היה[4] קו א״ב מצוי בפעל בלא תכלית לא היתה הנקדה המתנועעת יוצאת ממנו לעולם עד תכלית אותו זמן. וכבר התבאר שיצאה ממנו בתחלת אותו זמן. זה שקר. אם כן לא יהיה קו א״ב מצוי בפעל בלא תכלית ומש״ל.

[4.2 הערה תיאולוגית] ומי שירצה להבין מכאן מציאות האל בדרך קרוב, כי הוא מגביל תכליות מיוחדות ומוגבלות בפעל את המרחקים אחרי היותם בלתי מוגבלים בכח. והוא שאמר הב״ה לאיוב איפה הייתי ביסדי ארץ הגד אם ידעת בינה, מי שם ממדיה כי תדע או מי נטה עליה קו וגו׳, ואמר עד פה תבא ולא תוסיף ופה ישית בגאון גליך ושאר פסוקים כאלו.

---

1 בכתב היד: וכמו.

2 מילים מיותרות ששיכות למשפט הבא: ״קודם שירכב על ג״ד כמו שביארתי״. אפשר להבחין בסימני מחיקה מעל מילים אלה.

3 בכתב היד: כעת.

4 בכתב היד: היתה.

[5. תנועה בגיאומטריה: טעויות מקריות (המשך)]

[5.1 מדידת הקף המעגל על ידי גלגול] ומהמקרים האחרונים לחכמת השעור המונעים את ציור התנועה [112א] מהיות התחלה לה מאותו צד שהם אחרונים לה הוא מה שחשבו קצת חכמים כי מצאו קו ישר שוה למקיף העגלה בדרך זו: שאם יגלגל אדם עגולה והיא כתושבת גוף האסטונא על קו ישר עד שהנקדה ממקיף העגלה שהחלה לנגוע את הקו הישר תשוב ליגע בו פעם שנית, יהיה הקו הישר שבין שתי הנגיעות אלו שוה אל הקו המקיף בעגולה, כי כל נקדה אשר במקיף העגולה נוגעת לנקדה שהיא חברתה מהקו הישר אשר עליו היה הגלגול. ויהיה זה כמו הרכבת כל הקשת על כל הקו הישר, ולזה תהיה שוה לו. ונ״ל שאלו היה הקו מחובר מנקודות היה דעת זה אמתי אך אחר שאינו כן הוא דעת בטל.

[5.2 רציפות תנועת הגלגול] ולבאר בטולו שנשים שתי עגולות א״ב ד״ה על מרכז ג׳ בשטח אחד ובגוף אחד, ונוציא מנקדות א׳ ד׳ שני קוי אח״ל ד״מ הישרים נוגעין העגולות במקיפיהם. וכשיתגלגל מקיף העגלה א״ב על קו א״ל עד שתשוב נקדת א׳ ממקיף העגלה ליפול על נקדת ל׳ מקו א״ל, תשוב נקדת ד׳[1] ממקיף עגלת ד״ה ליפול על נקדת מ׳[2] מקו ד״מ, וישוב מרכז ג׳ על מרכז כ׳[3] וקו גד״א על קו כמ״ל. ונשליך נקדת ב׳[4] על מקיף עגלת א״ב באי זה מקום שנרצה ממנו ונדביק קו גה״ב הישר. אומר כי כשתפגוש נקדת ב׳ את קו א״ל על נקדת ח׳ ממנו תפגוש נקדת ה׳ את קו ד״מ על נקדת ט׳ ממנו.

מופתו נשים נקדת ז׳ מרכז העגולות בהיות נקדת ב׳ על נקדת ח׳ ונדביק קו ח״ז. אם כן יפול קו ג״ב[5] על קו ז״ח וקו [112ב] ג״ה על קו ז״ט. ולפי שהיה א״ח נוגע שני עגולי א״ב נ״ח יהיו שתי זויות גא״ח זח״א נצבות וקוי א״ג ז״ח נכחיים ושוין זה לזה, וכן שני קוי ג״ד ז״ט נכחיים ושוין זה לזה. ויהיה הקו הישר המדביק בין שני נקודות ד׳ ט׳ נכוחי לכל אחד מקוי ג״ז א״ח. אם כן זוית טד״ג נצבת. אבל זוית מד״ג נצבת. אם כן נקדת ט׳ היא בקו ד״מ. אבל נקדת ה׳ נפלה על נקדת ט׳, אם כן נקדת ה׳ תפגוש בקו ד״מ. וכן יתבאר שבפגוש נקדת ה׳ בקו ד״מ על נקדת ט׳ תפגוש נקדת ב׳ בקו א״ל על נקדת[6] ח׳. אם כן

[1] האות אינה ברורה.

[2] מילה מיותרת: ליפול.

[3] בכתב היד: ה׳. תיקון לפי גלוסקינה.

[4] האות אינה ברורה.

[5] בכתב היד: גה״ב.

[6] אות מיותרת: כ׳ או ס׳.

כשתתגלגל שוה[1] קשת מאחת שתי העגולות על קו ישר, תתגלגל קשת דומה לה בעגולה אחרת על קו ישר שוה לקו ישר האחר.

[5.3 תנועת הגלגול אינה יכולה לשמש למדידת הקף המעגל (פרדוקס הגלגל)] וידוע שאם התנועעו שתי העגולות על מרכז ג׳ בהיותו קיים במקום אחד יעבור כל מקיף עגלת ד״ה על נקדת ד׳ שבקו ד״מ. אך אם התנועע מרכז ג׳ עד נקדת כ׳ בזמן שהתנועעו שתי העגולות סבוב אחד תתחדש משתי אלו התנועות תנועה אחרת מורכבת מהם היא הנקרא גלגול.[2] ויתגלגל מקיף עגלת א״ב על קו א״ל ומקיף עגולת ד״ה על קו ד״מ. ולו היה קו א״ל שוה למקיף עגלת א״ב כמד נגיעתם זה בזה בתנועת הגלגול היה, קו ד״מ עוד שוה למקיף עגולת ד״ה כמד[3] נגיעתם זה בזה בתנועת הגלגול, והיו מקיפי שתי העגולות שוין זה לזה והוא שקר. וכן אם התנועע מרכז ג׳ עד נקדת ז׳ בזמן שהתנועעו שתי העגולות סבוב אחד יתגלגל כל מקיף עגלת א״ב על [113א] קו א״ח וכל[4] מקיף עגלת ד״ה על קו ד״ט. ואלו היה הגלגול משער את הקו הישר ומשוה אותו בקו המקיף בעגלה המתגלגלת עליו כמו שאמרו אותם אנשים היה קו א״ח הקטן שוה לקו א״ל הגדול וזה שקר.

[5.4 גלגול מגדיר התאמה חד-חד-ערכית בין מעגל לקטע ישר, אבל התאמה כזו אינה גוררת שויון] וכדי להשלים ציור הגלגול איך הוא שלא לטעות בו:

[כיוון ראשון] נשליך נקדת ח׳ בקו א״ל הישר.

[קיום] אומר שתִּפְגְּשֶׁנָּה נקדה ממקיף עגולת א״ב בגלגול הנזכר. **מופתו** שנוציא עמוד ז״ח. ולפי ששטח א״ז נכחי הצלעות, יהיה ז״ח שוה לקו ג״א. ולפי שמרכז ג׳ מתנועע בקו ג״ז הישר יפגוש בנקודה ז׳, והקו המדביק ממרכז ז׳ למקום פגישת קו א״ל במקיף העגולה יהיה עמוד אל קו א״ל ושוה לקו ג״א. אם כן הוא יהיה קו ז״ח. אם כן נקדת ח׳ היא במקיף העגולה.

[יחידות] ואומר שלא תעבור נקדה אחרת ממקיף העגולה על נקדת ח׳ שבקו א״ל בגלגול הנזכר. וזה שאלו תעבור נקדת נ׳ עוד עליה, אם כן כשתסוב קשת א״נ על נקדת ח׳ שבקו א״ל לא התנועע המרכז מנקדת ז׳ וזה[5] לא יהיה גלגול.

---

[1] ייתכן שהמלה מיותרת.

[2] שתי המילים האחרונות אינן ברורות.

[3] בכתב היד: כגון; מעל השורה תיקון: ״כמיד״ או ״כמוד״. התיקון שלנו הוא לפי ההופעה של ״כמד״ במשפט הקודם.

[4] בכתב היד: ועל.

[5] שלוש המילים האחרונות אינן ברורות. הקריאה לפי גלוסקינה.

[כיוון שני] ונשליך נקדת ב׳ על מקיף העגלה.

[קיום] אומר[1] שתִּפְגְּשֶׁנָּה נקדה מקו א״ל בגלגול הנזכר. וזה כי בהיות מרכז ג׳ קיים ויתנועע העגול עליו תעבור נקדת ב׳ על נקדת א׳. [אבל אם מרכז ג׳ אינו קיים אלא נע על קו ג״כ] ונשים קו ח״א נוגע את העגלה על נקדת א׳ ומתנועע עמה על קו א״ל בתנועת מרכז ג׳ על קו ג״כ. וכשתגיע נקדת ב׳ לנקדת א׳ שבקו ל״א[2] תגיע נקדת א׳ שבקו ח״א לנקדת ח׳. אם כן תגיע נקדת ב׳ את נקדת ח׳ שבקו א״ל.

[יחידות] ואומר שלא תגע נקדה אחרת מקו א״ל את נקדת ב׳. שאלו כן, תִגָּעֶנָּה עוד [113ב] נקדת ע׳ אם אפשר. אם כן כשתנועע נקדת ב׳ על קו ע״ח שהיא תנועה ישרה אין שם תנועה סבובית וזה לא יהיה גלגול.

אם כן כל נקדה שבקו הישר אשר עליו היה הגלגול נוגעת בנקדה חברה לה ממקיף העגולה וכן כל נקדה שבמקיף העגולה נוגעת בגלגול את נקדה שהיא חברתה מהקו הישר. וזהו מה שהטעה את האנשים ההם לחשוב שהקו הישר שוה למקיף העגלה. כי חשבו זה כמו הרכבת הקשת על הקו הישר בהרכבת כל נקדותיה על כל נקדותיו וכל אחת על חברתה. ולא ידעו שהדבר הוא כך בין בהיות הקו הישר ארוך בין בהיותו קצר. ולא יהיה לעולם הקו המקיף בעגלה שוה לקו הישר שעליו יהיה הגלגול עד[3] היות תנועת המרכז הישרה שוה באותו זמן לתנועת הנקדה שעל מקיף העגלה שהיא תנועה סבובית, ושתהיה כל אחת משתי תנועות אלו מדובקת ושוה בחלקיה. וידוע שאם לא יהיה קו ישר שוה למקיף העגלה לא תהיה התנועה השוה בחלקיה על קו ישר שוה באותו זמן לתנועה שוה בחלקיה על מקיף עגלה, לפי שהמרחק והזמן והתנועה מתערכים ומשוערים קצתם בקצתם כמבואר בשמע טבעי. ולזה[4] היתה תולדת אותו הקש שלהם הקדמה[5] אליו והוא בטל על פי חכמת ההגיון.

[5.5 לא ניתן למדוד את הקף המעגל באמצעות חוט] וממה שצריך לידע שלא נעלם מהחכמים מה שעושין בני אדם שכורכין חוטין ישרים דקים ממשי או מפשתן וזולתם סביב העגולה המושגמת ומודדין את מקיפה בקו ישר. אלא רצו [החכמים] לדעת אם הענין הוא כך בעצמו, [לא] כמו שנמצא אצל ההרגש [114א] בעל הקירוב, כי אין ההרגש מספיק לכך ועם הקירוב שיש בו. שהרי כל שני קוים בלתי שוים זה

[1] שתי המילים האחרונות אינן ברורות. הקריאה לפי גלוסקינה.

[2] בכתב היד: ס״א.

[3] אות מיותרת: ת׳.

[4] אות מיותרת: ת׳.

[5] המלה אינה ברורה.

בזה יכול אתה לחתכם לחלקים קטנים מאד עד שימצאם אצל ההרגש מדוברים בערכם זה אצל זה, ועם שאפשר שלא[1] יהיה ערך מדובר ביניהם, וכי מיני הערך האלם אין תכלית למספרים כמבואר במאמר[2] י׳ מאקלידס. ואדם יכול ליקח גדרי כל המנינים קצתם בדקדוק[3] אמתי וקצתם בקירוב.

ועוד שיתבאר שנמצא קירוב[4] מטיל ספק גדול באותו החוט שכורכין סביב העגלה. וזה שנשים אותו החוט כשהוא ישר קו א״ב והוא גוף, ונשים עביו בשעור קוי ג״ד ה״ז ח״ט הישרים העומדים על ארכו בזויות נצבות. אם כן כשיתעקם גוף א״ב לצד גה״ח יתקרבו נקדות גא״ח והנקדות שבאותו צד אצל המרכז זו את זו, ויתרחקו נקודות דז״ט וכל הנקודות שבאותו צד אצל המקיף זו מזו. ויהיו הקוים המתעקמים [5] לצד המרכז שבו נקדות גה״ח יותר[6] קצרים ממה שהיו בהיותם ישרים והקוים כלם המתעקמים בו לצד דז״ט יהיו יותר ארוכים ממה שהיו בהיותם ישרים, לפי שהמקיף גדול מהמוקף. וכשלא יהיו הקוים ההם המתעקמים אף אחד מהם עומדים על שעורם המיוחד ביושר ועקום לא יתכן השוים ביניהם. ועם מה שזכרתי מדברי הפילוסופים שאין הקו העקום ולא השטח העקום חוזרים לעולם להיות ישרים, ואולי שהם אמרו כך בעד הקירובים שזכרנו לנקדות גה״ח זו לזו[7] [114ב] והרחוקים לנקדות דז״ט זו מזו. וכבר נדבר בזה.

[5.6 הערה מסכמת] וג״כ בשער הראשון הזכרתי שהידיעה הכללית שמצא השכל לתת ערך בין העגולה ובין השטח ישר הקוים אינה מספקת לבעל חכמת השעור להגביל אותו שטח ישר הקוים. ועם שאולי יאמר אומר שאין השכל יכול להמציאו במציאות אשר בפעל אעפ״י שהמציאו במציאות האמצעי בין הכח והפעל, כמו שאינו יכול להמציא קו[8] מחולק לחלקים אין תכלית למספרים במציאות אשר בפעל אעפ״י שהמציאו במציאות האמצעי בין הכח והפעל. וזה עקר בקשתי וכוונתי בספר זה להגביל אותו שטח ישר הקוים במציאות אשר בפועל כשאר השטחים והשעורים הנמצאים לחכמי השעור. אבל מה שכתב איש נצרי שמו קנפאנו והתפאר ברבוע העגולה אינו כלום כפי הדקדוק שאנו מבקשים, וזה לפי שהוא סמך על הקירוב שביאר ארשמידס במקיף העגולה שהוא שוה לג׳ דמיוני קטרה ושביעי פחות

[1] המלה כתובה פעמים.

[2] אות מיותרת: ת׳.

[3] המלה כתובה פעמים.

[4] המילה אינה מתאימה.

[5] מילה מיותרת: כי.

[6] בכתב היד: היותר.

[7] אות מיותרת: ת׳ ועוד שני סימנים לא ברורים; כנראה להשלמת שורה.

[8] מלה מיותרת ומחוקה בכתב היד: משלש.

מעשירי.[1] וזה ראוי לחכמת התשבורת מצד שהיא מעשית ולקוחה בקירוב ואינו ראוי לחכמת השעור מהצד שלקוחה בדקדוק אמתי.

[1] בכתב היד : משריו.

השער השלישי. בסגולות השעורים ישרי הקוים והשטחים המועילות בחכמה זו

[**0. הגדרות**] אמר אלפונשו אני מזכיר בתחילת שער זה השמות המוסכמים הראויים אצלי להשתמש בם שלא בדרך שהשתמשו בם בשאר ספרים. וזה כדי [115א] להבין הדברים יותר בנקל, ועם שאשתדל להציע הדברים בכדי שלא נצטרך בספר זה לשום ספר אחר מחכמת השעור בלתי ספר אקלידס בעקרים לבדו.

ואומר [א] שאני קורא קשת הזוית הישרת הקוים את הקשת מן העגלה שתרכב עליה הזוית ההיא בהיותה על מקיפה. [ב] ואקרא יתר הזוית הישרת הקוים ליתר הקשת ההוא שתרכב הזוית עליה, והוא כפלי[1] במשען שלה הנקרא ג׳יב אלקוס בערב כמו שנקרא כפלו ותר[2] אלקוס. [ג] ואקרא רבוי קו ישר בקו ישר את המשלש ישר הקוים נצב הזוית שיקיפו בה אותם שני קוים. [ד] ואקרא רבוי קו ישר בקו ישר ובזוית מונחת את המשלש ישר הקוים שיקיפו שני הקוים ההם בזוית ההיא שבו. [ה] ואקרא רבוי קו ישר[3] ובזוית מונחת והנקבץ בקו ישר את המוצק ישר השטחים שתושבתו רבוי שני הקוים זה בזה וגבהו הקו השלישי. כמו שקוראים [ג*] רבוע קו ישר [בקו ישר] את התבנית נצב ארבע זויות ששני מרחקיו הם אותם שני הקוים, וכן קוראים [ה*] רבוע קו ישר בשטח ישר את האצטונא הנצבת שתושבתה היא אותו שטח ישר וגבהו אותו קו ישר. [ו] ונקרא שטח מיושר את השטח ישר הקוים והוא אשר מקיפים בו קוים ישרים. [ז] וכן נקרא מוגשם מיושר את המושגם[4] שיקיפו בו שטחים מיושרים.

ומעתה אזכור הסגולות.

[**1**] **סגולה ראשונה** כשישוו [115ב] שתי צלעות ממשלש ישר הקוים לשתי צלעות מן משלש אחר ישר הקוים כל צלע לחברתה יהיה ערך המשלש אחד אל המשלש השני כערך יתר הזוית[5] שיקיפו בה שני צלעות ההן מהמשלש האחר ליתר הזוית[6] שיקיפו בה שני צלעי משלש השני. **כיצד**? נשים שני משולשי אב״ג אב״ד ישרי הקוים ושתי צלעות א״ב א״ג מהמשלש האחד שוות לשתי צלעות א״ב א״ד מהמשלש

---

[1] תיקון מעל השורה. בכתב היד : קשת.

[2] בכתב היד : נטר.

[3] מילה שככל הנראה מיותרת : וביתר.

[4] בטקסט : מושגמים. מתוקן מעל השורה ל-ם׳.

[5] בכתב היד : הזויות.

[6] בכתב היד : הזויות.

השני כל צלע לחברתה. ונרשום על מרכז א׳ ובמרחק א״ג א״ד הגדולים והשוין זה בזה עגלת גד״ה. ונמשיך א״ב עד מקיף העגולה לנקדת ה׳. ונוציא מן ד׳ עמוד ד״ח על א״ה והוא משען זוית בא״ד וכפלו הוא יתר שלה. וכן נוציא מן ג׳ עמוד ג״ז על א״ה והוא משען זוית בא״ג וכפלו הוא....

[המשך סגולה א', סגולות ב'-ט' ותחילת סגולה י' חסרים]

[10 סגולה י׳][1]

[118א] כשיהיו שתי הצלעות המתנגדות בתבנית בעל ד׳ צלעות ישרות בלתי נכחיות זו לזו ויצאו שני אלכסוני התבנית מתחתכים על נקדה יהיה רבוי אותן[2] שתי הצלעות זו בזו ובזוית[3] שיפגשו עליה כמותר המשלש שיחדשו[4] שני האלכסונים עם הגדולה שבשתי הצלעות הנשארות על המשלש שחדשו עם הצלע הקטנה שבשתיהן. **כיצד**? נשים אבג״ד ישר הקוים בעל ארבע צלעות א״ב ג״ד המתנגדות בלתי נכחיות זו לזו, וכשימשכו יפגשו על נקדת ז׳. ונוציא אלכסוני א״ג ב״ד מתחתכים על נקדת ה׳. אומר שרבוי א״ב בג״ד ובזוית ז׳ כמותר משלש אה״ד על משלש בה״ג. **המופת** שרבוי א״ז בז״ד ובזוית ז׳ הוא משלש אז״ד ורבוי ב״ז[5] בז״ד ובזוית ז׳ הוא משלש בז״ד. ישאר רבוי א״ב בז״ד ובזוית ז׳ משלש אב״ד. וכן יהיה רבוי א״ב בז״ג ובזוית ז׳[6] כמשלש אב״ג. אם כן רבוי א״ב בג״ד שהוא מותר רבוי א״ב בז״ד[7] על רבוי א״ב בז״ג כמותר משלש אב״ד על משלש אב״ג שהוא כמותר משלש אה״ד על משלש בה״ג. ומשל.

[11] סגולה י״א כשיצאו שני קוים ישרים משתי קצוות קטר העגול מתחתכים וחותכים את מקיף חצי העגול[8] יהיה רבוי החלק מהקו האחד אשר מקצה הקטר עד מקום פגישתו את הקו השני בחלק ממנו שהוא יתר בעגולה עם רבוי החלק מהקו השני אשר מקצה הקטר עד מקום פגישתו [118ב] את הקו הראשון בחלק ממנו שהוא יתר העגלה שוה יחד לרבוי קטר העגלה בעצמה. **כיצד**? נשים קו אב קטר

---

[1] הכותרת חסרה אבל הטקסט של סגולה 10 נראה שלם.

[2] בכתב היד: אותו. תיקון גלוסקינה.

[3] בכתב היד: ובזויות.

[4] בכתב היד: ״שישנשו״, מתוקן מעל השורה.

[5] בכתב היד: ה״ז.

[6] בכתב היד: ג׳.

[7] בכתב היד: בא״ד.

[8] טקסט מוכפל: את מקיף חצי העגול.

עגלת אבג״ד ונוציא קוי א״ה ב״ה[1] מתחתכים בנקדת ה׳ וחותכין את מקיף העגול בשתי נקדות ג׳ ד׳. אומר שרבוי[2] א״ג בא״ה עם רבוי[3] ב״ד בב״ה שוה יחד לרבוי א״ב בעצמו. **המופת** נוציא מן ה׳ עמוד ה״ז על הקטר ונדביק א״ד ג״ב אם כן משלשי בה״ז[4] אד״ב דומים, וזוית בה״ז שוה לזוית בא״ד אשר יתרה הוא ב״ד. אם כן רבוי ב״ה ביתר זוית בה״ז שהוא ב״ד כרבוי ב״ז ביתר זוית ז׳ הנצבה שהיא קטר א״ב. וכן יתבאר שרבוי א״ה ביתר זוית אה״ז שהוא א״ג כרבוי א״ז בקטר א״ב. אם כן רבוי א״ג בא״ה עם רבוי ב״ד בב״ה יחד כרבוי קטר א״ב בעצמו ומשל. וסגלה זו היא יותר כוללת מהמבואר במאמר א׳ מספר העקרים לאקלידס בתבנית מ״ז באמרו שכל משלש נצב הזוית יהיה רבוי יתר[5] זוית הנצבה בעצמו שוה לרבוי שני הצלעות הנשארות כל אחת בעצמה יחד[6].

[12] **סגלה י״ב** [משפט תלמי] כשיהיה תבנית בעל ד׳ צלעות ישרות בעגול יהיה רבוי שתי צלעותיו המתנגדות זו בזו יחד כרבוי שני[7] אלכסוניו זה בזה. **כיצד?** נשים תבנית אבג״ד בתוך עגול א״ב ג״ד ונוציא א״ג ב״ד. אומר שרבוי א״ב בג״ד וא״ד בב״ג יחד כרבוי א״ג בב״ד. וכבר הזכיר בטלמיוס המופת בזה. והוא שנשים זוית הג״ד [שוה לזוית אג״ב, וזוית בא״ג] שוה לזוית בד״ג[8] [119ב] ישאר זוית גה״ד שוה לזוית אב״ג. אם כן רבוי ג״ד ביתר זוית דג״ה שהוא א״ב כרבוי ד״ה ביתר זוית ה׳ שהוא א״ג. ולפי שזוית אג״ד שוה לזוית הג״ב וכן זוית גא״ד שוה לזוית הב״ג ישאר זוית גד״א שוה לזוית בה״ג. אם כן רבוי ב״ג ביתר זוית בג״ה שהוא א״ד כרבוי ב״ה ביתר זוית ה׳ שהוא א״ג וכבר היה רבוי א״ב בג״ד כרבוי[9] א״ג בה״ד אם כן רבוי א״ד בג״ב וא״ב בג״ד יחד כרבוי א״ג בב״ד.

[13] **סגלה י״ג** רבוי שתי צלעי משלש ישר הקוים זו בזו כרבוי העמוד היוצא מביניהם על[10] התושבת בקטר העגול המקיף במשלש. **כיצד?** נשים משלש אב״ג ישר הקוים בתוך עגלה ונוציא מנקדת ב׳ עמוד ב״ד על תושבת א״ג. אומר שרבוי א״ב בב״ג כרבוי עמוד ב״ד[11] בקטר העגלה. כי לפי שהיה משלש אב״ד

---

1 בכתב היד ה״ד.

2 טקסט מיותר: א״ה.

3 טקסט מיותר: ב״כ .

4 בכתב היד: אה״ז.

5 בכתב היד: יותר

6 בכתב היד: יותר

7 טקסט מוכפל: שני.

8 בכתב היד: הד״ג.

9 טקסט מיותר: א״ב .

10 שתי המילים לא ברורות. הקריאה לפי גלוסקינה.

11 בכתב היד: ב״ג . תיקון לפי גלוסקינה.

ישר הקוים יהיה רבוי א״ב ביתר זוית א׳ שעל[1] תושבת והוא ב״ג כרבוי ב״ד ביתר זוית ד׳ שעל התושבת והוא הקטר ומש״ל.

[14] **סגלה י״ד** כל משלש ישר הקוים נצב הזוית יהיה רבוי חצי מותר שני קוים המקיפים בנצבה על הצלע הנשארה[2] בעצמו ובצלע הנשארת כחצי רבוי שני הקוים זה בזה שהוא חצי המשלש. **כיצד?** נשים משלש אב״ג נצב זוית ב׳. אומר שרבוי חצי מותר אב ב״ג יחד על יתר א״ג בעצמו וביתר א״ג כחצי משלש אב״ג. **המופת** נרשום תוך המשלש עגול [3] על מרכז ז׳ נוגע לצלעי המשלש על נקדות חט״כ ונדביק קוי ז״א ז״ב [ז״ג] ז״ח ז״ט ז״כ. [119ב] אם כן קוי ז״א ז״ב ז״ג חולקין את זוית המשלש לחצאין חציין. ולפי תהיה זוית זכ״ח[4] נצבת וזוית ב׳[5] נצבה אם כן ח״ב שוה אל ח״ז. וכן ב״ט שוה אל ז״ט, וג״ח אל ג״כ, וכן א״ט שוה אל א״כ. אם כן מותר א״ב בב״ג יחד על א״ג ככפל ז״ח. אבל רבוי ז״כ בכל קוי המשלש כמו המשלש שהוא רבוי א״ב בב״ג. אם כן רבוי ז״ח בכפלו שהוא ב״ח ב״ט ובכפל א״ג שהוא א״ט ח״ג [6] יחד כמו המשלש. אם כן רבוי ז״ח בעצמו ובא״ג כחצי משלש אב״ג ומש״ל.

[15] **סגלה ט״ו** כל משלש ישר הקוים נצב הזוית מקיף בעגול יהיה רבוי שני חלקי הקו המיתר את הנצבה זה בזה אשר חלקתהו בהם משושת העגול שוה לחצי המשלש. **כיצד?** נשים משלש אב״ג ישר הקוים נצב חזוית מקיף בעגול כח״ט, ונקדת כ׳ משושת העגול בקו א״ג יתר זוית נצבה.[7] אומר שרבוי א״כ בכ״ג[8] כחצי המשלש. **מופתו** נדביק הקוים כמו בצורה שלפני זו. אם כן רבוי א״ב בב״ג כרבוי א״ט בב״ט ובג״ח עם רבוי ב״ט בב״ח ובח״ג. אבל רבוי א״ט בב״ח כא״ט בב״ט. אם כן רבוי א״ט בב״ט ובג״ח עם רבוי ב״ט בב״ח ובח״ג כמשולש אב״ג. ונשליך[9] רבוי ב״ט בב״ח ובח״ג ובא״ט שהוא כחצי משלש אב״ג. ישאר א״ט בג״ח, כלומר א״כ בכ״ג כחצי משלש אב״ג ומש״ל.

---

[1] טקסט מוכפל: שעל.

[2] בכתב היד: הנשאר

[3] טקסט מיותר: כחצי.

[4] בטקסט זכ״ח. גלוסקינה מתקנת לזכ״א. נראה מהקשר שיותר מתאים זח״ב.

[5] בכתב היד: ח׳.

[6] מלה לא ברורה.

[7] שלוש המילים האחרונות אינן ברורות.

[8] המלה אינה ברורה.

[9] בכתב היד: ונשלים. תיקון גלוסקינה.

[16] סגלה י״ו נשים משלש בה״ד נצב זוית ד׳ תוך עגול א״ב ד״ה שמרכזו ז׳ ושני קטריו המתחתכין בזויות נצבות א״ג[1] ב״ה. [120א] ונרשום עד[2] נקדת א׳ ובמרחק א״ב גזרת אהכ״ב. אומר שמרכז העגול שיקיפהו משלש בהד נופל בקשת הכ״ב. **המופת** שקשת הכ״ב היא רביע מקיף עגלתה. ונחתוך זוית הב״ד על חציה בקו ב״ח חותך קשת בכ״ה על כ׳. אם כן קשת כ״ה דומה לקשת ה״ח שהוא חצי קשת ה״ד. וישאר קשת כ״ב דומה לחצי קשת ב״ד. אם כן כשנדביק ה״כ ונמשיכהו אל ט׳ בקשת ב״ד יהיה חותך את קשת ב״ד על חציה ועל כן יחתוך את זוית בה״ד לחציה. ולפי׳ תהיה נקדת כ׳ מרכז עגול שיקיפהו משלש בה״ד ונוגע לצלעותיו.

[17] סגלה י״ז כשיהיו שני משלשים ישרי הקוים נצבי הזויות ומתדמים בערכי צלעיהם יהיה רבוי שתי הצלעות המייתרות את הזויות הנצבות זו בזו כרבוי שתי הצלעות הנשארות במשלש האחד בשתי צלעות הנשארות במשלש האחר כל אחת בחברתה הדומה לה יחד. **כיצד?** נשים שני משלשים אב״ג דה״ז ישרי הקוים ונצבי זויות ג׳ ז׳ ומתדמים: א״ג דומה אל ד״ז וג״ב דומה אל ז״ה. אומר שרבוי א״ב בד״ה כרבוי א״ג בד״ז וב״ג בה״ז יחד. **מופתו** לפי שערך א״ב אל ד״ה כערך[3] א״ג אל ד״ז, יהיה ערך מרובע קו א״ב למרובע קו ד״ה כערך מרובע א״ג למרובע קו ד״ז, וערך רבוי א״ב בד״ה לרבוי א״ג בד״ז כערך מרובע א״ב למרובע א״ג. וכן ערך רבוי א״ב בד״ה לרבוי ב״ג בה״ז כערך מרובע קו א״ב למרובע ב״ג.[4] אבל מרובע קו א״ב שוה למרובעי שני קוי א״ג ב״ג יחד. אם כן רבוי א״ב בד״ה שוה לרבוי א״ג בד״ז וב״ג בה״ז יחד ומש״ל.

[18] סגלה י״ח כשיחתך מקיף [120ב] העגלה בחלקים שוים ויצאו יתרים מדביקים בין כל שתי נקדות מהם נכחיים ליתר אחד מאותם החלקים, יהיה רבוי יתר זה בעצמו ובשאר כל היתרים מהם יחד שוה לרבוי קטר העגלה בעצמו. **כיצד?** נשים יתר א״ב[5] ג״ד ה״ז[6] וחבריהם בעגול אבד״ז נכחיים, וקשתות א״ב ב״ד ד״ז ז״נ [נ״מ] מ״ה[7] כלן שוות זו לזו. אומר שרבוי יתר א״ב בעצמו וביתרים כלן שהם[8] ג״ד ה״ז

---

[1] גלוסקינה מתקנת לא״ד.

[2] בכתב היד: עד.

[3] מילה מוכפלת.

[4] בכתב היד: ״קוי א״ג ב״ג יחד״. שבוש שנובע כנראה מבלבול עם השורה הבאה.

[5] בכתב היד: א״ג. תיקון גלוסקינה.

[6] בכתב היד: ד״ז. תיקון גלוסקינה.

[7] בכתב היד: ד״ה. תיקון גלוסקינה.

[8] מיותר. כפל א״ב בעצמו כבר נזכר בתחילת המשפט.

וחבריהן שוין לרבוי קטר העגול בעצמו. **מופתו** נדביק יתרי ג״ד ד״ז א״ד ג״ז ונוציא מנקדת ב׳ קטר ב״מ חותך את יתרי א״ד ג״ז ה״נ[1] בזויות נצבות על שתי נקדות ח׳ כ׳ [פ׳]. ונוציא מנקדת ד׳ עמוד ד״ט על ג״ז. ולפי שרבוי שני צלעי המשלש זו כרבוי העמוד היוצא מביניהן על התושבת בקטר העגול המקיף באותו המשלש, כמו שביארנו בסגלת י״ג משער זה, יהיה רבוי א״ב בב״ד שוה לרבוי קטר העגול בעמוד [ב״ח וכן יהיה רבוי ג״ד בד״ז שוה לרבוי קטר העגול בעמוד[2]] ד״ט השוה אל ח״כ[3]. אם כן רבוי א״ב ג״ד יחד בב״ד שוה לרבוי קטר העגול בקו ב״כ. וכן יתבאר שרבוי א״ב בעצמו ובכל היתרים הנכחיים לו שהם ג״ד ה״ז וחבריהן שוה לרבוי קטר העגול בעצמו שהוא קטר העגול ב״מ ומש״ל.

[19] **סגולה י״ט** כשיהיה מותר שעור א״ב הראשון על שעור ג״ד השני שוה למותר ד״ה השלישי על ב״ז הרביעי יהיה א״ב הראשון עם ב״ז הרביעי יחד שוה אל ג״ד השני עם ד״ה השלישי יחד. **מופתו** נשים ח״ב מן א״ב הוא מותר א״ב על ג״ד, וכן נשים ד״ט מן ד״ה הוא מותר ד״ה על ב״ז. אם כן ח״ב ד״ט שוין זה לזה וא״ח שוה אל ג״ד. אם כן א״ב שוה אל ג״ט וכן ט״ה שוה אל ב״ז. אם כן א״ב ב״ז יחד שוה אל[4] ג״ד ד״ה יחד ומש״ל.

[20] **סגלה כ׳** כשיחתך יתר הזוית הנצבה במשלש נצב הזוית לשני חלקים מתחלפים [121א] בנקדה ויצא קו ישר מאותה נקדת עד הזוית הנצבה, יהיה ערך מותר רבוי החלק הגדול בעצמו על רבוי הקו היוצא בעצמו למותר רבוי הקו היוצא בעצמו על רבוי החלק הקטן בעצמו כערך רבוי הצלע מהמקיפות בנצבה שאצל החלק הגדול בעצמו לרבוי הצלע הנשארת שאצל החלק הקטן[5] בעצמו. **כיצד** נשים משלש אב״ג נצב זוית ב׳ ונחתוך יתר א״ג לשני חלקים בלתי שוין זה לזה בנקדת ד׳, וחלק א״ד גדול מחלק ד״ג. ונדביק ב״ד. אומר שערך מותר רבוי א״ד בעצמו על רבוי ב״ד בעצמו למותר[6] רבוי ב״ד בעצמו על רבוי ד״ג[7] בעצמו כערך רבוי א״ב בעצמו על רבוי ב״ג בעצמו. **מופתו** נוציא מנקדת ד׳ עמודי ד״ז ד״ה.[8] ויהיו משלשי אז״ד אב״ג דה״ג דומים. ולזה יהיה ערך מותר רבוי א״ז בעצמו על רבוי ד״ה בעצמו למותר רבוי

---

[1] בכתב היד: ב״נ.

[2] השמטה בכתב היד. תיקון גלוסקינה.

[3] בכתב היד: א״ט. תיקון גלוסקינה.

[4] טקסט מיותר: ש״ו.

[5] המלה מופיעה בשוליים.

[6] המלה מופיעה בשוליים. בכתב היד מילה נוספת מיותרת: ״על״.

[7] בכתב היד: ב״ג.

[8] בכתב היד: ז״ה. תיקון גלוסקינה.

ז״ד בעצמו ולרבוי ה״ג בעצמו כערך רבוי א״ז בעצמו לרבוי ז״ד בעצמו (השוה לערך רבוי א״ב בעצמו לרבוי ב״ג בעצמו). אבל מותר רבוי א״ז בעצמו על רבוי ד״ה בעצמו (השוה אל ז״ב בעצמו) שוה למותר רבוי א״ד[1] בעצמו על רבוי ב״ד בעצמו. ומותר רבוי ז״ד בעצמו (השוה אל [רבוי] ב״ה[2] בעצמו) על רבוי ה״ג בעצמו שוה למותר רבוי ב״ד בעצמו על רבוי ד״ג בעצמו. אם כן ערך מותר רבוי א״ד בעצמו על רבוי ב״ד בעצמו למותר רבוי ב״ד בעצמו על רבוי ד״ג בעצמו כערך רבוי א״ב בעצמו על רבוי ב״ג[3] בעצמו ומש״ל. ולכן אם יהיו א״ב ב״ג שוין זה לזה יהיו רבויי א״ד ד״ג כל אחד בעצמו כפל ב״ד בעצמו כמבואר בסגלת י״ט.

[121ב] **[21] סגלה כ״א** כל קו שנחלק לשני חלקים בלתי שוין זה לזה יהיה רבוי כל חלק מהם בעצמו יחד שוה לכפל רבויים זה בזה עם רבוי מותר חלק הגדול על החלק הקטן בעצמו. **כיצד**? נשים קו א״ב מחולק לחציו בנקדת ג׳ ולשני חלקים מתחלפים בנקדת ד׳ וא״ד גדול מד״ב. אומר שרבוי א״ד ד״ב[4] כל אחד בעצמו שוין יחד לכפל רבויים זה בזה עם ארבעה פעמים ברבוי ג״ד בעצמו. **המופת** נרשום על מרכז ג׳ חצי עגול אה״ב ונקים עמודי ה״ג ד״ז מעל קטר א״ב, ונדביק קוי ג״ז ה״ד ז״ב[5] ז״א. אם כן מהמבואר בסגלת כ׳ יהיו רבוי א״ד ד״ב כל אחד בעצמו יחד שוין לכפל רבוי ד״ה בעצמו וזה שוה לכפל רבוי ה״ג ג״ד כל אחד בעצמו, וזה שוה לכפל רבוי ד״ז בעצמו[6] עם ד׳ פעמים ברבוי ג״ד בעצמו. אבל כפל רבוי ד״ז בעצמו שוה לכפל רבוי א״ד בד״ב. אם כן רבוי א״ד ד״ב כל אחד בעצמו שוין יחד לכפל רבויים זה בזה עם ד׳ פעמים ברבוי ג״ד בעצמו[7] שוה לרבוי מותר א״ד על ד״ב בעצמו ומש״ל.

[22] סגלה כ״ב כל משלש נצב הזוית שיהיו שתי צלעותיו המקיפות בנצבה בלתי שוות זו לזו יהיה רבוי הצלע השלישית בעצמה שוה לכפל המשלש עם רבוי מותר שבין שתי הצלעות המקיפות בנצבה בעצמו. **כיצד?** נשים משלש הג״ד נצב זוית ג׳ וצלע ה״ג[8] גדולה מצלע ג״ד, ונמשיך ג״ד אל צד ד׳ עד שישוה ג״ב [122א] להה״ג. אומר שרבוי ה״ד בעצמו שוה לכפל משלש הג״ד עם רבוי ד״ב[9] בעצמו. **המופת** נרשום על

[1] בכתב היד: א״ז. תיקון גלוסקינה.
[2] בכתב היד: ד״ה. תיקון גלוסקינה.
[3] בכתב היד: ב״ד: תיקון גלוסקינה.
[4] בכתב היד: ג״ב. תיקון גלוסקינה.
[5] בכתב היד ד״ב. תיקון גלוסקינה.
[6] טקסט כפול: וזה שוה לכפל רבוי ה״ג ג״ד כל אחד בעצמו וזה שוה לכפל רבוי ד״ז בעצמו.
[7] טקסט מיותר: שוה לרבוי מותר וד׳ פעמים ברבוי ג״ד בעצמו.
[8] בכתב היד: ה״ד. תיקון גלוסקינה.
[9] האותיות מטושטשות. קריאה לפי גלוסקינה.

מרכז ג׳ חצי עגול אה״ב קטרו אג״ב ונדביק ה״א ה״ב. אם כן מהמבואר בסגלת כ׳ יהיה כפל רבוי ה״ד בעצמו שוה לרבוי ב״ד ד״א[1] כל אחד בעצמו יחד. אבל רבוי ד״א בעצמו שוה לרבוי ג״ד ג״א[2] כל אחד בעצמו עם כפל רבוי ג״ד בג״א, השוה לרבוי ה״ד בעצמו עם כפל משלש הג״ד. וישאר כפל משלש הג״ד עם רבוי ד״ב בעצמו שוה לרבוי ה״ד בעצמו ומש״ל.

[23] סגלה כ״ג נרצה לרשום תבנית ירחי שוה לשטח ישר הקוים.

[בניה ראשונה] נרשום עגול אבג״ד על מרכז ה׳ וקטריו מתחתכים בזויות נצבות א״ג ב״ד. ונרשום על נקדת א׳ ובמרחק א״ב גזרת אבז״ד. ויחודש בין שתי קשתות בג״ד בז״ד תבנית בגז״ד הירחי. אומר שהוא שוה למשלש אב״ד ישר הקויים. וזה שגזרת הג״ד[3] רובע העגול וכן גזרת הב״ג וערך גזרת אבז״ד לגזרת הב״ג הדומה לה כערך מרובע קו א״ב למרובע קו ב״ה. אבל מרובע קו א״ב כפל מרובע קו ב״ה. אם כן גזרת אב״ד כפל גזרת הב״ג שהיא חצי עגול בג״ד. ונשליך חתכת בזד״ה המשותפת, וישאר תבנית בזד״ג הירחי שוה למשלש אב״ד ישר הקוים. ויש כאן תבנית שהר[4] שוה לשטח ישר הקוים.

[בניה שניה] נשים קו א״ב הישר שיהיה רבויו בעצמו שוה לג׳ פעמים לרבוי קו ב״ג בעצמו. ונשים רבוי קו [א״ג] בעצמו שוה לרבוי קו ב״ג בעצמו עם רבוי א״ב בג״ב. ונקים מג׳ [122ב] קוים אלו משלש אב״ג כמו שהתבאר, ונרשום עליו עגולה[5] אבג״ד מרכזה נקדת ה׳. ולפי שקו א״ג גדול מן יתר ג״ב תהיה קשת אד״ג גדולה מקשת ג״ב ונחתוך ממנה קשת אד שוה[6] לקשת ב״ג. ונדבק (צ״ל נדביק) קוי א״ד ד״ג ה״א ה״ד ה״ג ה״ב ב״ד אם כן צלעי משלש אב״ג שוות לצלעי משולש אד״ב כל אחת לחברתה, וכן א״ד[7] ב״ד שוות זו לזו. ולפי שתבנית אבג״ד בעל ד׳ צלעות בתוך עגלה יהיה רבוי קוי א״ג[8] בקו ב״ד, כלומ׳ רבוי א״ג בעצמו, שוה לרבוי א״ד בג״ב שהוא רבוי ג״ב בעצמו עם רבוי א״ב בג״ב או ג״ד. ולפי יהיה ג״ב שוה אל ג״ד בעצמו אם כן רבוי א״ב בעצמו שוה לרבויי יתרי א״ד ד״ג ג״ב כל אחד בעצמו יחד. ונקים על א״ב

[1] בכתב היד: ד״ה. תיקון גלוסקינה.

[2] בכתב היד: ד״א. תיקון גלוסקינה.

[3] בכתב היד: אג״ד. תיקון גלוסקינה.

[4] המלה ״סהר״ או ״שהר״ מופיעה בשירת ספרד. תתכן גם הקריאה ״אחר״, כפי שקוראת גלוסקינה.

[5] בכתב היד: משלש. תיקון גלוסקינה.

[6] טקסט מיותר: שוה.

[7] גלוסקינה מתקנת לא״ג.

[8] בכתב היד: א״ב. תיקון גלוסקינה.

משלש אב״ז דומה למשלש אה״ד[1] נרשום על מרכז ז׳[2] גזרת זאח״ב, ולכן גזרת זאח״ב, שוה לגזרת האד״ב וחתכה אח״ב שוה למשלש חתכות א״ד ד״ג[3] ג״ב. אם כן תבנית אדגב״ח הירחי שוה לתבנית אדג״ב ישר הקוים ומש״ל.

[24] סגלה כ״ד כשיהיו ד׳ שעורים ערך ראשון לשני מחובר מערך הראשון לשלישי ומערך הרביעי לשני והיו הראשון והשני משתוים, יהיו השלישי והרביעי משתוים ג״כ. **כיצד**? נשים ערך א׳ אל ב׳ מחובר מערך א׳ אל ג׳ ומערך ד׳ אל ב׳ השוה לו. אומר ששעוריי ג״ד משתוים. **המופת** נדמה ערך ג׳ אל ה׳ כערך ד׳ אל ב׳.[4] אם כן [123א] ערך א׳ אל ה׳ כערך א׳ אל ב׳ השוה אל א׳. ולפי שערך ג׳ אל ה׳ כערך ד׳ אל ה׳, כערך ד׳ אל ב׳ השוה אל ה׳, יהיו ג׳ ד׳ שוין ומש״ל.

[25] סגלה כ״ה כשיהיו ששה שעורים ערך השני לראשון מחובר מערך השלישי לרביעי ומערך החמישי לששי והיו הראשון והשני שוין, יהיה ערך השלישי לרביעי כערך הששי לחמישי. **כיצד**? נשים ערך א׳ אל ב׳ מחובר מערך ג׳ אל ד׳ ומערך ה׳ אל ו׳, והיו א׳ ב׳ שוין. אומר שערך ג׳ אל ד׳ כערך ו׳ אל ה׳. **המופת** נדמה ערך א׳ אל ז׳ כערך ג׳ אל ד׳,[5] וכן נשים ערך ח׳ אל ב׳ כערך ה׳ אל ו׳. אם כן מהמבואר בכ״ד [6] יהיו ז׳ ח׳ שוין. אם כן ערך א׳ אל ז׳ השוה לערך ג׳ אל ד׳ הוא כערך ב׳ אל ח׳ שהוא כערך ו׳ אל ה׳ ומש״ל.

[26] סגלה כ״ו כשיהיו שני שעורים מונחים מסוג אחד, וכל אשר לוקחו לראשון חלקים כמנין החלקים שלוקחו לשעור השני היה כל חלק מהראשונים מתערך עם חבירו מהחלקים השניים בערך שוה, וכל שהיו החלקים הראשונים יחד מוסיפין על השעור הראשון היו החלקים השניים מוסיפין גם כן על השעור השני, וכל שהיו החלקים הראשונים יחד פותחין[7] משעור הראשון היו החלקים השניים ג״כ פותחין מהשעור השני, יהיה ערך השעור הראשון אל השעור השני כערך המונח אשר מאחד מחלקיו הנלקחים[8] [מהשעור הראשון] לו לחברו[9] מהנלקחים לשעור השני. **כיצד**? נשים שני שעורי [123ב] א״ב ג״ד מסוג אחד ונקח משעור א״ב חלקי ז״נ נ״מ מ״ל ל״ח, וכל חלק מהראשונים מתערך לחבירו מהשניים בערך

[1] בכתב היד אה״ז. תיקון גלוסקינה.

[2] בכתב היד: ג. תיקון גלוסקינה.

[3] בכתב היד ד״ב. תיקון גלוסקינה.

[4] טקסט מיותר: אם כן ערך א׳ אל ב׳ המחובר מיחס א׳ אל ג׳ ומיחס ד׳ אל ב׳ השוה לערך א׳ אל ב׳.

[5] טקסט מיותר: ומערך ה׳ אל ו׳ והיו א׳ ב׳ שוין אומר שערך.

[6] טקסט מיותר: ד׳.

[7] שיכול אותיות, כאן ובהמשך וכן בסגולה כ״ז.

[8] המלה מופיעה פעמיים.

[9] שתי המלים מופיעות פעמיים.

שוה: ז״נ לט״פ ונ״מ לפ״ע ומ״ל לע״ס ול״ח לס״כ, וכל שהיו חלקי ז״ח מוסיפין על שעור א״ב היו חלקי ט״כ[1] גם כן מוסיפין על שעור ג״ד, וכל שהיו חלקי ז״ח חוסרין משעור א״ב היו חלקי ט״כ חוסרין גם כן משעור ג״ד. אומר שערך ח״ב אל ג״ד כערך שעור ז״נ לשעור ט״פ. **מופתו** שאם לא כן נשים ערך א״ב לג״ד גדול מערך ז״נ לט״פ אם אפשר, ונשים ה״ב קטן מן א״ב שיהיה ערכו אל ג״ד בלתי קטן מערך ז״נ לט״פ. ונקח חלקי ק״צ צ״ו מוסיפין על שעור ה״ב ובלתי מוסיפין על שעור א״ב. ונקח ג״כ חלקי ת״ש ש״ר על מנין חלקי ק״צ צ״ו. ויהיה הערך ת״ש אל ק״צ וש״ר אל צ״ו כערך המונח מן ט״פ אל ז״נ. אם כן בקבוץ יהיה ערך ת״ר אל ק״ו כערך ט״פ אל ז״נ, ובחלוף ערך ק״ו אל ת״ר כערך ז״נ אל ט״פ. ולפי שחלקי ק״ו בלתי מוסיפין על שעור א״ב יהיו גם כן חלקי ת״ר בלתי מוסיפין על שעור ג״ד. אם כן יהיה ערך ק״ו הגדול מן ה״ב[2] אל ת״ר שאינו גדול מג״ד כערך ז״נ אל ט״פ. ולזה יהיה ערך ה״ב אל ג״ד קטן מערך ז״נ אל ט״פ. וכבר שמנוהו בלתי קטן ממנו זה שקר. אם כן אין ערך א״ב אל ג״ד גדול מערך ז״נ אל ט״פ וכבר יתבאר שאינו קטן ממנו ולכן יהיה שוה לו.

[27] סגלה כ״ז כשיהיו ארבעה שעורים ולוקחו לראשון ולשלישי חלקים על מספר שוה והיה ערך השעור הראשון לכל אחד מהחלקים הלקוחים לו כערך [124א] השעור השלישי לחברו מהלקוחים לו, וכל שהחלקים הראשונים יוסיפו על השעור השני היו גם כן החלקים האחרים מוסיפין על שעור הרביעי, וכל שהחלקים הראשונים[3] פותחין מהשעור השני היו גם כן החלקים האחרים פותחים מהשעור הרביעי, יהיה ערך השעור הראשון אל השעור השני כערך השעור השלישי אל השעור הרביעי. **כיצד?** נשים ד׳ שעורי א׳ ב׳ ג׳ ד׳[4] ונקח שעורי[5] ה״ז ז״ח ח״ט מסוג שעור א׳,[6] וכן נקח שעורי ל״מ מ״נ נ״ס כמנין שעורי ה״ט [מסוג שעור ג׳][7], וערך שעור א׳ לכל אחד משעורי ה״ז ז״ח ח״ט כערך שעור ג׳[8] לחברו בשעורי ל״מ מ״נ נ״ס, וכל שהוסיפו שעורי ה״ט אל שעור ב׳ יוסיפו ג״כ שעורי ל״ס על שעור ד׳ וכל שחסרו שעורי ה״ט מן ב׳ חסרו ג״כ שעורי ל״ס מן ד׳. אומר שערך א׳ אל ב׳ כערך ג׳ אל ד׳. **מופתו** שאם לא כן נשים ערך ב׳ אל א׳ גדול מן ערך ד׳ אל ג׳ אם אפשר, ונשים כ׳ קטן מן ב׳ שיהיה ערכו אל א׳ בלתי קטן מערך ד׳ אל ג׳,

---

[1] בכתב היד: ס״כ. תיקון גלוסקינה.

[2] בכתב היד ז״נ. תיקון גלוסקינה.

[3] טקסט מיותר: יוסיפו על השעור השני היו גם כן החלקים האחרים מוסיפין.

[4] בכתב היד: א״ב ג״ד. זה בד״כ הסימון לשני קטעים אולם כאן זו אי תשומת לב של המחבר או טעות של המעתיק.

[5] אות מיותרת: י׳. כאן ושוב בשטרה הבאה שנזכר קטע ה״ז.

[6] בכתב היד: ב׳. תיקון גלוסקינה.

[7] תוספת של גלוסקינה.

[8] בכתב היד ב׳. תיקון גלוסקינה .

ונקח חלקי ע״פ צ״פ מוסיפין על שעור כ׳ ובלתי מוסיפין על שעור ב׳. ונקח ג״כ חלקי ק״ר ר״ש על מנין חלקי ע״צ בעניין שיהיה ערך א׳ אל ע״פ כערך ג׳ אל ק״ר וערך א׳ אל פ״צ כערך ג׳ אל ר״ש. אם כן בקבוץ יהיה ערך[1] א׳ אל ע״צ כלו כערך ג׳ אל ק״ש כלו. ולפי שחלקי ע״צ בלתי מוסיפין על שעור ב׳[2] יהיו גם כן חלקי ק״ש בלתי מוסיפין על שעור ד׳, וחלקי ע״צ מוסיפין על כ׳, וחלקי ק״ש בלתי מוסיפין על ד׳. אם כן ערך [124ב] ע״צ אל כ׳[3] גדול מערך ק״ש אל ד׳ וערך כ׳ אל א׳ בלתי קטן מערך ד׳ אל ג׳. אם כן ערך ע״צ אל א׳ גדול מערך ק״ש אל ג׳, וכבר היו שוין זה לזה, זה שקר. אם כן אין ערך ב׳ אל א׳ גדול מערך ד׳ אל ג׳, וכן יתבאר שאינו קטן ממנו. אם כן הוא שוה לו ומש״ל.

[28] סגלה כ״ח כשיהיה שטח חצי עגלה נוטה על שטח ישר ויצאו ממקיף העגלה עמודים על השטח הישר, יהיה ערך העמודים ההם זה לזה כערך משעני אותן קשתות שיצאו מקצותיהן אותם העמודים זה לזה. **כיצד**? נשים שטח חצי עגול אבג״ד קטרו א״ד נוטה על שטח ישר, ומנקדות ב׳ ג׳ ייצאו עמודי ב״ח ג״ט על השטח הישר. ומשעני ב״ה ג״ז על קטר א״ד. אומר שערך עמוד ב״ח[4] לעמוד ג״ט כערך משען ב״ה למשען ג״ז. **המופת** נדביק קוי ה״ח ז״ט. אם כן שני עמודי ב״ח ג״ט נכחיים וכן משעני ב״ה גז נכחיים, ואין הקוים בשטח אחד. ולזה זוית חב״ה שוה לזוית טג״ז כמו שבאר אקלידס׳ וזויות[5] ח׳[6] ט׳ נצבות. אם כן משלשי בה״ח גז״ט דומים, ערך ב״ה אל ג״ז כערך ב״ח אל ג״ט ומש״ל.

[29 סגלה] כ״ט נרצה למצא יציאת שני קוים האחד ישר והשני כפוף שיהיה ביניהן בתחלת יציאתן מרחק מה, וכל אשר יתארכו שני הקוים יחסר המרחק שבין שתי קצותיהן ויקרב האחד מהם לאחר ולא יחתכו[7] אפילו יצאו לאין תכלית. **כיצד!** נשים שני קוי א״ב ב״ג מקיפין בזוית ב׳ נצבה ונשליך נקדת ד׳ בקו א״ב כפי מה הן בין א״ב הן חוץ מנקדת ב׳. ונניע [125א] נקדת ב׳ שבקו ב״ד על קו ב״ג לצד נקדת ג׳, ויתנועע עמה קו ב״ד בענין שתהיה נקדת ד׳ ממנו מתנגדת לנקדת א׳, כלומר שתהיינה תמיד ג׳ נקדות א׳ ד׳ ב׳ בקו אחד ישר. ותרשום נקדת ד׳ בתנועה זו[8] קו דז״ח הכפוף, הנקרא אצלנו הקו הפרוץ. אומר ששני קוי ב״ג דז״ח כל אשר יתארכו לצד ג״ח יחסר המרחק בין קצותיהן ולא יפגשו לעולם. **המופת**

[1] טקסט מיותר: אנ.
[2] בכתב היד: כ. תיקון גלוסקינה.
[3] בכתב היד: ב. תיקון שלנו.
[4] בכתב היד: ב״ט. תיקון גלוסקינה.
[5] בכתב היד: זוית.
[6] בכתב היד: ה׳.
[7] בכתב היד: ייתכן. תיקון גלוסקינה.
[8] שתי מילים לא קריאות.

נוציא שני קוי אח״ג אז״ה הישרים. אם כן יהיו שלשת קוי ב״ד ז״ה ח״ג שוין זה לזה. **ונוציא** מנקדות ז״ח שני עמודי ז״ט ח״כ על קו ב״ג. ולפי [i] שמרובע קו ז״ה גדול ממרובע קו [ז״כ גדול מן] ז״ט, ולפי [ii]ששני משלשי זט״ה כח״ג נצבי הזויות [iii] ויתרי ז״ה ח״ג שוין [iv] וזוית זה״ט גדולה מזוית כג״ח,[1] יהיה עמוד ז״ט גדול מעמוד ח״כ. וכן יתבאר שכל העמודים היוצאין מקו דז״ח הפרוץ על קו ב״ג הישר, כל הקרוב מהם לקו א״ב גדול מאשר הוא רחוק ממנו. אם כן שני קוי ב״ג דז״ח כל אשר ימשכו לצד ג״ח יחסר המרחק בין קצותיהן ולא יפגשו לעולם ומש״ל. והכלי המוכן לרשום בו את הקו הפרוץ הוא נקל להמצא לבעלי המלאכה המעשית והוא מועיל מאד בחכמה זו.

[30] סגלה ל׳ לחלק כל זוית ישרת הקוים לג׳ חלקים שוים. **כיצד?** נשים זוית אב״ג הישרת הקוים, ונקים ב״ד עמוד על א״ב, ונשים ב״ג באיזה שעור שנרצה, ונמשיך א״ב לאין תכלית לצד ב׳. ונשים המשרטטת על נקדת ג׳ בענין שתפגוש שני קוי א״ב ב״ד על שתי נקודת ד׳ ה׳, ושיהא קו ד״ה כפל ב״ג, וזה יהיה ברשום [125ב] קו פרוץ כפי הנזכר. ותהיה זוית דה״ב[2] שליש זוית אב״ג המונחת. **המופת** לפי שקו ד״ה ממשלש בד״ה הוא יתר זוית דב״ה[3] הנצבה וכשנחתכהו על נקדת ז׳ ונדביק ב״ז יהיה ה״ז שוה אל ז״ב, והוא שוה אל ד״ז. אם כן זוית בז״ג השוה לזוית[4] זב״ה עם זוית זה״ב. ותהיה זוית דה״ב שליש שתים זויות זה״ב זג״ב יחד השוות לזוית אב״ג המונחת ומש״ל.

[31] סגלה ל״א נרצה למצא שני קוים ישרים אמצעיים בערך בין שני קוים ישרים אחרים שאינם שוים זה לזה. **כיצד?** נשים קו א״ב הקטן וג״ה הגדול, ונרשום חצי עגול קטרו ג״ה הגדול ומרכזו ב, ונקדת ד על חצי אב על הקטר. ונוציא עמוד ד״ז על הקטר למקיף העגול ונמשיך א״ד לצד ח׳ עד שיהיה א״ח שוה אל א״ב. ונדביק ח״ז ונוציא מנקדת ב׳ קו ב״ל נכחי אל ח״ז ולאין תכלית. ונוציא מנקדת ז׳ קו ז״מ אל קו גה״מ פוגש את קו ב״ל על נקדת ל׳, בענין שיהיה קו ל״מ שוה לחצי הקטר. ויהיה זה ברשום הקו הפרוץ אומר שערך א״ב אל ז״ל כערך ז״ל[5] אל ב״מ וכערך ב״מ אל ג״ה[6]. **מופתו** נמשיך ז״מ עד היות ל״נ

[1] בכתב היד: כח״ג. תיקון גלוסקינה.

[2] בכתב היד: דח״ב. תיקון גלוסקינה.

[3] בכתב היד: בד״ה. תיקון גלוסקינה.

[4] אי בהירות בכתב היד.

[5] בכתב היד: ז״מ. תיקון גלוסקינה.

[6] בכתב היד: ג״ה. תיקון גלוסקינה.

שוה אל[1] הקטר. ולפי שיתרון רבוי מ״ז בעצמו על רבוי ב״ז בעצמו כרבוי מ״ב בעצמו ובכפל ב״ד[2] והוא גם[3] כרבוי ז״ל בעצמו ובכפל ל״מ, יהיה רבוי א״מ במ״ב כרבוי נ״ז בז״ל, ויהיה ערך א״מ אל נ״ז כערך ז״ל[4] אל מ״ב, וערך א״ב[5] אל ז״ל כערך חצי מ״ב אל ל״מ השוה לערך מ״ב אל ל״נ. וכבר היה ערך א״ב אל [126א] ז״ל וכערך א״מ אל נ״ז השוה לערך ל״ז אל מ״ב.[6] אם כן ערך[7] א״ב אל ז״ל כערך ז״ל אל מ״ב וכערך מ״ב אל ל״נ השוה אל הקטר ומש״ל.

[32] [126ב] **סגלה ל״ב** נרצה לעשות מושגם ישר הקוים שוה למושגם מונח ישר הקוים ודומה למושגם שני מונח ישר הקוים. **כיצד?** נשים מושגם שתושבתו שטח א"ג[8] וגבהו[9] ד״ב ונשים מושגם שני שתושבתו שטח ה״ח[10] וגבהו ט״ז, ונרצה למצוא מושגם שלישי שוה אל הראשון ודומה אל השני. נקים על שטח ה״ח מושגם אחר שגבהו ז״כ שיהיה שוה אל המושגם הראשון ונוציא בין קוי ז״כ ז״ט שני קוי ע׳ ל״נ מתערכים ביניהן ערך ז״כ אל ע׳ כערך ע׳ אל ל״נ וכערך ל״נ אל ט״ז. ונשים ערך ל״נ הידוע אל ט״ז הידוע כערך נ״מ שאינו ידוע אל ה״ז[11] הידוע. ונקים על[12] קו נ״מ שטח מ״ס[13] דומה לשטח ה״ח, והוא יהיה תושבת המושגם השלישי הדרוש וגבהו ל״נ.[14] **מופתו** לפי שערך ז״כ אל ל״נ כערך ל״נ אל ט״ז שנוי השוה לערך נ״מ אל ה״ז שנוי, השוה לערך שטח מ״ס לשטח ה״ח, יהיה רבוי ז״כ בשטח ה״ח כרבוי ל״נ בשטח מ״ס. ויהיו שני מושגמים שוים ולזה המושגם השלישי שוה אל הראשון, ולפי שערך ל״נ אל ט״ז כערך נ״מ אל ה״ז יהיה המושגם השלישי דומה לשני .

[33] סגלה ל״ג נרצה לצייר תנועה ישרה מדובקת ותמידית הלוך ושוב על קו ישר בעל תכלית מבלי השקט בין הליכה ושיבה. **כיצד?** נשים קו א״ב הישר ונרשום עגול אז״ב עליו מרכזו ג׳, ונרשום על קו

---

[1] שתי המלים האחרונות לא ברורות.

[2] בכתב היד: ב״ל. תיקון גלוסקינה.

[3] בכתב היד: ״עם״. גלוסקינה מציעה: עוד.

[4] בכתב היד: ז״נ. תיקון גלוסקינה.

[5] בכתב היד: א״ל. תיקון שלנו.

[6] בכתב היד: ו׳ במקום ז׳ והסדר הפוך: מ״ב אל ל״ו.

[7] מילה לא ברורה.

[8] גלוסקינה מתקנת כאן לא״ב.

[9] מלה מיותרת: שטח.

[10] בכתב היד: ה״ז. גלוסקינה אינה מתקנת.

[11] בכתב היד: ח״ז. תיקון **שלנו**

[12] בכתב היד: אל. תיקון גלוסקינה.

[13] בכתב היד: מ״ט. תיקון גלוסקינה.

[14] טקסט חזרתי מיותר: **מופתו** לפי שערך ז״כ אל ל״נ הידוע אל ט״ז הידוע כערך נ״מ שאינו ידוע אל ח״ז הידוע, ונקים על קו נ״מ שטח מ״ט דומה לשטח ה״ח והוא יהיה תושבת המושגם השלישי הדרוש וגבהו ל״נ.

א״ג עוד עגול דח״ג נוגע עגול אז״ב. ויהיו שלש אותיות אה״ד בנקדת הנגיעה, ונסכים שנסוב מרכז הכוכב שהוא ד'[1] בעגול דח״ג מנקדת ד' עד נקדת ח' בתנועה שוה בחלקיה. ותסוב נקדת ה' שהיא נגיעת שתי העגולות בתנועה שוה בחלקיה מנקדת א' לנקדת ז' להפך תנועת ד'[2], כענין שתהיה תנועת ד' כפל תנועת ה'[3]. ר״ל שכשתשלים נקדת ד' את כל עגול דח״ג שתי פעמים תשלים נקדת ה' את כל עגול אז״ב פעם אחת. ונדביק שני קוי ג״ז ג״ח הישרים. ולפי שקשת ד״ח דומה לכפל קשת א״ז, וזוית דג״ח במקיף העגול רוכבת על קשת ד״ח, וזוית אג״ז במרכז העגול רוכבת על קשת א״ד, תהיה זוית דג״ח שוה לזוית אג״ז. ולפיכך כשירכב קו ד״ג על קו ג"ז[4] ירכב קו ח״ג על קו א״ג[5] ותפול נקדת ח' שהוא מקום מרכז הכוכב בקו א״ג על נקדת ט'. נמצא שחלק הכוכב מנקדת א' עד נקדת ט'. וכן יתבאר כשישלים הכוכב כל עגולתו פעם אחת, ויגיע נקדת א' לנקדת ב' ויהיה הכוכב בנקדת ב', וכן יחזור מנקדת ב' לנקדת א', מבלי שיטה מקו א״ב לשום צד מהצדדין. ואומר שלא ישקוט הכוכב שום זמן לא בנקדת א' ולא בנקדת ב'.

מופתו שאלו היה שוקט ילך הככב בזמן ההוא בעגלתו שום [127א] קשת, לפי שתמיד הוא מתנועע על העגול. ונסכים שהלך מנקדת ד' עד נקדת כ'[6]. ונדביק קו כ״ג. ואם היתה נקדת כ' נופלת על נקדת א' היה קו כ״ג שוה לקו א״ג והוא שקר. אם כן לא ישקוט הכוכב בנקדת א'. הנה מצאנו התנועה הישרה הוזרת מבלי שום השקט בין הליכה ושיבה. **[מסקנה]** ונראה מזה שהסבה בהשקט הנמצא בין שתי תנועות המתהפכות אינה הקצה והתכלית המצויה בפועל במרחק שהתנועה עליו כמו שחשב ארסטוטו' והנמשכים לדעתו, לפי שאותה הקצה והתכלית לא היתה מפעל המתנועע כדי שישקוט ברשומה, אבל היא רשומה במרחק. ואפשר שיעבור עליה הגוף המתנועע מצד שהוא מתנועע מבלי שישקוט בה. וכ״ש אם היה המתנועע נמשך בתנועתו לתנועה אחרת סבובית ומדובקת או לכח[7] שומר את הדבוק כמו שהוא בשני גלגלי נגה וככב חמה. וכבר כתב אבו עאלי בן רצואן שמה שאמרו שהתנועות המתהפכות יש השקט בינותן אינו אמת תמיד, וכי אין ההשקט המצוי בין תנועות הדבוק כי אם שינוח הכח המניע מיגיעו.

---

1 בכתב היד : ה'.

2 בכתב היד : ה'.

3 בכתב היד להפך : תנועת ה' כפל תנועת ד' .

4 בכתב היד : ב״ז. תיקון גלוסקינה.

5 האותיות לא ברורות.

6 בכתב היד : ג.

7 בכתב היד : לכת.

השער הרביעי לבאר דרך החלק הגוף לשטחים והשטח אל הקוים כמו שאמר אפלטון ודרך שעוריהן ומדידתן אלו באלו, ובו י״ו יסודות.

[1] היסוד הראשון כשהיו שני שטחים בעלי ד׳ צלעות ישרות נצבי הד׳ זויות [127ב] על תושבת אחת או על שתי תושבות שוות זו לזו, יהיה ערך השטח האחד [לאחר] כערך מה שיפול בתוך השטח האחד מכל העמודים אין תכלית למספרם היוצאין על התושבת ומחלקין אותה ואת שני השטחים בחלקים שוים - הם העמודים הנרשמים עליו בתנועה הישרה השוה בחלקיה שיתנועע הקו העמוד (כמו שזכרתי בשער הראשון) - אל כל מה שיפול מכל אותן העמודים בתוך השני. וכן בשעורי האצטוונות המתדמים והמוצרפים המתדמים שגבהם אחד, שיהיה ערך הגדול שבהם אל הקטן כערך השטחים הנכחיים אין תכלית למספרם החותכים את השעור הגדול בחלקים שוים אל כל מה שיפול מהם בשעור הקטן. וכמו כן באצטוונות העגלות שירשמום השטחים נצבי הזויות בתנועתם סביב חצם, יהיה הערך האחת אל השנית כערך מרובעי כל העמודים היוצאין בשטח האחד על החץ שלו למרובעי כל העמודין היוצאין בשטח השני על החץ שלו.

כיצד? נשים שני שטחי אבג״ד אבה״ז בעל ד׳ צלעות ישרות נצבי הזויות על תושבת א״ב. אומר שערך שטח אבג״ד [לשטח אבה״ז כערך] כל העמודים אין תכלית למספרם הרשומים בתנועה הישרה השוה בחלקיה בתוך שטח אבג״ד אל כל מה שיפל מאותם עמודים בתוך שטח אבה״ז.

מופתו, נחלק תושבת א״ב לחלקים שוין באיזה מספר שנרצה בקוים חמ״כ טנ״ל וחבריהם. אם כן ערך שטח [128א] א״כ לשטח א״מ כערך קוי א״ד ח״כ יחד לקוי א״ז ח״ז [ח״מ??] יחד. וכן ערך שטח ח״ל לשטח ח״נ כערך קו ט״ל לקו ט״נ. וכן ערך שטח

[הטקסט נקטע באמצע המשפט]

Glossaries

A.1. Hebrew-English Glossary

Verbal forms are listed under the three root letters. Other words (including participle forms) are listed as they appear in the text. An asterisk refers to the paragraph in which the term is explained.

R. Glasner, A. Baraness, *Alfonso's Rectifying the Curved*,
Sources and Studies in the History of Mathematics and Physical Sciences,
https://doi.org/10.1007/978-3-319-77303-2

אלכסון	diagonal III.10, III.12
אלם	irrational I.5.4, *II.5.5
אמצעי	intermediate I.1.3–4, I.4.1–2, II.5–6
אצטונא/אסטונא/אסטוונא	cylinder I.1.4, I.2.1, I.5.4, II.5.1 IV.1
אצטונא מעוגלת/אסטונא עגולה	cylinder 0, I.1.2, 4, I.2.4, I.5.4, IV.1
אצטונה נצבת	right circular cylinder I.5.1,
	right general cylinder III.0
גוף האסטונא	cylinder i.5.4, ii.5.1
מוצק האסטונה	cone i.5.1 (in the text אצטולא), I.5.4
אפשרות	capacity II.1.7
	possibility i.1.2
בטל	false *II.1.2, II.2.2, II.5.1, 4
גבול	limit ii.1.4, ii.1.7
גבל (v.)	
מגביל/להגביל	construct *I.2.5, I.5.4, II.5.6
	limit I.1.3, II.1.4, 7, II,4.2
	see also מוגבל, נגבל
גדר	definition II.2.4
	r oot *II.5.5
גוף	body 0, I.3.2–5, I.4.3–4, I.5.4, II.1.7, II.5.2, 5, III.33, IV.0
גזרה	statement
גזרה תנאית מדובקת	conditional conjunctive statement I.1.3
גלגל	sphere I.3.4
	orb *I.5.3, III.33
גלגל המזלות	zodiac I.2.1, I.4.1–2
גלגל חצי היום	meridian I.2.1, I.4.1
גלגל (v.)	
התגלגל	roll II.5.1–3
גרם	*gerem* (left untranslated) I.4.3
גרמים, גרמים ראשוניים	primary *geramim* *I.3.2, 4, I.4.3
דבק (v.)	
נדביק/להדביק	join I.3.5, II.1.7, II.2.1–3, II.3.1–5, II.5.2, 4, III.11, 14–16, 18, 20–22, 28, 30–31, 33
נדבק	joined ii.3.4
	see also קו מדביק, מדובק
דמה (v.)	assume II.1.5, II.2.1, II.4.1, III.24–25
	see also התדמות, מתדמה

דמות	likeness 0, I.5.4
דקדוק	precision II.5.5–6
היולי	*hiuli* (left untranslated) *I.3.2–4
הצטרפות	relation I.4.2
הקדמה	
	principle I.3,3
	postulate II.2.1–4, II.3.0, 6
	lemma *ii. 2.3–4
	premise I.1.3–4, I.2.3–4, I.5.2, II.1.2, II.2.2,
	premise (of a syllogism) II.5.4
הקדמה תנאית	conditional premise I.1.3, I.2.3
הקדמה תנאית מדובקת	conjunctive conditional premise I.1.3
הקדמה תנאית מתפרקת	disjunctive conditional premise I.2.4
הקדמה קטנה	minor premise I.1.3–4, I.2.3
הקדמה גדולה	major premise I.1.3–4
	see also גזרה תנאית מדובקת
הקש	syllogism *I.1.3–4, I.2.3, II.5.4
	argument, proof ii.2.2 –4, II.3.0
הקש הסבוב	circular argument, ii.2.2, ii.3.0
הקש מופתי	demonstrative proof II.3.0
הקש התנאי	conjunctive conditional syllogism I.1.4
	see also הקדמה, תולדה, גזרה
הרגש	perception II.5.5
הרכבה	superposition/ superposing * i.1 .1, I.2.1–4
הרכבה עצמית	essential superposition I.2.1
הרכבה שוקטת	static superposition I.2.1
הרכבה שזכרנו/הרגלנו לציירה	superposition that we are used to imagine I.2.2–3, I.3.5
הרכבה שכלית	mental superposition I.2.3–4, I.3.1, I.3.5, I.4.2, I.5.4
הרכבה עליונה	superior superposition i.2.2 –3
הרכבה בתנועה	superposition by motion i.4.1
הרכבת שעורים, מרחקים	superposition of distances, magnitudes I.2.1–2, I.4.2, I.5.1–2, II.1.3
הרכבת קשת על קו ישר	superposition of an arc on a line II.5.1, 4
	see also רכב
השבה	transforming, 0, i.5.4
השקט	a state of rest iii.33
	s ee also שוקט
התדמות	likeness i.4.2
	see also דמה
התחלה	first principle 0, I.1.3 , I.3.5, I.5.1–4, II.0, II.2.0, II.3.3, 6, II.5.1

זוית	angle I.5.4, II.2.1, 3, II.3.1–6, II.4.1, II.5.1, III.0, 10-16, 28–30, 33
זוית ישרת קוים	rectilinear angle i.5.4 , III.30
זוית מקשתות עגולות	angle between arcs i.5.4
זוית חדה	acute angle i.2.1
זוית נרוחת	obtuse angle i.2.1
זוית מתנגדת	opposite angles II.3.1
זוית מתחלפת	alternate angle II.3.3
זוית נצבת/נצבה	right angle I.5.1, many more.
	see also משלש, קשת, יתר
זיוף	sophistical argument I.1.1
זמן	time I.4.1–2, 4, I.5.3, II.4.1, II.5.3–4, III.33
חבר/חברה	corresponding I.2.1, II.2.1, II.3.1, II.5.1, 4, III.1, 17–18, 23, 26–7, IV.1
חוט	thread ii.5.5
חכם	sage i.1.1, 3, i.3.3
	expert i.2.3, i.3.5
	scholar i.4.3, ii.5.5
חכמה	science, discipline, I.2.3, I.3.1, 5, I.5.2, II.0, II.1.6, II.3.6, III.0, 29
חכמת השעור	theoretical geometry 0, I.1.3, I.2.5, I.3.3 I.5.1–4, II.0, II.1.1, 6–7, II.3.7, II.5.6, III.0
חכמת התשבורת	computational geometry i.5.3, ii.5.6
בעלי החכמה	geometers i.5.4
חלוק/ חלוקה	division 0, I.3.5, I.4.1–2, 4, I.5.3–4, II.1.7
חמר	matter II.1.6–7
חסר (v.)	fall short II.1.3, III.27, III.29
חץ	axis i.5.1 *IV.1
חתך (v.)	cut I.2.3, I.5.3, II.3.1–6, II.5.5, III.11, 16, 18, 20, 23, 30, IV.1
חתך על חצי	bisect II.3.6 , III.16
מתחתך	intersect II.1.5, II.3.1, II.3.6, III.10–11, 16, 23, 29
חתכה	segment of a circle III.23
טבעת	
טבעת כדורית	spherical annulus 0
טבעת רבועה	square annulus 0
טבעת עגולה	circular annulus 0
טעות	error II.1.2, II.1.4
יושר	
(לנוע) על יושר	(move) in a straight line I.5.1, II.3.3, II.4.1
(להיות) על יושר נקודה	(be) in a straight extension of a point *II.1.1, II.4.1

יחס	ratio *iii.24 see Introduction Section 7, 2.
יצא (v.)	
יצא	produced (a line) i.5.2 separate (point, from a line) II.4.1, III.29 drawn II.2.4, II.3.1–2, 5–6, III.10–11, 18, 20, 28
הוציא, נוציא	draw e.g. II.2.3, II.3.5–6, II.4.1, II.5.2, 4, III.1, 10–13, 18, 20, 29, 31–2
ישר	see קו, שטח
יתר (in a right-angled triangle)	hypotenuse II.3.2–3, 5–6, III.14, 20
יתר הזוית הנצבה	hypotenuse III.20
הקו/צלע המייתר את הנצבה	hypotenuse III.15, III.17
יתר הזוית	side subtended by the angle II.3.2, 5 , III.0, 1, 11–13, 1
(in a circle)	chord i.3.5 , iii.0 –1, 11, 18, 23
יתר (v.)	*III.15, 17
יתרון	excess iii.31
כח	potency/ potentiality I.3.4, I.4.1–2, I.5.6
בכח	potential/potentially i.3.5, i.4.1–2, I.5.1, 3, II.1.7, II.4.2
כפל	twice ii.3.6, iii.0–1, 14, 20–23, 30–31, 33
לקח (v.)	take iii.26–7
מדד (v.)	measure ii. 5.5,
מדידה	measure 0, I.3.4–5, *I.5.3, IV.1 see also שעור
מדובק	continuous ii.3.5, i.5.3, ii.1.2, ii.5.4, iii.33 in logic: conjunctive II.1.3 see also הקדמה
מדובר	rational I.5.4, II.5.5
מוגבל	limited I.3.4, I.4.3–4, II.1.1, 7, II.4.2 constructible *I.2.5, I.5.4 see also גבל (v.)
מוגשם מיושר	rectified solid *III.0, see also מושגם
מונח	substratum I.3.2, 4 given I.2.5, I.4.3, I.5.1, II.3.4, III.0, 26, 30, 32.
מופת	demonstration, proof *I.1.1, I.1.4, I.2.3–4, I.5.2, II.1.1-2, 5, 7, II.2.2

מופתו או המופת (as a title)	II.3.1-6, II.5.2, II.5.4, III.10-33.
מופתי	demonstrative.3.0
מופת עצמי	essential demonstration I.2.3
מופת חותך/נחתך	decisive demonstration I.1.1
מופת אמתי	genuine demonstration I.1.1
מופת החלוף	*per impossibile* demonstration I.2.4
	see also ברהן/ברהאן in the Arabic glossary below
מוצק	pyramid (in Plato's cosmology) I.3.2-4
מוצק ישר השטחים	pyramid III.0
מוצק האסטוונה	cone I.1.4, I.5.1, I.5.4
מוצק עגול, מוצק העגול	cone 0, I.2.4
מוצרף	prism (?) *IV.1
מוקדם	logic: antecedent I.1.3, I.2.3
מורכב	made of I.3.3, I.4.2, II.5.3
מושגם	solid, polyhedron II.5.5, *III.0, III.32 see also מוגשם
מותר	excess. III.10, 14, 19-22
מחובר	composed of I.3.5, II.1.2, II.5.1 see also ערך מחובר
מחוגה	compass *I.2.5
מיושר	see שטח, מוגשם
מישור	plane I.5.4
מנין	number of II.5.5, III.26-7
מצא (v.)	*I.0
מצא	find e.g. *I.0, I.1.4, I.5.4, III.29, III.31
נמצא	there exists, there is I.1.3-4, I.3.5, II.1.6
המציא	realize II.5.6
מציאות	existence I.1.3, I.3.4, I.4.1-4, II.1.2-3,7, II.4.2, II.5.6
מציאות בכח	potential existence I.4.1, II.5.6,
מציאות בפעל	actual existence I.2.5, I.4.1, II.5.6
מציאות בתנועה	existence in motion I.4.1
מציאות אמצעי	intermediate existence I.4.1-2, II.5.6
מקיף	circumference , surface area I.1.2-4, I.3.4, I.5.4, II.1.4, II.5.1-6, III.1, 11, 13, 15, 18, 28 see also עגלה/עגול, נקף
מקרה	accident 0, I.5.2, II.0, II.1.1, II.5.1 attribute II.1.7
במקרה	accidentally II.1.6
מקרה אחרון	secondary accident II.0, II.5.1
מרובע	square I.1.2

מרובע קו	square (on a line) *iii.0, 17, 23, 29, iv.1
מרחק	distance *i.4.1-2, ii.2.3 -4, II.4.1, III.1, 29
משיחה	mensuration i.0
משך (v.)	
ימשך, נמשך המשיך	to be extended i.5.2, ii.1.5, ii.3.6, iii.10, 29 II.1.7, II.3.1-4, II.4.1, III.1, 10 see also נמשך
משולח	has no qualities (simpliciter) ii.1.7
משלש/משולש	triangle (משלש 117 times, משולש 14 times)
משלש ישר הקוים	rectilinear triangle ii.3.5, iii.0-1, 13-15
משלש נצב הזויות	right-angled triangle ii.3.2, iii.0, 11, 14-5, 20, 22
משלש שוה שוקיים	isosceles i.5.1, ii.3.2
משער	geometer i.1.3, i.2.3 see also שער (v.)
משרטטת	drawing instrument *i.2.5, iii. 30.
משׁשת העגול	tangent point iii.15
משתוה	
משתוה/שוה בחלקיו	equally-divided 0, *I.4.2, I.5.3, II.5.4, III.33, IV.1 see also שוה בחלקיו, מתדמה בחלקיו
מתדמה	similar i.4.2, iii.17, iv.1
מתדמה בחלקיו	similarly divided i.4.2
מתחלף	different i.1.3, i.2.1, i.5.4, ii.3.3 unequal iii.20-21 see also זוית, תנועה
מתחלק	divisible, divided i.3.3, 5, i.4.2-4
מתחתך	see חתך (v.)
מתנגד	opposite iii.1,10,12, 29
זוית מתנגדת	vertical angle iii.1
מתנועע	moved [body] i.4.2, ii.1.4-7, ii.2.2, II.4.1, II.5.4, III.33
מה שממנו יתנועע/התנועע	*terminus ex quo* II.1.2
מה שאליו יתנועע/התנועע	*terminus ad quem* II.1.2
מתעקם	bent ii.5.5
מתפרק	dissectible i.5.3
הקדמה תנאית מתפרקת	logic: disjunctive conditional premise i .2.4
נגבל	determined i.5.3, ii.1.4 see also גבל (v.)

נגע (v.) — t ouch I.2.1, I.4.4, II.5.1-4, III.14, 16, 33

נושא — substratum ii.1.7

נכחי — parallel ii.1.1, ii.2.1-4, ii.4.1, ii.5.2, 4, iii.10, 18, 28, 31
 נכחי הצלעות — parallelogram ii.5.4

נמנע — impossible i.1.1, i.4.3, ii.1.1, *2, 4, 7

נמשך — follower i.1.1, i.3.1, iii.33
subsequent upon iii.33
in logic: consequent i.1.3
see also משך (v.)

נצב — see שטח , זוית , משלש

נקבץ — expanded [area by a line] *iii.0.
see also קבוץ

נקף (v.)
 מקיף/מוקף — encompassing/encompassed ii.1.4, ii.5.5,
 מקיף ב / מקיף עם — enclose ii.2.4, ii.3.5

סבב (v.) — r otate II.4.1, II.5.4, III.33

סבה — cause i.3.2-3, iii.33
 סבה עצמית — essential cause i.2.2

סבוב — rotation i.5.1, ii.5.3
revolution. i.5.3

סבובי — circular i.5.1, ii.1.1, ii.5.4, iii.33

סגלה/סגולה — property, used to denote proposition 0, iii.1,11-31. * introduction section 7

סובב — curved (?) ii.3.5 (meaning not certain)

סוג — genus i.1.3, i.5.4, ii.3.6, iii.26-7
 סוג קרוב — close genus I.1.3

עגולה/עגלה/עגול — circle (many occurrences)
 מקיף העגול/ העגלה — circumference of a circle i.1.4, i.2.4, ii.5.1-2, 4, 6 III.1, 11, 28, 31, 33
 עגול ישר, עגלה ישרה — plane circlei.1.4

עיון — inquiry i.3.1, 5, i.5.3, ii.1

עמוד — perpendicular ii.2.3-4, ii.3.2, 4-6, ii.4.1, ii.5.4, III.1,11,13, 18, 20-1, 28-31, IV.1

עצמי — essential, see סבה, הרכבה, מופת

עקום	curved ii.5.5
	see קו, שטח, see also מתעקם
ערך	ratio (many occurrences)
ערך מדובר	rational ratio ii.5.5
ערך אלם	irrational ratio ii.5.5
ערך מחובר	compound ratio *iii.24- 5
ערך שנוי	duplicate ratio *iii .32
ערך שוה	same quantity, equal quantity *i.4.2, i.5.3.
עשוי בתוכו	inscribed ii.1.3
עשוי חוצה לו	circumscribing ii.1.3
עת	time interval i.4.4
פגש (v.)	meet i.5.2, ii.1.5, ii.2.4, ii.3.6, ii.4.1, ii.5.4, III.10, 29
פותח (presumably anagram of פוחת)	fall short of III.26-7
פעל	actuality i.3.4, i .4.1-2, I.5.6
	effect iii.3 3
בפעל	in act, actually I.2.5, I.3.5, I.4.1-2, II.1.7,II.4.1-2, II.5.6
פרוץ	see קו
צודק	logic: true i.2.3
צורה	diagram *II.3.2, III.15
ציור	imagining 0, i.2.4, *.i.4.2,i.5.1, 4, ii.1.6
ציור תנועה	of motion 0, II.5.1-3, II.0, II.1.1, II.2.0, 3, II.5.1
ציור הרכבה	of superposition I.5.4, II.5.1
ציור חלוקים שוים	of equal divisions
ציור קוים	of lines II.1.6-7
ציור הגלגול	of rolling II.5.4
ציור שכלי	imagining by the intellect ii.0, ii.1.1-2, 4
צייר (v.)	imagine i.2.2, *i.4.2, i.5.1, ii.1.6, iii.3 3
צלע	side (many occurrences)
צלעות נוכחיות	parallel sides ii.2.1
צלעות מתנגדות	opposite sides ii.3.1, iii.10
צלעות מקיפות בנצבה	sides which enclos e a right angle III.22
	see also תבנית, שטח
קבוץ	summing *iii.26-7, see also נקבץ
קו	lin e (many occurrences)

קו ישר	straight line i.1.1,4, i.2.1, 4-5, many more
קו מעוגל	circular line i.1. 1
קו קשתי	arched line i.1. 1
קו עקום	curved line ii.5.5
קו מקיף בעגולה	circumference i.2.1, i.5.4 ,
קו מדובק	continuous line ii.1. 2
קו מדביק	joining line ii.2.2-3, ii.3.4, ii.5.2, 4
קו בעל תכלית	finite line ii.1.7, iii.33
קו בלא תכלית	infinite line ii.1. 7
קו עמוד	perpendicular ii .2.4, IV.1
קו פרוץ	split line (conchoid) iii. *29-31
קום (v.)	
הקים/נקים	construct ii.3.2, iii.23 , III.32 erect (a perpendicular) II.2.1, II.3.2, 4, II.4.1, III.21, 30
קטר	diameter i.1.4, i.2.4, i.5.1,6, iii.11, 13, 16, 18,21-3, 28, 31
קיים	fixed ii.5.3-4
קרוב/קירוב	approximation i.5.3, ii.4.2, ii.5. 5
קצה	extremity, boundary i.2.3, ii.1.1,4-5,iii.11, 28-9, 33
קצה ותכלית	extremity and end *ii.1.1, III.33
קצה וגבול	extremity and limit ii.1.4
קצוב	determined *II.1.7
קרוב	see קירוב
קשה	rigid i.4. 4
קשת	arc i.2.1, 4,i.3.5, i.4.1-2, i.5.4, ii.5.1-2, 4, III.0, 16, 18, 23, 28, 33
קשתי	arched i.1.1 .
קשת הזוית	arc of an angle iii. 0
ראוי	appropriate, proper i.3.2-4, i.4.3.i.5.4, ii.1.6 can qualify 0, i.1.3, i .4.2, II.0
ראיה	proof (*demonstratio quia*) (*I.1.1), I.1.4, I.2.3 see also *dalil* in the Arabic glossary below
ראש	vertex (of a triangle) ii.3.1- 2
רבוי	amount i.3.2- 3 expansion *III.0, many more times in Part III
רבוע	product i.1.4, i.2.4,* iii.0
רבע (v.)	to square i.1.2-3, ii.5.6
רכב (v.)	superpose, * I.1.1-2, I.2.1-4, I.3.2-3, 5, I.4.1-2, 4, I.5.1-2, 4, II.1.5, II.3.2, 4-5, II.4.1, II.5.1, 3-4, III.0, 33 see also הרכבה

רקות/ריקות	void i.4. -34, II.1.1
רקיע	surface (of a sphere) i.5.4
רשום	effect iii.33
רשם (v.)	describe (by motion) i.5.1, ii.2.1-2, iii.1, 14, 16, 21-23, 31, IV.1 mark (a point in a diagram) ii.1.3,ii.3.6
שברים	sections 0, *i.5.3
שוה	equal (many occurrences)
שוה בחלקיו	equally divided ii.5.4, iii.33 see also משתוה
שטח	area (many occurrences) surface, plane 0, I.3.2-5, I.4.4, III.28, IV.1
שטח ישר	plane surface/plane area i.2.1, iii.28
שטח ישר הקוים	rectilinear area i.1.1, 3-4, i.2.3, 5, i.5.4, ii.5.6, III.0, 23
שטח מיושר	rectified area iii.0
שטח נכחי הצלעות	parallelogrami i.5.
שטח עגול	circular area i.1.1
שטח קשתי	arched area i.1.1
שטח עקום	curved area ii.5.5
שטח בעל ד׳ צלעות	quadrilateral ii.3.1-2
שטח בעל ד׳ צלעות נצב הד׳ זויות	four-sided right-angled area (= rectangle) IV.1
See also תבנית	
שום (v.)	
נשים	consider ii.3.1-6, iii.1, 10-33, iv.1 assume ii.1.5, ii.2.3, ii.5.2, iii.30
שכל	intellect i.2.5, i.3.1, i.4.2, ii.5.6
שלך (v.)	
נשליך	pick (a point on a line) *ii.5.2, 4, iii.29 discard iii. 15, 23
שנוי	see ערך שנוי
שעור	magnitude (many occurrences)
חכמת השעור	see חכמה
שער(v.)	e valuate 0, i.3.4-5, i.4.2, *i.5.3, ii.5.3, iv .0
שעור ומדידה	Evaluating and measuring , *I.3.4-5, *I.5.3
שער (Arabic *maqāla*)	part (of a book, literally “gate”) i.0, ii.0, iii.0, 18
שוקט	fixed i.5.1, ii.1.5, static i.2.1, i.4.2 at rest iii.33

שקר		false i.1.1,3, 7, ii.2.3, ii.3.2, 6, ii.4.1, ii.5.3-4, III.26-7, 33 see also נמנע, בטל
תבנית (Arabic *shakl*)		figure proposition, (greek *theorema*) I.5.3, II.2.2-3, II.3.0-7, III.11, see also introduction, section 7 plane figure (greek *schema*) I.1.3, I.2.3-4, I.3.5, III.23
	תבנית ישר קוים	rectilinear figure (=polygon) i.3.5
	תבנית בעל ד׳ צלעות ישרות	four sided rectilinear figure (=quadrilateral) III.10, 12, 23,
	תבנית נצב ארבע זויות	right four-angles figure (=rectangle) II.2.1, II.3.2, III.0
	תבנית ירחי	lunar figure i.2.3, iii.23
תולדה		logic: consequence (of a syllogism) i.1.3, ii.5.4
תוספת		extension ii.1.7
תושבת		base of a cylinder I.5.1, II.5.1, III.0 of a triangle I.5.1, III.13, 18 of a pyramid III.0 of a polyhedron III,32 of a rectangle IV.1
תכלית		end (many occurrences)
	בעל תכלית	finite e.g. i.4.1, i.5.3
	בלא תכלית, בלתי בעל תכלית, בלי תכלית	infinite e.g. i.4.3, i.5.3
	עד אין תכלית, בלי תכלית, לאין תכלית	indefinitely, infinitely e.g. i.3.5, i.4.1-4, i.5.2-4
	בתכלית האמת	truest i.3.4
	תכלית קצובה	determined boundary *ii.1.7 see also קצה ותכלית
תנאי		condition i.2.2, see also הקדמה, הקש
תנועה		motion (many occurrences)
	תנועה פשוטה	simple motion i.4.2
	תנועה ישרה	linear motion i.5.1, ii.2.2, ii.5.4, iii.33, iv.1
	תנועה סבובית	circular motion i.5.1, II.1.1, II.5.4, III.33
	תנועה מדובקת	continuous motion iii.33
	תנועה תמידית	perpetual motion iii.33
	שוה/משתוה בחלקיה	equally divided *i.4.2, i.5.3
	מתחלפת בחלקיה	unequally divided *i.4.2 see שוה/ משתוה/מתחלף
תפיסה		charge i.3.1-2
תשבורת		see חכמת התשבורת, שברים

A.2. List of Verbs Frequently Used in Proofs and Constructions (in the First Person Plural, as they Appear in the Text)

נדביק	join, see דבק
נדמה	imagine, assume, see דמה
נוציא	draw, produce, depart, see יצא
נחתוך	cut, see חתך
נחתוך לשני חצאים	bisect, , see חתך
נמשיך	extend, see משך
נעביר	pass (a parallel line through a point), see עבר
נקח	take, see לקח
נקים	construct, erect (a perpendicular) see קום
נרכיב	superpose, see רכב
נרשום	describe, mark, see רשם
נשים	consider, assume, see שום
נשליך	pick (a point on a line), discard

A.3. Arabic-English Glossary (Spelling as it Appears in thc Tcxt)

ברהן/ברהאן	برهان	*demonstratio propter quid*	I.2.3
ברהן אלכלף		*demonstratio per impossible*	I.2.4
ג׳יב אלקוס	جيب القوس	sine iii. 0	
דליל	دليل	*demonstratio quia*	I.2.4
ותר אלקוס	وتر القوس	twice the sine	III.0

References

Primary Sources

Alexander Aphrodisias

Le commentaire d'Alexandre d'Aphroddise aux 'Second analytiques d'Aristote, edited by P. Moraux (Berlin 1979).

Alexander Pseudo

Commentary on Aristotle's *Sophistici Elenchi,* in S. Ebbesen, *Commentators and Commentaries on Aristotle's Sophistici Elenchi*, (Leiden, 1981).

Alfonso de Valladolid

Meyasher `Aqov edition, Russian translation and notes by G. Gluskina (Moscva, 1983).
Mostrador de Justicia, see Szpiech 2006.
Teshuva la-Meharef, ms. Parma Palatina 2440, Di Rossi 533, IMHM 13444 (Institute of Microfilmed Hebrew Manuscripts).

R. Glasner, A. Baraness, *Alfonso's Rectifying the Curved*,
Sources and Studies in the History of Mathematics and Physical Sciences,
https://doi.org/10.1007/978-3-319-77303-2

Al-Ghazali

Maqāṣid al-Falāsifa, Hebrew translation by Yehuda ben Shlomo Nathan, Paris BNF ms. héb 904, IMHM 26860 fol. 42b20-43a5.

Intenciones de los filósofos, Spanish translation, introduction and notes by M. Alonso (Barcelona, 1963).

Al-Nayrizi

Euclidis Elementa ex interpretatione al-Hadschdschadschii cum commentariis al-Narizii. Arabic and Latin translations, edited by R.O. Besthorn and J.L. Heiberg: Codex Leidensis 399,1 (Kopenhagen, 1897–1905, repr. 1997).

The Commentary of al-Nayrizi on Book I of Euclid's Elements of Geometry, translation by A. Lo Bello (Boston and Leiden, 2003)

The Commentary of al-Nayrizi on Books II-IV of Euclid's Elements of Geometry, translation by A. Lo Bello (Boston and Leiden, 2009)

Gerard of Cremona's Translation of the Commentary of al-Nayrizi on Book I of Euclid's Elements of Geometry, translated by A. Lo Bello (Boston and Leiden, 2003).

Aristotle

New *Clarendon Aristotle Series* when available, otherwise J. Barnes (ed.) *The Complete Works of Aristotle* (Oxford, 1985).

Aristotle, Pseudo, *Mechanica*, translation by W.S. Hett, Loeb classics edition.

Archimedes

The Two Books of the Sphere and the Cylinder, translation and commentary by R. Netz (Cambridge, 2009).

The Measurement of the Circle, in T.L. Heath, *The Works of Archimeded* (Cambridge, 1897).

Medieval Hebrew translations, see Lévy, 2011.

Bradwardine Thomas

Tractatus de Proportionibus, edition and translation by H. Lamar Crosby (Madison, 1955).

Cardano Girolamo

Opus novum de proportionibus, Online version available through Archimedes project: http://archimedes.mpiwg-berlin.mpg.de/cgi-bin/toc/toc.cgi?dir=carda_propo_015_la_1570;step=thumb.

Duns Scotus John

Ioannis Duns Scoti doctoris subtilis ac Mariani Ordinis Fratrum Minorum Commentaria oxoniensia ad IV libros magistri Sententiarus, vol. II.

Euclid

Elements

Greek and Latin: *Elementa*, ed. I.L. Heiberg, Leipzig 1883.
English translation: *The Thirteen Books of Euclid's Elements*, translated with introduction and commentary by Sir T.L. Heath (Cambridge, 1925).
Arabic translation by Yiṣḥaq-Thābit: Rabat, Hassaniya ms. 56, fol. 142–143; Paris BNF ms. héb. 1381 (Arabic in Hebrew letters).
Hebrew translation by Moshe Ibn Tibbon: Paris private collection ms. IMHM 39116; Oxford Bodl. ms. Mich. 405, N. 2007, IMHM 19287.
Hebrew translation by Ya'aqov ben Makhir, Oxford ms. Hunt. 16 N. 2005, IMHM 19290.
Hebrew Geometrical Compendium (HGC), Mantova, Comunita Israelitica ms. ebr. 2, IMHM 783.
Latin translation by Campanus: *Campanus of Novara and Euclid's Elements Edition*, introduction and notes by H.L.L. Busard (München, 2005)

Fibonacci

Fibonacci's De Practica Geometrie, edited by Barnabas Hughes (New York, 2008).

Heron

Opera vol. l, edited by L. Nix and W. Schmidt, Leipzig 1900.

Hilbert D.

The Foundations of Geometry, translated by E. J. Townsend (La salle Illinois, 1950).

Ibn al-Haytham

Ibn al-Haytham's Commentary on the Premises of Euclid's Elements, Arabic edition, English translation and commentary by B. Hooper-Sude (Princeton, dissertation, 1974).

Ibn Rushd (Averroes)

The Epistle on the Possibility of Conjunction with the Active Intellect by Ibn Rushd with the Commentary of Moses Narboni, edited and translated by Kalman P. Bland (New York, 1982).
Middle Commentary on Aristotle's Posterior Analytics, in *Paraphrase de la logique d'Aristote*, text arabe inédit par G. Jéhamy (Beyrouth, 1982) 3 vols.
Hebrew translation by Ya'aqov Anatoly, Paris BNF ms. héb 932/1, IMHM 30916.
Long Commentary on Aristotle's Posterior Analytics
Hebrew translation by Qalonimus ben Qalonimes, Parma Biblioteca Palatina ms. ebreo 3022/2 De Rosi 285, IMHM 13751; Oxford Bodl. ms. N. 1365/1, IMHM 22390.
*Middle Commentary on Aristotle's Sophistical Refu*tations, in *Paraphrase de la logique d'Aristote*, text arabe inédit par G. Jéhamy (Beyrouth, 1982).
Hebrew translation by Qalonimus ben Qalonimes, Paris BNF ms. héb 933, IMHM 30917.
Middle Commentary on Aristotle's Physics
Hebrew translation by Qalonimus ben Qalonimus, Hamburg Staats- und Universitaetsbibliothek, MS. Cod. hebr. 264.
Averroes on the Principles of Nature: the Middle Commentary on Aristotle's Physics I-II, a critical edition and English translation of books I-II by S. Harvey. Dissertation (Harvard, 1977).
Middle Commentary on Aristotle's De caelo
Arabic: *Talkhīṣ al-Sama' wa-l-'alam* by G. al-'Alawi (Fās, 1984).
Hebrew translation by Shlomo Ibn Aiub, Berlin Staatsbibliothek ms. Catalog Steinschneider 212.2, IMHM 1769,
Middle Commentary on Aristotle's De generatione et corruptione
Averroes on Aristotle's *De generatione et corruptione*: Middle Commentary and Epitome, trans., from the original Arabic and the Hebrew and Latin versions, Samuel Kurland (Cambridge Mass., 1958).
Hebrew translation by Qalonomus ben Qalonimus, *Commentarium medium et epitome in Aristotelis De generatione et corruptione libros,* edited by Samuel Kurland (Cambridge Mass., 1958).

Maimonides

Introduction to Logic, in the Hebrew Version of Moshe Ibn Tibbon. Eds. L. Roth and D.H., Baneth (Jerusalem, 1987).

Oresme Nicole

Tractatus de commensurabilitate vel incommensurabilitate motum celi, edited and translated by E. Grant in *Nicole Oresme and the Kinematics of Circular Motion*. (Madison, 1971).
De Proportionibus Proportionum and Ad Pauca Respicientes, edited and translated with introductions and notes by E. Grant (Madison, 1966).
The Questiones de spera of Nicole Oresme, Latin text with English translation, commentary, and variants by Garrett Droppers (University of Wisconsin dissertation, 1966).

Pappus of Alexandria

Book 4 of the *Collection*, edition, translation and commentary by Heike Sefrin-Weis (London Dodrecht, 2010).

Proclus

In primum euclidis elementorum librum commentarii ed. G. Friedlein (Leipzig, 1873).
English translation: *A Commentary on the First Book of Euclid's Elements*, translated with introduction and notes by G.R. Morrow (Princeton, 1970).

Ptolemy

Almagest, English translation: translated and annotated by G.J. Toomer (Princeton, 1998).
Greek text of the first book in: Premier livre de la composition mathematique de Claud, ed. N. Halma (Paris, 1816).
Hebrew translation by Ya'aqov Anatoly, Paris BNF ms. héb 1019, online: http://gallica.bnf.fr/ark:/12148/btv1b9064456z/f10.image.r=H%C3%A9breu.langEN.swf

Russel B.

The Principles of Mathematics (Cambridge, 1903).

Saccheri Gerolamo

Euclid Vindicated from Every Blemish, edited and annotated by V. De Risi, translated by G.B. Halsted and L. Allegri (Dordrecht, 2014).

Simplicius

Simplicii in Aristotelum Physicorum, liber quattour priores commentaria, ed. H. Diels, 1882.

Secondary Sources

Abattouy, M. 2001. "Nutaf min al-Hiyal: a Partial Arabic Version of Pseudo-Aristotle's Problemata Mechanica", *Early Science and Medicine*, 6, 96–122.

Aradi N. 2015. *The Arithmetic of the Jews in the Middle Ages*. Jerusalem, Hebrew University, Ph.D dissertation.

Assis Y.T. 2012. "On the Language and Script of the Jews of Spain as an Expression of Their Religious and Cultural Identity", *Peamim* 132, 57–116 (Heb).

Baer I.F. 1958. "The Qabbalistic Doctrine in the Christological Teaching of Abner of Burgos", *Tarbiz* 27, 278–289.

Baer I.F. 1959. *A History of the Jews in Christian Spain*. Tel Aviv (Heb).

Baraness 2016. "Abner of Burgos, Sefer Meyasher 'Aqov", in V. Katz et al. *Sourcebook in the Mathematics of Medieval Europe and North Africa*, Princeton, 345–353.

Baraness 2018, *A mathematical commentary on Sefer Meyasher 'Aqov by Alfonso*, Jerusalem, Ph.D dissertation.

Barker P. and Heidarzadeh T. 2016. "Copernicus, the Tusi Couple and East-West Exchange in the Fifteenth Century", in M. A. Granada, P. J. Boner and D. Tessicini (eds.), *Unifying Heaven and Earth: Essays in the History of Early Modern Cosmology*, Barcelona, 19–57.

Becker O. "Eudoxus-Studien: I: Eine voreudoxische Proportionenlehre und ihre Spuren bei Aristoteles und Euklid," *Quellen und Studien zur Geschichte der Mathematik, Astronomie und Phyik* B. II (1933), 311–330.

Blåsjö, V. 2014. "A Critique of the Arguments for Maragha Influence on Copernicus," *Journal for the History of Astronomy*, 45: 183–195.

Bobzien S. 2002. "The Development of Modus Ponens in Antiquity: From Aristotle to the 2nd Century AD". *Phronesis* 47, 359–394.

Bos H.J.M. 2001. *Redefining Geometrical Exactness*, New York and Berlin.

Brentjes S. 1996. "Remarks about the Proof Sketches in Euclid's *Elements*, Book I as Transmitted by Ms Paris, B. N., Fonds Latin 10257," In M. Folkerts (Ed.), *Mathematische Probleme im Mittelalter: der lateinische und arabische Sprachbereich*, Wiesbaden, 115–137.

Brentjes S. 2001a. "Two comments on Euclid's *Elements*? On the relation between the Arabic text attributed to al-Nayrızı and the Latin text ascribed to Anaritius", *Centaurus* 43, 17–55.

Brentjes S. 2001b. "Observations on Hermann of Carinthia's Version of the *Elements* and its Relation to the Arabic Transmission", *Science in Context* 14, 39–84.

Brentjes S. 2018. "*Who Translated Euclid's Elements into Arabic?*", in J. Hämeen-Anttila and I. Lindstedt (eds.), *Translation and Transmission*. Münster.

Chabás J. and Goldstein B.R. 2000. "Astronomy in the Iberian Peninsula: Abraham Zacut and the Transition from Manuscript to Print", *Transactions of the American Philosophical Society* New Series 90, 1–196.

Clagett M. 1964. *Archimedes in the Middle Ages: The Arabo-Latin Tradition*. Madison.

Corry L. 2016. "Distributivity-like Results in the Medieval Traditions of Euclid's *Elements*: Between Geometry and Arithmetic", *Mathesis* Ser. V, Vol. 1.

Crozet P. 2009 "Thabit ibn Qurra et la composition des rapports", *Arabic Sciences and Philosophy* 14, 175–211, reprinted in R. Rashed (ed.) 2009, 391–536.

Dales R.C. "Henry of Harclay on the Infinite", *Journal of the History of Ideas*, 45, 295–301.

De Risi V. 2014. See Primary sources, Saccheri.

Del Barco J. 2015. *The Late Medieval Hebrew Book in the Western Mediterannean*. Leiden.

Drabkin I.E. 1950. "Aristotle's Wheel: Notes on the History of a Paradox". *Osiris* 9 162–198.

Eisenmann E. 2009, "The Role of the Medium and the Natural Place in the Explanation of the Natural Motion in Rabbi Moses ben Yehuda's Thought", *Tarbitz* 78, 399–419 (Heb.).

Elior O. 2012. "Isaac Abravanel's Rejection of Corporeal Form", *Aleph* 12, 367–402.

Elior O. forthcoming. *Euclid's Elements, Books I-II, in Hebrew Garb: Critical Editions of the Translation by Moses Ibn Tibbon and the Translation Ascribed to Rabbi Jacob with an Introduction and Glossary.*

Engel E. 1992. "Abraham ben Mordecai Farissol: Sephardic Tradition of Book Making in Northern Italy of the Renaissance Period", *Jewish art* 18, 149–167.

Engroff, J.W. 1980. *The Arabic Tradition of Euclid's Elements: Book V.* Cambridge Mass., Harvard University Ph.D. dissertation.

Feldhay R. and Ragep F.J. 2017. *Before Copernicus: The Cultures and Contexts of Scientific Learning in the Fifteenth Century.* Montreal.

Fowler D.H. 1981. "Anthyphairetic Ratio and Eudoxan Proportion", *Archive for History of Exact Sciences,* 24, 69–72

Fowler D.H. 1982. "Book II of Euclid's *Elements* and a pre-Eudoxan theory of ratio, Part 2: Sides and diameters", *Archive for History of Exact Sciences,* 26, 193–209.

Frank R.M. 1967. "Al-*ma'nà*: *Some Reflections* on the Technical Meanings of the Term in the Kalâm and Its Use in the Physics of Mu'ammar", *Journal of the American Oriental Society* 87, 248–259.

Freudenthal Gad 1988. "Maimonides' Guide of the Perplexed and the Transmission of the Mathematical Tract 'On Two Asymptotic Lines' in the Arabic, Latin and Hebrew Medieval Traditions", *Vivarium* 26, 113–140.

Freudenthal Gad, 2005. "Two notes on *Sefer Meyasher `aqov* by Alfonso, alias Abner of Burgos", in Gad Freudenthal, *Science in the Medieval Hebrew and Arabic Traditions,* Ashgate, chapter IX. Originally published in 1991 in Herbew in *Qiriat Sefer* 63, 984–6.

Freudenthal Gad and Zonta M. 2012. "Avicenna Among Medieval Jews: The Reception of Avicenna's Philosophical, Scientific and Medical Writings in Jewish Cultures, East and West", Arabic Sciences and Philosophy 22, 217–287.

Gandz S. 1929. "The Terminology of Multiplication in Arabic and Hebrew Sources", *Hebrew Union College Annual* 6, 247–262.

Geoffroy M. and Steel C. 2001. 'La tradition latine'. In M. Geoffroy and C. Steel (eds.), *Averroès La Béatitude de l'âme: Éditions, traductions et études.* Paris 2001, 83–129.

Gershenzon S. 1984. *A Study of Teshuvot La-Meharef by Abner of Burgos.* Jewish Theological Seminary, Ph.D. dissertation.

Glasner R. 2009. *Averroes' Physics.* Oxford.

Glasner R. 2013. "Hebrew Translations in Medieval Christian Spain: Alfonso of Valladolid Translating Archimedes?", *Aleph* 13, 185–199.

Glasner R. 2019. "The Hebrew Geometrical Compendium:", *Aleph* 19, 201–268.

Gluskina G. 1974. "On the Unpublished Medieval Treatise *Meyasher 'Aqov*, in the British Museum" *Palestinskii Sbornik* 25, 152 (Russian).

Gluskina G. 1978. "On the Authorship of the Mathematical Treatise *Meyasher 'Aqov,*" *Palestinskii Sbornik* 26, 66–75 (Russian).

Goddu A. 2010. *Copernicus and the Aristotelian Tradition.* Leiden.

Goldstein B.R. 2019. "Preliminary Remarks on Medieval Hebrew Trigonometric Tables". *Aleph* 19, 131–136.

Gordon L.S. 2016. "Matter, Place, and Being from a Scholastic Point of View: A Bypass to the Psycho-Physical Problem?" *Philosophy and Theology* 28, 101–140.

Grant E 1966. See Primary sources, Nicole Oresme.

Grant E. 1971. See Primary sources, Nicole Oresme.

Grant E. 1974. *A Source Book in Medieval Science.* Cambridge Mass.

Grant E 1981. *Much Ado about Nothing.* Cambridge.

Grant E. and Murdoch J.E. 1987. *Mathematics and its Applications to Science and Natural Philosophy in the Middle Ages,* Cambridge.

Grellard C. and Robert A. 2009. *Atomism in Late Medieval Philosophy and Theology.* Leiden.

Hacker J. 2015 "Jewish Book Owners and their Libraries in the Iberian Peninsula, Fourteenth-Fifteenth Centuries. In Del Barco, 70–104

Harari O. 2008, "Proclus' Account of Explanatory Demonstrations in Mathematics and its Context", *Archiv für Geschichte der Philosophie* 90, 137–164.

Harari O. 2012. "John Philoponus and the Conformity of Mathematical Proofs to Aristotelian Demonstration", in K. Chemla (ed.) *The History of Mathematical Proofs in Ancient Traditions.* Cambridge, 206–227.

Harvey S. 1977. See Primary sources, Ibn Rushd.

Harvey S. 2001. "Why Did Fourteenth-century Jews Turn to Al-Ghazali's Account of Natural Science?" *Jewish Quarterly Review* 91, 359–376.

Harvey S. 2003. "Did Alfarabi Read Plato's *Laws*", *Medioevo Rivisita di storia della filosofia medievale* 28, 51–68.

Heath T.L. 1897. See Primary sources, Archimedes.

Heath T.L. 1908/1956. See Primary sources, Euclid.

Heath T.L. 1921/1981. *A History of Greek Mathematics*. Oxford; reprinted New York 1981.

Heath T.L. 1949. Mathematics in Aristotle, Oxford.

Hecht J.L. 1993. *The Problem of Exchange between Isaac Polgar and Abner of Burgos/ Alfonso of Valladolid according to Parma manuscript 2440: Iggeret Theshuvat Apikoros and Teshuvot la-Meharef.* New York, NYU Ph. D. dissertation.

Hogendijk J. P. 1991. "The geometrical parts of the *Istikmal* of Yusuf al-Mu'taman Ibn Hud (11th century). An analytical table of contents". *Archives Internationales d'Histoire des Sciences* 41, 207–281.

Hogendijk J. P. 1992–4. "Four constructions of two mean proportionals between two lines in the Book of Perfection *Istikmal* of al-Mu'taman", *Journal for the History of Arabic Science* 10, 13–29.

Hogendijk J. P. 1994. "Mathematics in medieval Islamic Spain", *Proceedings of the International Congress of Mathematicians, August 3-11, 1994, Zürich.* Basel, vol. 2, 1568–1580.

Hogendijk J. P. 1995. "Al-Mu'taman ibn Hud, 11th-century king of Saragossa and brilliant mathematician". *Historia Mathematica*, 22, 1–18.

Hogendijk J.P. 2003. "Lost geometrical parts of the *Istikmal* of Yusuf al-Mu'taman ibn Hud (11th century) in the redaction of Ibn Sartaq (14th century): An analytical table of contents", *Archives internationales d'histoire des sciences* 53, 19–34.

Hogendijk J.P. and Langermann Y.T. 1984. "A Hitherto Unknown Hellenistic Treatise on the Regular Polyhedra", *Historia Mathematica* 11, 325–326.

Hughes B. See Primary sources, Fibbonacci.

Jaouiche K. 1986. *La théorie des parallèles en pays d'islam*. Paris.

Katz V., Imhausen A., Robson E., Dauben J.W., Plofker K., and Berggren J.L. (eds.) 2007. *The mathematics of Egypt, Mesopotamia, China, India, and Islam: a sourcebook*. Princeton.

Katz V., Folkerts M., Wagner R., and Berggren J.L. (eds.) 2016. *Sourcebook in the Mathematics of Medieval Europe and North Africa.* Princeton.

Keyt D. 1961. "Aristotle on Plato's Receptacle", *The American Journal of Philology*, 82, 291–300.

Klatzkin J. 1928–1933. *Thesaurus Philosophicus*. Berlin.

Kleiner I. 1989. "Evolution of the Concept of Function: A Brief Survey", *The College Mathematics Journal* 20, 282–300,

Knorr W.R. 1978. "Archimedes and the pre-Euclidean Proportion Theory", *Archives Internationales d'histoire des sciences* 28, 183–244.

Knorr W.R. 1986. *The Ancient Tradition of Geometrical Problems*. Toronto.

Knorr W.R. 1987. "The Medieval Tradition of Archimedes' Sphere and Cylinder", in Grant and Murdoch, 3–42.

Knorr W.R. 1989. *Textual studies in Ancient and Medieval Geometry*. Boston.

Koningsveld P. S. Van 1992. "Andalusian-Arabic Manuscripts from Christian Spain: A Comparative Intercultural Approach", *Israel Oriental Studies* 12, 75–110.

Kren, Claudia. "The Rolling Device of Nasir al-Dīn al-Ṭūsī in the *De spera* of Nicole Oresme." *Isis* 62 (1971): 490–498.

Kretzmann N. 1982. *Infinity and Continuity in Ancient and Medieval Thought*, Ithaca.

Kretzmann N., Kenney A. and Pinborg J.(eds.) 1982. *The Cambridge History of Later Medieval Philosophy*, Cambridge.

Lachterman D.R. 1989. *The Ethics of Geometry*, New York and London.

Langermann Y.T. 1984. "'Sefer Uqlidis' by an Anonymous Author", *Kiryat Sefer* 54, 74 (Heb.).

Langermann Y.T. 1988. "The Scientific Writings of Mordekhai Finzi" *Italia* 7, 7–44.

Langermann Y.T. 1996a. "Medieval Hebrew Texts on the Quadrature of the Lune", *Historia Mathematica* 23, 31–53.

Langermann Y.T. 1996b "Arabic Writing in Hebrew manuscripts: A Preliminary Re-listing", in *Arabic Sciences and Philosophy* 6, 137–160.

Langermann Y.T. 2009. "Quies media: A Lively Problem on the Agenda of Post-Avicennian Physics", in M. Mazak and N. Ozkaya (eds.) *International Ibn Sina Symposium Papers*, 53–67. Istanbul.

Lévy T. 1992. "Gersonide, le pseudo-Ṭūsī, et le postulat des parallèles: Les mathématiques en hébreu et leurs sources arabes", *Arabic Sciences and Philosophy* 2, 39–82.

Lévy T. 1997a: "Une version hébraïque inédite des *Eléments* d'Euclide", *in* D. Jacquart (éd.), *Les voies de la science grecque,* Genève, 181–239.

Lévy T. 1997b : "Les *Eléments* d'Euclide en hébreu (XIIIe-XVIe siècles)", *in* M. Aouad, A. Elamrani-Jamal, and A. Hasnawi (eds.), *Perspectives arabes et médiévales sur la tradition scientifique et philosophique grecque*, Paris/Louvain, 79–94.

Lévy T. 2005. "Le manuscrit hébreu Munich 36 et ses marginalia: un témoin de l'histoire textuelle des *Eléments* d'Euclide au moyen age", in D. Jacquart et C. Burnett (éd.), *Scientia in margine. Etudes sur les marginalia dans les manuscrits scientifiques du Moyen Age à la Renaissance.* Genève, 103–116.

Lévy T. 2011. "La mesure du cercle d'Archimède au moyen age: Le témoinage des textes hébreux", in R. Fontaine, R. Glasner, R. Leicht, G. Velri, *Studies in the History of Culture and Science.* Leiden, 103–137.

Lloyd G. "The Alleged Fallacy of Hippocrated of Chios", *Apeiron* 20, 103–128.

Lo Bello 2003, see Primary sources, al-Nairizi.

Lo Bello 2009, see Primary sources, al-Nairizi.

Luria S. 1933. "Die Infinitismaltheorie der antiken Atomisten". *Quellen und Studien zur Geschichte der Mathematik, Astronomie und Physik* 2, 106–185.

Maier A. 1982. "The achievements of late Scholastic natural philosophy", in *On the Threshold of Exact Science: Selected Writings of Anneliese Maier.*

Maor E. 1998. *Trigonometrical Delights*. Princeton.

Mcginnis J. 2010. "Avicennan Infinity: A Select History of the Infinite through Avicenna". *Documenti e Studi*, issue dedicated to Avicenna and Averroes 21, 199–222. Online: http://www.umsl.edu/~philo/People/Faculty/McGinnis%20Works/Avicennan%20infinity.pdf

McGinnis J. 2006. "Avoiding the Void: Avicenna on the Impossibility of Circular Motion in a Void". Online: http://www.umsl.edu/~philo/People/Faculty/McGinnis%20Works/Avoiding%20the%20void.pdf

Meirinhos J. F. 2007. "Afonso de Dinis de Lisboa: percurso de um filósofo, médico, teólogo, tradutor e eclesiástico do século XIV", *Península, Revista de Estudos Ibéricos* 4, 47–64, http://repositorio-aberto.up.pt/bitstream/10216/8195/2/4193.pdf

Melechen N. 1999. *The Jews of Medieval Toledo: Their Economic and Social Contacts with Christians from 1150 to 1391.* New York, Ph.D. dissertation.

Mueller I. 1971. "Homogeneity in Eudoxus's Theory of Proportion", *Arabic Sciences and Philosophy* 7, 1–6.

Mueller I. 1981. *Philosophy of Mathematics and Deductive Structure in Euclid's Elements.* Cambridge Mass.

Mueller I. 1982. "Aristotle and the Quadrature of the Circle", in Kretzmann 146–164.

Murdoch J.E. 1957 *Geometry and the Continuum in the Fourteenth Century: A Philosophical Analysis of Thomas Bradwardine's tractatus de continuo.* University of Wisconsin, Ph.D dissertation.

Murdoch J.E. 1964. "Superposition, Congruence and Continuity in the Middle Ages", *Melanges Alexandre Koyre*, Paris, 416–441.

Murdoch J.E. 1969 "*Mathesis* in philosophiam scholasticam introducta: The Rise and Development of the Application of Mathematics in Fourteenth-Century Philosophy and Theology", *Arts Libéraux et Philosophie au Moyen Age. Actes du Quatrième Congrès International de Philosophie Médiévale*, 215–253.

Murdoch J.E. 1974. “Naissance et développement de l’atomisme du bas moyen age latin”, in *La science de la nature: théories et pratiques. Cahiers d’études médiévales* 2, 11–32.

Murdoch J.E. 1982. “Infinity and Continuity”, in Kretzman et al. 564–591.

Murdoch 1987. “Thomas Bradwardine: Mathematics and Continuity in the Fourteenth Century”, in Grant and Murdoch, 103–137.

Murdoch J.E. 2009, “ Beyond Aristotle: Indivisibles and Infinite Divisibility in the Later Middle Ages”, in Grellard and Robert, 15–38.

Netz R. 1999. *The Shaping of Deduction in Greek Mathematics.* Cambridge.

Netz R. 2004. See Primary sources, Archimedes.

Nolan L. *Primary and Secondary Qualities.* Oxford.

North J.D. 1987. “Coordinated and Categories: the Graphical Representation of Functions in Medieval Astronomy”, In Grant and Murdoch, 173–188

O’Brien D. 1984. Theories of Weight in the Ancient World: Plato Weight and Sensation. Leiden.

Palermino C.R. 2001. “Galileo’s and Gassendi’s Solutions to the *Rota Aristotelis* Paradox: a Bridge between Matter and Motion Theories”, in Luthy C., Murdoch J.E., Newman, W. R. (eds.) *Medieval and Early Modern Science, Late Medieval and Early Modern Corpuscular Matter Theories.* Leiden, 381–422.

Pedersen O. and Jones A. 2010 *A Survey of the Almagest, with annotation and a new commentary by A. Jones.* New York and Dordrecht.

Philip J.A. 1966. “The ‘Pythagorean’ Theory of the Derivation of Magnitudes”, Phoenix 20, 32–50.

Podkonski, R. 2006. “Al-Ghazali’s “Metaphysics” as a Source of Anti-atomistic Proofs in John Duns Scotus’s Sentences Commentary”, in A. Speer and L. Wegener (eds.) *Wissen über Grenzen. Arabisches Wissen und lateinisches Mittelalter* (*Miscellanea Mediaevalia* 33), 612–625.

Ponte, J. P. 1992. “The History of the Concept of Function and Some Educational Implications”. *The Mathematics Educator, 3*, 3–8. Online: http://www.educ.fc.ul.pt/docentes/jponte/ind_uk.htm

Ragep, F. J. 1993. *Nasir al-Din al-Tusi’s “Memoir on Astronomy,” Sources in the History of Mathematics and Physical Sciences*,12. 2 vols. Berlin New York.

Ragep, F. J. 2006. “The Two Versions of the Tūsī Couple”, in *From Deferent to Equant*: A Volume of Studies on the History of Science of the Ancient and Medieval Near East in Honor of E.S. Kennedy *(Annals of the New York Academy of Sciences* 500), 329–356.

Ragep F.J. 2017. “From Tūn to Toruń: The Twists and Turns of the Ṭūsī-Couple” in Feldhay and Ragep, 161–197.

Rashed M. *Alexandre d’Aphrodise, Commentaire perdu a` la “Physique” d’Aristote* (Livres IV–VIII) Berlin 2011.

Rashed R. 1993. “Al-Kindī’s Commentary on Archimedes’ ‘The Measurement of the Circle’”, *Arabic Sciences and Philosophy* 3, 7–53.

Rashed R. 1997a. *Histoire des sciences arabes*, vol. 3. Paris.

Rashed R. 1997b: “Les commencements des mathématiques archimédiennes en arabe: Banū Mūsā”, in A. Hasnawi, A. Elamrani-Jamal et M. Aouad (eds.) *Perspectives arabes et médiévales sur la tradition scientifique et philosophique grecque*, Paris/Louvain, 79–94.

Rashed R. 2000. *Ibrahim Ibn Sinān logique et géometrie au Xe sicele.* Leiden.

Rashed R. 2009. “Thabit ibn Qurra et l’art de la mesure”, in Rashed (ed.) *Thabit ibn Qurra. Science and Philosophy in Ninth-Century Baghdad,* Berlin/New York, 173–210.

Rashed R. 2012. *Founding Figures and Commentators in Arabic Mathematics.* Edited by N. El-Bizri. New York.

Rashed R. 2013a. *Ibn al-Haytham and Analytical Mathematics.* New York.

Rashed R. 2013b. *Ibn al-Haytham’s Theory of Conics, Geometrical Constructions and Practical Geometry.* New York.

Rashed R. 2015. *Angles et Grandeur.* Boston/Berlin.

Rashed R. and Houzel C. 2005, “Thābit Ibn Qurra et la théorie des paralléles”, *Arabic Sciences and Philosophy* 15, 9–55.

Rashed R. and Vahabzadeh B. 1999. *Al-Khayyām Mathématician.* Paris.

Rosnenfeld B. and Youschkevitch A.P. 1997. "Géometrie", in Rashed 1997a, 121–162.

Rossini P. 2018. "Giordano Bruno and Bonaventura Cavalieri's theories of indivisibles: a case of shared knowledge", *Intellectual History Review*, 28, 461–476.

Sabra A.I. 1968. "Thābit Ibn Qurra on Euclid's Parallels Postulate", *Journal of the Warburg and Courtauld Institutes*, 31, 12–32.

Sabra A.I. 1969. "Simplicius's Proof of Euclid's Parallels Postulate", *Journal of the Warburg and Courtauld Institutes*, 32, 1–24.

Sadik S, 2008, "Rabbi Ḥasdai Crescas' Critique of Aristotelian Science and the Lost Book of Abner of Burgos". *Tarbiz* 77, 133–155 (Heb.).

Sadik S. 2011. *Trinity and Determinism in the Thought of Abner of Burgos*, Ben Gurion University, Ph.D. dissertation (Heb.).

Sadik S. 2016a. "Abner de Burgos and the Transfer of Philosophical Knowledge between Judaism and Christianity". *Medieval Encounters* 22, 95–112.

Sadik S. 2016b, "Utilization of Maimonides Writing by Jewish Apostates", *Jewish Studies* 51, 31–68 (Heb.).

Sainz de la Maza Vicioso C. 1990. *Alfonso de Valladolid: edición y estudio del manuscrito "Lat. 6423" de la Biblioteca Apostólica Vaticana*, Universidad Complutense de Madrid, Ph.D dissertation.

Saito K. 1985. "Book II of Euclid's *Elements* in the Light of the Theory of Conic Sections", *Historia Scientiarum* 28, 31–60. Reprinted in Jean Christianidis (ed.), *Classics in the History of Greek Mathematics, Boston Studies in the Philosophy of Science*, Vol. 240, Dordrecht/Boston, 139–168.

Sarfatti G. B. 1968. *Mathematical Terminology in Hebrew Scientific Literature of the Middle Ages*. Jerusalem (Heb.).

Schreiber S.G. 2003. *Aristotle on False Reasoning*, New York.

Sela S. 2017. "The Ibn Ezra – Bates Astrological Connection and the Three Abrahams", *Mediterranea. International journal for the transfer of knowledge* 2, 163–186.

Sidoli N. 2004. *Ptolemy's Mathematical Approach: Applied Mathematics in the Second Century.* University of Toronto, Ph.D. dissertation.

Sirat C. 2009. "La version arabo-hébraïque médiévale du Grand Commentaire d'Averroès sur le *De anima* d'Aristote et Shem Tov b. Joseph b. Shem Tov", in J. Meirinhos and O. Weijers (eds.), *Florilegium mediaevale: Études offertes à Jacqueline Hamesse à l'occasion de son éméritat*. (*Textes et Études du Moyen Âge*, 50) Louvain, 541–561.

Sirat C. 2015. "Studia of Philosophy as Scribal Centers in Fifteenth-Century Iberia". In De Barco, 46–69.

Sirat C. and Geoffroy M. 2005. *L'original arabe du Grand Commentaire d'Averroès au De Anima d'Aristote*. Paris.

Smith D.E. 1925/1953, *History of Mathematics*. New York.

Sorabji R. 1990. *Aristotle Transformed: The Ancient Commentators and Their Influence*, Trowbridge.

Steel C. and Guldentops G. 1997. "An Unknown Treatise of Averroes against the Avicennians on the First Cause", in *Recherches de Théologie et Philosophie Médiévales* 64, 86–135.

Steinschneider M. 1893. *Die Hebräischen Übersetzungen des Mittelalters und die-Juden als Dolmetscher*. Berlin.

Sude-Hooper 1974, see Primary sources, *Ibn al-Haytham*.

Sylla E.D. 1997. "Thomas *Bradwardine's* "De continuo" and the Structure of Fourteenth-Century Learning" in E. Sylla and M. Mc Vaugh (eds.), *Texts and contexts in ancient and medieval science: studies on the occasion of John E. Murdoch's seventieth birthday*, Leiden. 148–186.

Szabo A. 1978. *The Beginnings of Greek Mathematics*. Dordrecht.

Szpiech R. 2006 *From Testimonia to Testimony. Thirteenth-Century Anti-Jewish Polemic and the Mostrador de Justicia of Abner of Burgos/Alfonso de Valladolid.* Yale University, Ph.D dissertation.

Szpiech R. 2010. "In Search of Ibn Sīnā's 'Oriental Philosophy' in Medieval Castile", *Arabic Sciences and Philosophy* 20, 185–206.

Thorup A. 1992. "A Pre-Euclidean Theory of Proportions", *Archive for history of exact sciences* 45, 1–16.
Toomer G. J. 2008. "Nicomedes". *Complete Dictionary of Scientific Biography*. Online: http://www.encyclopedia.com/doc/1G2-2830903168.html
Unguru S. 1974. "Pappus in the Thirteenth Century in the Latin West", *Archive for history of exact sciences* 13, 307–324.
Vahabzadeh B. 2002. "Al-Mahani's Commentary on the Concept of Ratio", *Arabic Sciences and Philosophy* 12, 9–52.
Vahabzadeh B 1994. "Two Commentaries on Euclid's Definition of Proportional Magnitudes", *Arabic Sciences and Philosophy* 4, 181–198.
Vahabzadeh B. 2013. "Le commentaire d-Ibn Mu'āḏ sur le concept de rapport", *Arabic Sciences and Philosophy* 23, 221–276.
Vajda G. 1958. "A Note on the Texts of Al-Ghazālī Quoted by Abner of Burgos", *Tarbiz* 28, 99–100
Verlinden C. 1953. "Italian Influence on Iberian Colonization", *The Hispanic American Historical Review* 33, 199–211.
Vitrac B. 2000. "Umar al Khayyam et Eutocius: Les antécédents grecs du troisième chapitre du commentaire 'Sur certaines prémisses problématiques du Livre d'Euclide'", *Farhang* 12, 51–105. Online: https://hal.archives-ouvertes.fr/hal-00174927
Vitrac B. 2002. "Umar al Khayyam et l'anthyphérèse: Étude du deuxième Livre de son commentaire Sur certaines prémisses problématiques du Livre d'Euclide". *Farhang* 14s, 137–192.
Vitrac B. 2005. "Quelques remarques sur l'usage du mouvement en géométrie dans la tradition euclidienne: de Platon et Aristote à Omar Khayyâm". Online: https://hal.archives-ouvertes.fr/hal-00174944/document
Vlastos G. 1995. "Minimal Parts in Epicurean Atomism", in G. Vlatos, *Studies in Greek Philosophy,* Princeton II, 285–314.
Wagner Robert J. 1983. "Euclid's Intended Interpretation of Superposition", *Historia Mathematica* 10, 63–89.
Wagner Roi. 2010. "The natures of numbers in and around Bombelli's *L'algebra*", *Archives for the History of Exact Sciences* 64, 485–523.
Wagner Roi 2013. "Mordekhai Finzi Translation of Maestro Dardi's Italian Algebra". In A. Fidora, H.J. Hames, Y. Schwartz (eds.) *Latin into Hebrew* vol. II: *Texts in Contexts*. Leiden. 195–221.
Wardy R. 1990. *The Chain of Change: A Study of Aristotle's Physics VII,* Cambridge.
Wartenberg I. 2007. *The Epistle of the Number by Isaac ben Solomon ben al-Ahdab (Sicily 14th century): A Episode of Hebrew Algebra.* Tel Aviv University and Université Paris 7, Ph.D dissertation.
Wolfson H.A. 1929. *Crescas' Critique of Aristotle*, Cambridge.
Wolfson H.A. 1973. "The Terms *Tasawwur* and *Tasdiq* in Arabic Philosophy and their Greek, Latin and Hebrew Equivalents", in I. Twersky (ed.) *Studies in the History of Philosophy and Religion*, Cambridge Mass., 478–492.
Wood R. 2009 "Indivisibles and infinites : Rufus on points", in C. Grellard and R. Aurélien, *Atomism in Late Medieval Philosophy and Theology,* Leiden, 9–39.
Young de G. 1984, "The Arabic textual Tradition of Euclid's Elements", *Historia Mathematica* 2, 147–160.
Young de G. 1992. "lshaq ibn Hunayn, Hunayn ibn Ishaq, and the Third Arabic Translation of Euclid's Elements", *Historia Mathematica* 19, 188–199.
Young de G. 2002 "Euclidean Geometry in Two Medieval Islamic Encyclopaedias", Al-Masāq, 14:1, 47–60.
Young de G. 2005 "Diagrams in the Arabic Euclidean tradition: a preliminary assessment", *Historia Mathematica* 32, 129–179.
Zimmermann F. W. 1986. "The Origins of the So Called *Theology of Aristotle*", in J. Kraye, W. F. Ryan, and C. B. Schmitt *Pseudo-Aristotle in the Middle Ages: The Theology and Other Texts*, London, 110–240.
Zupko J. 1993. "Nominalism Meets Indivisibilism", *Medieval Philosophy and Theology* 3, 158–185.

Index

R. Glasner, A. Baraness, *Alfonso's Rectifying the Curved*,
Sources and Studies in the History of Mathematics and Physical Sciences,
https://doi.org/10.1007/978-3-319-77303-2

CPSIA information can be obtained
at www.ICGtesting.com
Printed in the USA
BVHW011214170221
599677BV00033B/41